Transition-Metal-Containing Bifunctional Catalysts: Design and Catalytic Applications

Transition-Metal-Containing Bifunctional Catalysts: Design and Catalytic Applications

Guest Editors

Linda Zh Nikoshvili
Lioubov Kiwi-Minsker
Valentin Yu Doluda

MDPI

Basel • Beijing • Wuhan • Barcelona • Belgrade • Novi Sad • Cluj • Manchester

Guest Editors

Linda Zh Nikoshvili
Department of Biotechnology, Chemistry and Standardization
Tver State Technical University
Tver
Russia

Lioubov Kiwi-Minsker
Ecole Polytechnique Fédérale de Lausanne
GGRC-ISIC-EPFL
Lausanne
Switzerland

Valentin Yu Doluda
Department of Biotechnology, Chemistry and Standardization
Tver State Technical University
Tver
Russia

Editorial Office
MDPI AG
Grosspeteranlage 5
4052 Basel, Switzerland

This is a reprint of the Special Issue, published open access by the journal *Catalysts* (ISSN 2073-4344), freely accessible at: www.mdpi.com/journal/catalysts/special_issues/H5S1NN3491.

For citation purposes, cite each article independently as indicated on the article page online and using the guide below:

Lastname, A.A.; Lastname, B.B. Article Title. *Journal Name* **Year**, *Volume Number*, Page Range.

ISBN 978-3-7258-4074-8 (Hbk)
ISBN 978-3-7258-4073-1 (PDF)
https://doi.org/10.3390/books978-3-7258-4073-1

Contents

About the Editors

Linda Zh Nikoshvili

Linda Zh. Nikoshvili is an associate professor in the Department of Biotechnology, Chemistry, and Standardization (former Department of Biotechnology and Chemistry), Tver State Technical University (TSTU), Tver, Russia. She graduated from the Department of Biotechnology and Chemistry of TSTU (with honors) with an MS in 2006. She received her PhD in Catalysis in 2009. After one year, she obtained the position of associate professor at TSTU. The theme of her scientific works is connected with the development of new highly effective nanostructured polymeric systems for the hydrogenation of acetylene alcohols, nitro-compounds, biomass-derived compounds (levulinic acid), the oxidation of monosaccharides, and cross-coupling processes for the selective production of platform chemicals and intermediates for the synthesis of active pharmaceutical ingredients. She actively participates in international, all-Russian, and regional scientific conferences and symposia related to the synthesis and study of nanostructured catalytic systems and their applications in chemistry and chemical technology. For her achievements in the field of heterogeneous catalysis, she has received an "Honorary Diploma of the Ministry of Economic Development of the Tver Region" (2013), the award "Engineer of the Year" (2013), and an "Honorary Diploma of the Governor of the Tver Region" (2017). She has published more than 100 research articles, including several patents.

Lioubov Kiwi-Minsker

Lioubov Kiwi-Minsker is a professor emeritus at the Swiss Federal Institute of Technology Lausanne (EPFL), Lausanne, Switzerland. She recieved her PhD in 1982 in Physical and Colloidal Chemistry from Moscow University and passed habilitation in 1992 at Novosibirsk University in Physical Chemistry. She taught chemical thermodynamics, kinetics of heterogeneous reactions, and colloidal chemistry and carried out research in the field of heterogeneous catalysis for environmental protection. In 1994, she entered the EPFL, appointed to the Laboratory of Chemical Reaction Engineering. Her research activities continue to be in the field of heterogeneous catalysis and reactor technology, especially reactors with structured catalytic beds and micro-reactors. She received Late Innovation awards from "EniTechnology" (Eni Snamprogetti group, Italy) in 2002, from the Royal DSM Company in 2005, and "Doctor Honoris Causa" from the Lomonosov Moscow State Academy of Fine Chemical Technology (Russia) in 2010. She is currently a member of the EU Research Institute of Catalysis (ERIC) and a member of the Working Party on "Chemical Reaction Engineering" of the European Federation of Chemical Engineering (EFCE) and of the European Federation of Catalysis (EFCATS). She is a co-author of around 500 publications, including monographs, textbooks, patents, research articles, and reviews.

Valentin Yu Doluda

Valentin Yu. Doluda is full professor of the Department of Biotechnology, Chemistry, and Standardization, Tver State Technical University (TSTU), Tver, Russia. He graduated from TSTU in 2005 with a degree in Biotechnology. He holds the degree of Doctor of Chemical Sciences (2019), is a Candidate of Chemical Sciences (2008), and holds the academic title of associate professor (2012). Over the past five years, Doluda V. Yu has published more than 45 publications. Since, 2024 he was appointed Acting Head of the Department of Chemistry and Polymer Technology.

Doluda V. Yu was awarded Honorary diplomas of the Ministry of Education of the Tver Region (2012, 2018), Honorary Diploma of the Legislative Assembly of the Tver Region (2014), and Honorary Diploma of the Ministry of Education and Science of the Russian Federation (2016). The field of scientific interests of Doluda V. Yu includes different heterogeneously catalyzed processes, such as selective oxidation, hydrogenation, and various reactions of biomass processing, and obtaining synthetic liquid fuels.

catalysts MDPI

Editorial

Transition-Metal-Containing Bifunctional Catalysts: Design and Catalytic Applications

Linda Zh. Nikoshvili [1,*], Valentin Yu. Doluda [1] and Lioubov Kiwi-Minsker [2,*]

1 Department of Biotechnology, Chemistry and Standardization, Tver State Technical University, A.Nikitina Street, 22, 170026 Tver, Russia; doludav@yandex.ru
2 Ecole Polytechnique Fédérale de Lausanne, ISIC-FSB-EPFL, CH-1015 Lausanne, Switzerland
* Correspondence: nlinda@science.tver.ru (L.Z.N.); lioubov.kiwi-minsker@epfl.ch (L.K.-M.); Tel.: +7-904-005-7791 (L.Z.N.); +41-76-334-0861 (L.K.-M.)

Chemical processes catalyzed by transition metals are widely applied and well established in modern industry. Nevertheless, the scientific background and new approaches to a catalyst's synthesis are constantly developing. Research efforts are mainly directed towards an improvement in catalytic activity and decrease in by-products and pollutants. For heterogeneous catalysts, there are a lot of opportunities for their rational design in terms of morphology and creation of different active centers in the same catalyst [1]. The multifunctionality of catalytic systems became a research trend in recent decades. Bifunctional catalytic systems are a particular case of multifunctional catalysis and are a promising area of research [2–4], providing broad opportunities for carrying out simultaneously few catalytic processes in a single apparatus. In this way, modern principles of "green" chemistry and sustainable development of chemical production can be implemented.

Bifunctional catalysis is a broad concept represented by several types of systems:

- Bimetallic catalysts containing two different active metals or their oxides;
- Catalysts consisting of an active metal in combination with the Brønsted and/or Lewis acid sites;
- Biocatalytic systems that include metal nanoparticles (NPs) and/or acid sites.

The combination of different functions in a single catalyst reduces the duration of the multi-stage process and usually increases the overall yield of the target product. At the same time, the implementation of cascade or tandem processes in the same apparatus using multifunctional catalysts is associated with several difficulties. For example, optimal conditions for each simultaneously proceeding reaction may differ; intermediate products occurring at different active centers can affect each other; and finally, the diffusion of reagents between active centers should be considered.

Due to the importance and complexity of catalysis using bifunctional systems, this Special Issue, titled "Transition-Metal-Containing Bifunctional Catalysts: Design and Catalytic Applications", was organized. It consists of 11 contributions, including 2 reviews and 9 research papers. The aim of this Special Issue was to present and summarize recent advances in the development of bifunctional catalytic systems based on transition metals for different applications.

In the review by Nikoshvili and Matveeva (contribution 1), a variety of Pd-catalyzed tandem processes are described, which include not only truly tandem catalysis but also cascade and orthogonal catalytic reactions. The problem of defining tandem and related processes is discussed while emphasizing the role of distinct catalytic mechanisms and active species. Among the large number of homogeneous Pd-catalyzed tandem reactions presented, there are truly tandem processes, but also the related (one-pot, orthogonal, or domino) reactions. They were clearly distinguished, since in the referred scientific papers a double attribution was applied, being sometimes misleading. It is concluded by the authors that in the field of homogeneous tandem catalysis, a variety of conditions, possible additives, and reaction routes can be applied to provide the highest yields of desired products with the lowest environmental impact. In contrast, in the case of heterogeneous catalysis, the main direction of processes' development is an improvement in support properties, i.e., the use of supports bearing acid-base centers of Brønsted and Lewis. Such Pd-containing catalytic systems are usually based on oxidic supports and metal-organic frameworks and are often applied for the transformation of "platform chemicals" derived from plant biomass. Another approach used for obtaining bifunctional catalysts that is highlighted in the review is the combination of homogeneous and heterogeneous catalysis, e.g., immobilized metal complexes or enzymes. The combination of metal catalysis and biocatalysis is a promising alternative to traditional chemical routes of the synthesis of chiral compounds with high enantioselectivity. Thus, the review covers all areas of development regarding palladium-catalyzed tandem processes.

Another review (contribution 2) included in this Special Issue is the work of Stepacheva et al., which is devoted to the application of plant-biomass-derived materials as catalytic supports. Plant biomass is a renewable source used not only to produce fuels and "platform chemicals" but also for cheap and effective sorbents and catalytic supports. The authors of the review describe the current methods of biochar obtained from a variety of starting materials in detail. It is shown that acid-functionalized biochars produced by pyrolysis can be successfully used as catalysts for biodiesel production and for the transformation of carbohydrates into furfural, 5-hydroxymethylfurfural (5-HMF), and other valuable chemicals. Base-functionalized biochars are less used in biodiesel production due to their high saponification activity and sensitivity to water. Despite such disadvantages, it is noticed that base-functionalized biochars can provide high rates of transesterification under mild conditions. Moreover, authors show that biochars and biocarbons can be effective supports for a wide variety of transition metals (Mo, Co, Cu, Ni, Fe, Au, Ru, Pd) and be applied in heterogeneously catalyzed organic reactions and in electrocatalytic processes. This review can serve as a concise guide for the development of functionalized biocarbon-based material for catalytic applications.

The topic of carbon-supported transition metal catalysts is continued in contributions 3 and 4 of this Special Issue.

In the contribution 3 by Taran et al., the synthesis of Ru-containing catalysts supported on Sibunit® (commercial mesoporous carbon materials) and their application for reductive catalytic fractionation (RCF) of spruce wood is described. Sibunit® was functionalized with different oxygen-containing species, which allowed the researchers to prevent the sintering of ruthenium particles under high temperatures, being responsible for acidic properties. Thus, the series of bifunctional Ru/C catalysts was developed by varying the support acidity, catalyst grain size, content, and dispersion of Ru. The yield and composition of solid, liquid, and gaseous products of spruce wood RCF were carefully studied with various catalyst types, reaction times, and temperatures. Moreover, solid products were characterized with IR spectroscopy and XRD. Under optimal conditions (225 °C, 3 h) using the most acidic Ru/C, the liquid products enriched with phenolic compounds were synthesized, and the optimal trade-off between phenols (30.0 wt.%), polyols (18.6 wt.%), and solid products containing 84.4 wt.% of cellulose was achieved.

Kustov and Kalenchuk (contribution 4) reported a series of mono- and bimetallic Ni-Cr/C catalysts synthesized using Sibunite® as a support for the reaction of bicyclohexyl dehydrogenation applied in hydrogen storage systems. The catalysts' composition and morphology were characterized by the XPS, XRD, TPR, HRTEM, and electron diffraction methods. In contrast to the work of Taran et al. (contribution 3), the authors of contribution 4 highlighted the interactions between two active metals rather than the influence of the support. Despite the formation of a Cr-Ni solution and the change of nickel crystal lattice parameters, no obvious advantages for bimetallic Ni-Cr/C compared to monometallic systems were found, though the authors observed strong positive effect for the trimetallic Pt-Ni-Cr/C catalyst in an earlier study [5].

Among carbonaceous supports, polymers can serve as a promising alternative to chars. There is a big variety of functionalized polymeric materials for sorbents and catalytic supports [6]. Hypercrosslinked aromatic polymers are of special interest [7] due to their high porosity, chemical, and thermal stability.

Dutra et al. (contribution 5) developed bimetallic (Ru-Ni, Ru-Co, Ru-Cu) catalysts containing 3wt.% of Ru and 0.1 wt.% of the second metal supported on commercial hypercrosslinked polystyrene (HPS), which were tested in the reactions of transformation (hydrogenation, hydroformylation, oxidative cleavage, and deoxygenation) of canola fatty acid methyl esters (FAMEs). High values of FAME conversion (up to 99%) were attained, showing the prospect of HPS-based catalytic systems.

Though there are numerous carbonaceous materials available as supports, the synthesis of multifunctional catalytic systems is commonly based on different oxides and their mixtures. The series of contributions within this Special Issue were devoted to oxide catalysts (contributions 6–8) and to transition metals supported on oxides (contributions 9 and 10).

The group of scientists from Brazil in contribution 6 reported the synthesis of cerium oxide, molybdenum oxide, and mixed material $Ce_2(MoO_4)_3$ for the conversion of fructose to 5-HMF. It is well known that the main issue of this reaction is the formation of humins, which can be controlled by balancing acidity [8]. $Ce_2(MoO_4)_3$ revealed lower acidity compared to pure molybdenum oxide (see the data presented in contribution 6) that inhibited the formation humins to some extent (selectivity to 5-HMF increased from 14.5% up to 27.5% at a high fructose conversion). Among other byproducts, different acids (levulinic, acetic, formic, lactic), aldehydes (glyceraldehyde and pyruvaldehyde), and dihydroxyacetone were found. The highest selectivity to 5-HMF was found in the case of pure cerium oxide (70.2%); however, the conversion of fructose was about 50% (150 °C, 6 h). The stability and reuse of $Ce_2(MoO_4)_3$ was studied in four repeated runs and revealed a negligible variation in conversion and selectivity. Similarly, in contribution 7, the conversion of fructose was tested using tin, niobium, and tin-niobium oxides obtained with three different methods: coprecipitation (PPT), hydrothermal treatment using cetyltrimethylammonium bromide (CTAB), and Pechini treatment using glycerol (GLY). Independently of the synthesis method, mixed SnNb oxides were more active compared to the corresponding pure oxides, and they were highly selective due to having an appropriate balance between the Lewis and Brønsted acid sites. Among SnNb (GLY), SnNb (CTAB), and SnNb (PPT) materials, the highest fructose conversion (76.9% for 2 h and >90% for 6 h at 150 °C) and selectivity to 5-HMF (about 80%) were found for SnNb (CTAB), which makes SnNb oxides promising catalysts for the processes of biomass valorization.

Interesting work (contribution 8) devoted to oxide catalysts was presented by Mukhtarova et al., who studied the hydrodeoxygenation (HDO) of guaiacol (lignin-derived compound) to aromatics in the presence of MoOx and WOx generated in situ from corresponding carbonyls. The detailed characterization of the catalysts was carried out with XRD, XPS, HRTEM EDX, SEM, DRIFTS, and Raman spectroscopy methods. The influence of reaction conditions (temperature and H_2 pressure) on the conversion and selectivity of guaiacol to different products was studied, and the HDO pathways were proposed. Under the chosen reaction conditions (380 °C, 5 MPa of H_2, 6 h, solvent—dodecane), the benzene-toluene-xylenes fraction was obtained with high yield (96%) over WOx catalyst.

Shivtsov et al. in contribution 9 prepared a series of magnesium-aluminum oxides (Mg-Al-O_x) with different Mg/Al molar ratios (4:1, 2:1, 1:1, 1:2, and 1:4) with the aerogel method, which served as supports for Pd NPs. Chosen method of oxide synthesis allowed them to obtain materials with high specific surface areas (up to 600 m^2/g) for the development of bifunctional catalysts (1 wt.% of Pd)of tetradecahydrophenazine dehydrogenation with a high hydrogen evolution rate (252 mmol$\cdot g_{Met}{}^{-1}\cdot min^{-1}$). The authors of contribution 9 revealed that the most promising mixed oxides were characterized by significant concentrations of basic electron-donor sites responsible for the efficient stabilization of palladium NPs and alteration of the dehydrogenation mechanism.

Philippov et al. (contribution 10) reported transfer hydrogenation of biomass-like phenolic compounds such as guaiacol and dimethoxybenzenes (DMB) over novel Ni-SiO_2 and NiCu-SiO_2 catalysts under supercritical conditions at 250 °C using 2-PrOH as the only hydrogen donor. The role of copper in bimetallic catalytic systems was highlighted. Detailed kinetic analysis was carried out, showing the negative effect of Cu addition on the rate of guaiacol and 1,4-DMB conversion. However, in the case of 1,2-DMB and 1,3-DMB, Cu promoted their transformation to corresponding dimethoxycyclohexanes. Moreover, the addition of copper influenced the selectivity of the reaction products, which is an interesting result deserving further study.

The most exciting multifunctional catalytic systems are those that combine the features of transition metals, oxides, and carbonaceous materials. Thus, in contribution 11 presented by Sorokina et al., novel Cr-Zn/Ni composite materials based on an oxide support (SiO_2 or Al_2O_3) with embedded magnetic NPs and covered by a crosslinked pyridylphenylene polymer (PPP) layer were described. The developed nanocomposites were applied for CO_2 hydrogenation to methanol. The necessity of all the components was proven by the results of catalyst characterization and testing, as follows: PPP was found to prevent the aggregation of catalytically active species and enhanced the adsorption of CO_2 and H_2; magnetic NPs insured structural promotion, which was likely due to the formation of oxygen vacancies, and allowed for the fast and simple magnetic separation of the catalyst; the combination of the two metals (Cr-Zn and Cr-Ni) resulted in the creation of bimetallic interfaces. All these features allowed the authors of contribution 11 to obtain a highly active and stable catalytic system of methanol synthesis.

Finally, we are grateful to all the authors who contributed to this Special Issue. We express our sincere gratitude to the editorial team of *Catalysts* for their assistance and continuous support in the preparation of this Special Issue. Moreover, we would like to thank the reviewers for their valuable contributions.

Author Contributions: Conceptualization, L.Z.N. and L.K.-M.; writing—original draft preparation, L.Z.N.; writing—review and editing, V.Y.D. and L.K.-M. All authors have read and agreed to the published version of the manuscript.

Funding: This research was funded by the Russian Science Foundation, grant number 23-79-00009.

Conflicts of Interest: The authors declare no conflicts of interest.

List of Contributions:

1. Nikoshvili, L.Z.; Matveeva, V.G. Recent Progress in Pd-Catalyzed Tandem Processes. *Catalysts* **2023**, *13*, 1213. https://doi.org/10.3390/catal13081213.
2. Stepacheva, A.A.; Markova, M.E.; Lugovoy, Y.V.; Kosivtsov, Y.Y.; Matveeva, V.G.; Sulman, M.G. Plant-Biomass-Derived Carbon Materials as Catalyst Support, A Brief Review. *Catalysts* **2023**, *13*, 655. https://doi.org/10.3390/catal13040655.
3. Taran, O.P.; Miroshnikova, A.V.; Baryshnikov, S.V.; Kazachenko, A.S.; Skripnikov, A.M.; Sychev, V.V.; Malyar, Y.N.; Kuznetsov, B.N. Reductive Catalytic Fractionation of Spruce Wood over Ru/C Bifunctional Catalyst in the Medium of Ethanol and Molecular Hydrogen. *Catalysts* **2022**, *12*, 1384. https://doi.org/10.3390/catal12111384.
4. Kustov, L.M.; Kalenchuk, A.N. Effect of Cr on a Ni-Catalyst Supported on Sibunite in Bicyclohexyl Dehydrogenation in Hydrogen Storage Application. *Catalysts* **2022**, *12*, 1506. https://doi.org/10.3390/catal12121506.
5. Dutra, R.C.; Martins, T.V.S.; Rocha, D.d.G.; Meneghetti, M.R.; Meneghetti, S.M.P.; Sulman, M.G.; Matveeva, V.G.; Suarez, P.A.Z. Doped Ruthenium/Hypercrosslinked Polystyrene (HPS) Catalysts in the Modification of Fatty Acid Methyl Esters. *Catalysts* **2023**, *13*, 630. https://doi.org/10.3390/catal13030630.
6. Pryston, D.B.d.A.; Martins, T.V.d.S.; Vasconcelos Júnior, J.A.d.; Avelino, D.O.d.S.; Meneghetti, M.R.; Meneghetti, S.M.P. Investigation of CeO_2, MoO_3, and $Ce_2(MoO_4)_3$, Synthesized by the Pechini Method, as Catalysts for Fructose Conversion. *Catalysts* **2023**, *13*, 4. https://doi.org/10.3390/catal13010004.

7. Martins, T.V.d.S.; Pryston, D.B.d.A.; Meneghetti, S.M.P.; Meneghetti, M.R. Influence of Synthesis Methodology on the Properties and Catalytic Performance of Tin, Niobium, and Tin-Niobium Oxides in Fructose Conversion. *Catalysts* **2023**, *13*, 285. https://doi.org/10.3390/catal13020285.
8. Mukhtarova, M.; Golubeva, M.; Sadovnikov, A.; Maximov, A. Guaiacol to Aromatics: Efficient Transformation over In Situ-Generated Molybdenum and Tungsten Oxides. *Catalysts* **2023**, *13*, 263. https://doi.org/10.3390/catal13020263.
9. Shivtsov, D.M.; Koskin, A.P.; Stepanenko, S.A.; Ilyina, E.V.; Ayupov, A.B.; Bedilo, A.F.; Yakovlev, V.A. Hydrogen Production by N-Heterocycle Dehydrogenation over Pd Supported on Aerogel-Prepared Mg-Al Oxides. *Catalysts* **2023**, *13*, 334. https://doi.org/10.3390/catal13020334
10. Philippov, A.; Nesterov, N.; Martyanov, O. Transfer Hydrogenation of Biomass-Like Phenolic Compounds and 2-PrOH over Ni-Based Catalysts Prepared Using Supercritical Antisolvent Coprecipitation. *Catalysts* **2022**, *12*, 1655. https://doi.org/10.3390/catal12121655.
11. Sorokina, S.A.; Kuchkina, N.V.; Grigoriev, M.E.; Bykov, A.V.; Ratnikov, A.K.; Doluda, V.Y.; Sulman, M.G.; Shifrina, Z.B. Cr-Zn/Ni-Containing Nanocomposites as Effective Magnetically Recoverable Catalysts for CO_2 Hydrogenation to Methanol: The Role of Metal Doping and Polymer Co-Support. *Catalysts* **2023**, *13*, 1. https://doi.org/10.3390/catal13010001.

References

1. Vojvodic, A.; Nørskov, J.K. New Design Paradigm for Heterogeneous Catalysts. *Nat. Sci. Rev.* **2015**, *2*, 140. [CrossRef]
2. Chakraborty, D.; Mullangi, D.; Chandran, C.; Vaidhyanathan, R. Nanopores of a Covalent Organic Framework: A Customizable Vessel for Organocatalysis. *ACS Omega* **2022**, *7*, 15275. [CrossRef] [PubMed]
3. Rudolph, M.A.; Isbrücker, P.; Schomäcker, R. Bifunctional Catalysts for the Conversion of CO_2 intoValue-Added Products—Distance as a Design Parameter for New Catalysts. *Catal. Sci. Technol.* **2023**, *13*, 3469. [CrossRef]
4. Natarajan, S.; Manna, K. Bifunctional MOFs in Heterogeneous Catalysis. *ACS Org. Inorg. Au* **2024**, *4*, 59. [CrossRef] [PubMed]
5. Bogdan, V.I.; Kalenchuk, A.N.; Chernavsky, P.A.; Bogdan, T.V.; Mishanin, I.I.; Kustov, L.M. Synergistic Effect of Metal Components of the Low-loaded Pt-Ni-Cr/C Catalyst in the Bicyclohexyl Dehydrogenation Reaction. *Int. J. Hydrogen Energy* **2021**, *46*, 14532. [CrossRef]
6. Yuan, Y.; Yang, Y.; Zhu, G. Multifunctional Porous Aromatic Frameworks: State of the Art and Opportunities. *EnergyChem* **2020**, *2*, 100037. [CrossRef]
7. Huang, J.; Turner, S.R. Hypercrosslinked Polymers: A Review. *Polym. Rev.* **2017**, *58*, 1. [CrossRef]
8. Bounoukta, C.E.; Megías-Sayago, C.; Ivanova, S.; Ammari, F.; Centeno, M.A.; Odriozola, J.A. Pursuing Efficient Systems for Glucose Transformation to Levulinic acid: Homogeneous vs. Heterogeneous Catalysts and the Effect of Their Co-Action. *Fuel* **2022**, *318*, 123712. [CrossRef]

catalysts

MDPI

Review

Recent Progress in Pd-Catalyzed Tandem Processes

Linda Z. Nikoshvili * and Valentina G. Matveeva *

Department of Biotechnology, Chemistry and Standardization, Tver State Technical University, A. Nikitina Street, 22, 170026 Tver, Russia
* Correspondence: nlinda@science.tver.ru (L.Z.N.); matveeva@science.tver.ru (V.G.M.);
Tel.: +7-904-005-7791 (L.Z.N.); +7-905-601-0144 (V.G.M.)

Abstract: In recent years, Pd-containing catalytic systems for tandem processes have gained special attention due to their enhanced catalytic properties and their possibility of performing several reactions without the necessity of separating the intermediates. In this review, recent progress in Pd-catalyzed tandem processes is considered. Three types of catalytic systems are described: homogeneous catalysts (including immobilized Pd complexes); heterogeneous catalysts supported on oxides, MOFs, COFs, etc., with particular attention to the supports containing acid/base sites; and metal-enzyme catalysts for chemoenzymatic tandem processes applied in fine organic synthesis and biotechnology. For homogeneous Pd-catalyzed reactions, different tandem reactions were considered, i.e., cross-coupling, cyclization, carbonylation, isomerization, alkylation, arylation, etc.

Keywords: tandem processes; one-pot processes; palladium; bifunctional catalysts; bimetallic catalysts; chemoenzymatic catalysis

1. Introduction

Multiple catalytic processes in one reactor are of high importance due to the time–space economy and global tendency towards greener sustainable processes [1]. However, at present, there are many misinterpretations in the terminology used. As was mentioned by Camp et al. [2], such processes have different names: single-pot catalysis, one-pot catalysis, dominocatalysis, dual catalysis, tandem-catalysis, and multifaceted catalysis. Many papers use the term "tandem" for different processes proceeding one after another. Fogg and dos Santos [3] defined the one-pot catalytic processes that are not tandem catalysis as bicatalytic reactions, in which the second catalyst is added after the first one, which completes the reaction. When all the catalytic species are present in the reactor together, it can be defined as either domino or tandem catalysis. If all the reagents and catalysts are simultaneously present in the reactor, and the functionality formed in the previous step undergoes subsequent transformation, such a process can be classified as domino catalysis. It is important to emphasize that in many cases, domino catalysis is equated to the tandem catalysis, while Fogg and dos Santos [3] indentified of the domino/cascade catalysis as the multiple transformations proceeding via a single catalytic mechanism (Figure 1).

In the instance of tandem catalysis, the substrate is sequentially transformed through a number of mechanistically distinct mechanisms. Moreover, tandem catalysis can be divided in orthogonal, assisted, and auto-tandem catalysis [3]. The difference between auto-tandem and assisted tandem processes is that in the latter case, an additional compound is added in the reactor, switching the second mechanistically distinct reaction. As was mentioned by different authors [4–6], auto-tandem catalysis is an effective method for generating complex molecules from basic starting materials since it can support several mechanistically distinct reactions in a single reactor. When sequential C–C bond-forming processes take place in the reactions, the adoption of the auto-tandem catalysis technique has a very significant effect [6]. In contrast to auto-tandem catalysis, a linear synthetic method would not be able to produce unstable or sensitive organic intermediates in situ. Although auto-tandem processes have many benefits, it might be challenging to control them. Moreover, interference from side reactions might make them more difficult, particularly when the optimal conditions for certain catalytic cycles differ from one another [4].

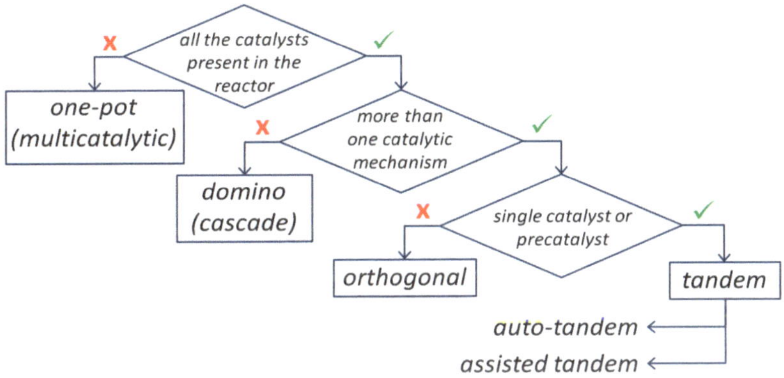

Figure 1. Distinguish between one-pot, domino, and tandem processes (based on [3]).

Tandem catalysis is a powerful tool for the production of a wide range of organic compounds by means of a variety of synthetic transformations, among which are arylation, alkylation, cyclization, cycloaddition, carbonylation, cross-coupling, amination, isomerisation, and other processes. Different metals such as Co [7], Cu [8], Ag [9,10], Au [11], Ni [12], Ru [13], Pt [14,15], and Pd [16] in the composition of both homogeneous [17,18] and heterogeneous [19] systems can catalyze tandem reactions.

Moreover, bimetallic catalytic systems (Cu-Zn [20], Cu-Ni [21], Pd-Cu [22,23], Au-Pd [24,25], Ir-Pd [25], Pd-Ru [26,27], Pd-Rh [28], Ag-Pd [29], Pd-Ni [30], Au-Cu [31], Cu-Ag [10,32], Cu-Bi [10], etc.) can be also used.

Among the transition metals applied for tandem processes, palladium is one of the most widely used. Thus, this review is devoted to the recent advances (for the last five years) in tandem reactions catalyzed by Pd complexes and nanoparticles (NPs), including those catalyzed by bimetallic systems and metal-enzyme catalysts.

2. Pd-Catalyzed Homogeneous Tandem Catalysis

At present, homogeneous Pd-catalyzed processes, including the tandem ones, are the most widespread in fine organic synthesis for the production of active pharmaceutical ingredients (APIs) and polymers.

2.1. Coupling Tandem Processes

Palladium is known as a catalyst for the cross-coupling processes [33,34]. Both sequential and iterative Pd-catalyzed reactions can be carried out in tandem cross-coupling reactions, and the order of C–C bond forms can be regulated by either the attenuated leaving groups of the multireactive substrate or certain catalyst/ligand combinations [35–38]. Both the nucleophilic and electrophilic sites may be coupled in a specific manner. Additionally, in addition to halogen and metal leaving groups, the C–H bond can be used as an appealing leaving group [35].

In the catalytic cycle of cross-coupling reactions, palladium constantly changes its oxidation state from Pd(0) to Pd(II) and vice versa, hence the process can be catalyzed by any form of the metal [39,40]. Lamb et al. [41] employed Pd(II) to catalyze in the one-pot process the dehydrogenation of 2,2-disubstituted cyclopentane-1,3-diones and the oxidative Heck coupling (Figure 2). It was postulated that Pd(II) was an active form, which underwent partial decomposition during the reaction if the ligand loading was unsufficient.

However, it was difficult to create the ideal one-pot conditions since the optimal conditions for the dehydrogenation stage were inappropriate for the oxidative Heck process.

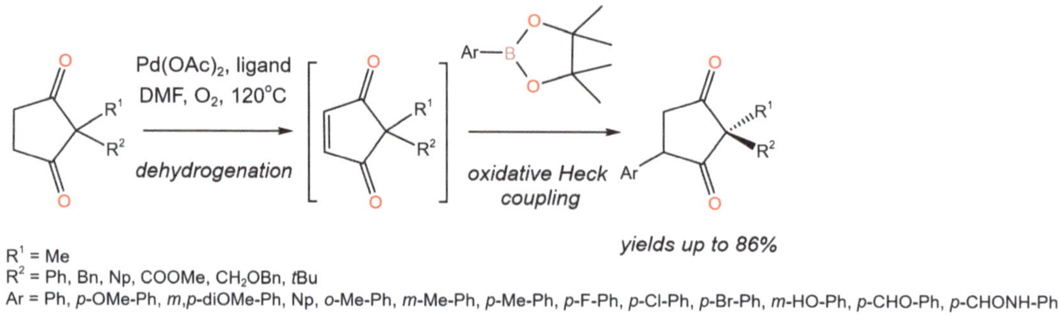

Figure 2. Scheme of the dehydrogenation of 2,2-disubstituted cyclopentane-1,3-diones and the oxidative Heck coupling in a one-pot process [41].

By the example of Np-substituted cyclopentane-1,3-dione, it was shown that due to the presence of unligated Pd, which was likely formed during the aerobic dehydrogenation stage, the established one-pot reaction allowed for the achievement of a reasonable yield of 60% at moderate enantioselectivity at 74:26 e.r. In order to study the telescoped reaction, the reaction mixture was filtered to remove any unligated Pd before the cross-coupling partner was added. Such treatment enabled the e.r. to be comparable to the separate oxidative Heck step (88:12 vs. 90:10 e.r.) with agood 70% yield [41].

The improved synthesis of indolines via tandem decarboxylative amination/Heck/annulation reaction was carried out by Wang et al. [42] (Figure 3).

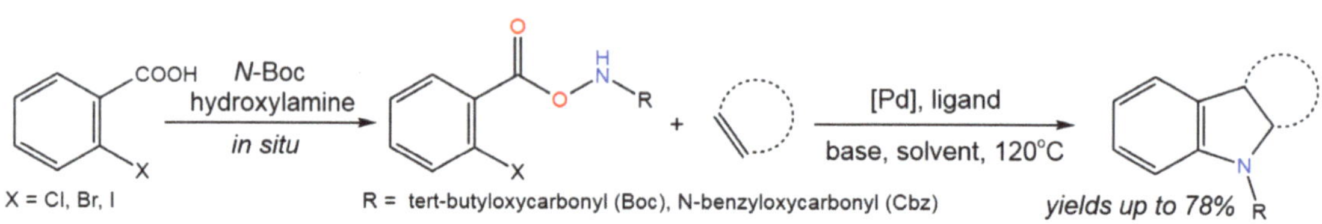

Figure 3. Scheme of the synthesis of indolines via tandem decarboxylative amination/Heck/annulation reaction [42].

Different palladium compounds ($Pd(OAc)_2$, $Pd(TFA)_2$,$Pd_2(dba)_3$, $PdCl_2$) and ligands ((t-Bu)$_3$P·BF_4, Ph3P, etc.) were used. $PdCl_2$ and (t-Bu)$_3$P·BF_4 revealed the best performance. Among the bases (Cs_2CO_3, K_2CO_3, t-BuOK, DBU, and Et_3N), Cs_2CO_3 was optimum. Moreover, different solvents were used (benzene, toluene, xylene, MeCN, $CHCl_3$, and DMF), and benzene was the most preferable one. Toluene was the second in efficiency allowing up to 70% yield of target product. In general, depending on the halogen, protecting group, substituent, and nature of the second substrate, as well as on the reaction conditions, up to 78% yield of the target product was achieved [42]. The reaction mechanism was proposed to proceed via formation of the intermediate of an intramolecular decarboxylation process. The cyclization product was then synthesized by reacting this intermediate with norbornadiene using a tandem Heck-type reaction and nitrogen nucleophilic route. The reaction followed a similar tandem Heck/intramolecular Tsuji–Trost pathway when the 1,3-diene substrate was used [42].

Song et al. [43] developed the synthesis of 2-(1-phenylvinyl)-indoles via the novel Pd(0)-catalyzed intermolecular coupling reaction of 2-*gem*-dibromovinylanilines and *N*-tosylhydrazones (Figure 4). Reaction conditions were optimized, ($PdCl_2(PPh_3)_2$ was chosen as a catalyst, and tBuOLi was chosen as a base (the other bases were DMAP, tBuOK,

TEA, $CsCO_3$, and CsF)). Using 2-*gem*-dibromovinylanilines as a substrate, the scope of *N*-tosylhydrazones was studied. The reaction products were obtained in moderate to good yields (from 42% up to 94%), depending on the substituents: electrondonating (*p*-OMe and -OCH_2O-) or electronwithdrawing (NO_2 and CN). The substitutions on 2-*gem*-dibromovinylaniline were also studied (-CO_2Me, Br, F, Ns, Ph, 4-F-Ph, 4-Cl-Ph, 4-OMe-Ph, 4-*t*Bu-Ph, 4-Me-Ph, *N*-Ac, *N*-Ms), giving good yields (>80% in most of cases).

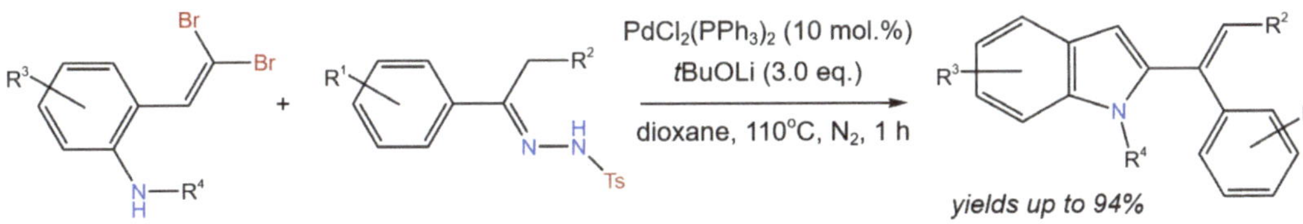

R^1 = H, 3-OMe, 4-OMe, 3,4-(OCH_2O), 4-NO_2, 4-CN, 4-F, 4-Cl, 4-Br, 3-Br
R^2 = H, Me, Et, Ph
R^3 = H, 6-CO_2Me, 5-Br, 6-F, 5-Ph, 5-(*p*-F-Ph), 5-(*p*-Cl-Ph), 5-(*p*-OMe-Ph), 5-(*p*-*t*Bu-Ph), 5-(*p*-Me-Ph) (position is indicated in the indole residue)
R^4 = Ts, Ac, Ms, Ns

Figure 4. Scheme of the tandem reaction of 2-*gem*-dibromovinylanilines and *N*-tosylhydrazones for the synthesis of 2-(1-phenylvinyl)-indoles [43].

The reaction mechanism was also proposed, which included two catalytic cycles, one of which was started from Pd(0) and included oxidative addition, deprotonation, and reductive elimination, resulting in theformation of 2-bromo-1-tosyl-1*H*-indole. The latter entered the second cycle that also started from an oxidative addition to Pd(0) with the formation of the Pd(II)complex. Subsequently, alkylpalladium species were formed with carbene intermediate generated from the diazo compound. Finally, β-H elimination allowed for the synthesis of the desired product and the regeneration of Pd(0) [43].

In situ-produced vinyl fluorosulfate intermediates were coupled with electron-deficient olefin partners (Figure 5) by Revathi et al. [44] to create a protocol for the one-pot conversion of alcohols to 1,3-dienes. The use of DMSO as a solvent was mandatory, due to the involvement in the reaction mechanism. Broad substrate scopes of alcohols were examined for the SO_2F_2-mediated dehydrative cross-coupling. Different starting compounds containing both electron-donating groups (alkyl, aryl, trifluoromethoxy, methoxy, ether, thioether) and electron-withdrawing groups (nitro, trifluoromethyl) including halogens were investigated. Corresponding products were synthesized with the yields varying from 40% up to 99%. Moreover, the substrate scope of electron-deficient olefin coupling partners was studied. The best results were obtained with acrylamide, which provided the corresponding product at a 82% yield, and ethyl vinyl ketone, which produced the corresponding 1,3-diene at a 88% yield [44].

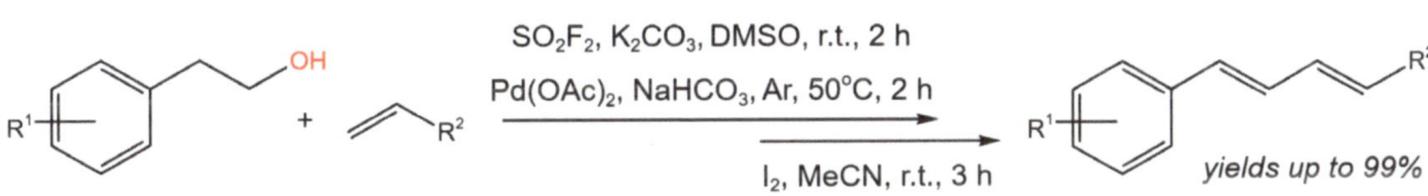

R^1 = H, 4-Ph, 2-Me, 3-Me, 4-Me, 2,5-diMe, 4-OMe, 3,5-diOMe, 3,4,5-triOMe, 4-*t*Bu, 4-CF_3, 3-OCF_3, 4-OCF_3, 2-OBn, 3-OBn, 4-SMe, 4-NO_2, 4-F, 3-Cl, 4-Cl, 3,4-diCl, 2-Br, 3-Br, 4-Br, 4-I, 2,3-benzo, 3,4-benzo
R^2 = COOMe, COOEt, COOPh, $COOCH_2Ph$, COEt, $CONH_2$

Figure 5. Scheme of dehydrative cross-coupling for the conversion of alcohols to 1,3-dienes [44].

The C–C/C–N bond formation in the reaction between *N*-methyl benzamide and arylboronic acid was catalyzed by the dinuclear palladium(II) complex, resulting in the direct synthesis of phenanthridinone (Figure 6) [45]. A broad range of phenanthridinones was synthesized in good to excellent yields (from 79% up to 96%) using 1 mol.% of the catalyst and a variety of substrates, which also included condensed aromatics and heterocycles (in the case of arylboronic acid).

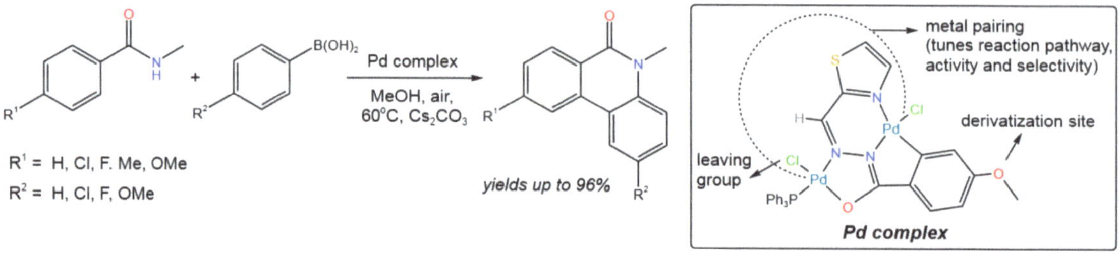

Figure 6. Scheme of the synthesis of phenanthridinones [45].

It was proposed that the reaction proceeds via *ortho*-arylation [45]. First, *N*-methyl arylamide was coordinated with the catalyst along with the *ortho*-C–H bond activation. Then, the transmetalation reaction occurred with participation of an arylboronic acid, leading to the formation of an *ortho*-arylated intermediate, which subsequently resulted in the seven-membered palladacycle. In this process, one of the two Pd(II) nuclei was transformed to Pd(0). Finally, the reductive elimination led to the desired product, while Pd(0) was oxidized by air.

The Pd pincer complex supported by 2,6-bis(pyrrolyl)pyridine ligands was shown [46] to be an efficient catalytic system for one-pot tandem Heck alkynylation (copper-free Sonogashira coupling)/cyclization reaction) (Figure 7). It was shown that a wide range of benzofuran derivatives can be obtained at 90 °C using 0.1 mol.% of catalyst loading for 10 h with high yields (up to 96%).

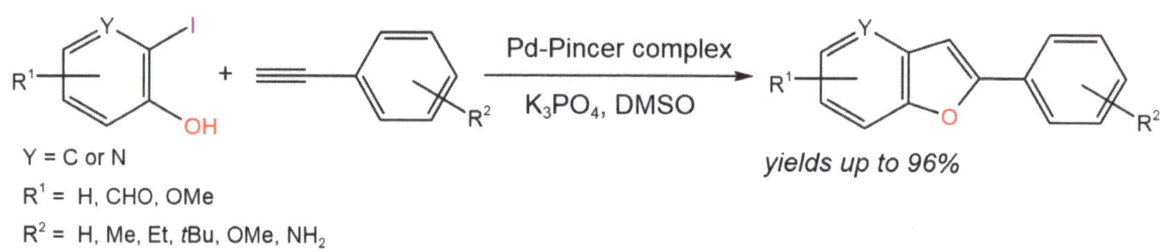

Figure 7. Scheme of the synthesis of benzofuran derivatives via the Heck alkynylation/cyclization reaction [46].

The reaction mechanism included two cycles (Heck alkynylation cycle and intramolecular cyclization cycle) associated with the product of Heck alkynylation, which underwent the cyclization process. It is noteworthy that the first cycle was likely initiated by Pd(II)/Pd(IV) transformations instead of the most common Pd(0)/Pd(II) chemistry [46].

In a similar way, microwave-assisted, domino [Pd]-catalyzed Heck cyclization followed by intermolecular Sonogashira coupling was carried out by Karu and Gedu [47] for the synthesis of substituted dihydrobenzofurans with yields of up to 99% (Figure 8). The reaction was proposed to proceed via a single catalytic cycle including classical Pd(0)/Pd(II) transformations. Initiated by Pd(0), the cyclization took place first, followed by Sonogashira coupling.

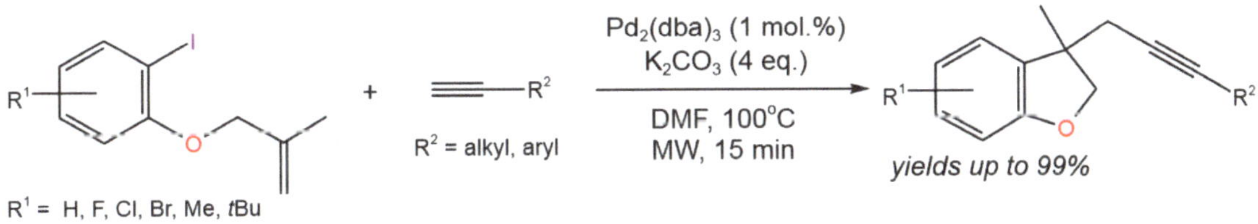

Figure 8. Scheme of the synthesis of substituted dihydrobenzofurans via Heck cyclization followed by Sonogashira coupling [47].

Such a kind of substituted heterocycles, containing a triple carbon–carbon bond, can be a subject of further transformations. For example, Ho et al. [48] developed the one-pot process for the dearomatising spirocyclization/cross-coupling of indole/pyrrole derivatives (Figure 9). In the proposed protocol the use of $Pd(PPh_3)_4$or *trans*-PdBr(*N*-succinimide)$(PPh_3)_2$ allowed up to 97% yields of target products. Palladium complexes were proposed to play a dual role: (i) π-acid to activating the alkyne towards dearomatising spirocyclization, and (ii) cross-coupling catalysts.

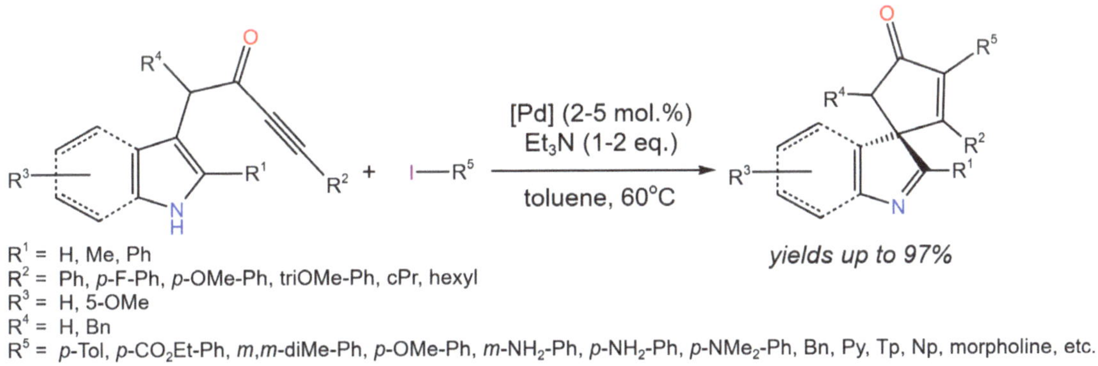

Figure 9. One-pot dearomatising spirocyclisation/cross-coupling [48].

Moreover, a variety of the spirocyclic pyrroline-based compounds, which are important scaffolds in drug development, could be produced with the yields up to 80% via Pd-catalyzed tandem Narasaka–Heck/C(sp^3 or sp^2)-H activation reaction [49] with γ,δ-unsaturated oxime esters as starting materials via the five-membered spiro-palladacycle intermediate.

Functionalized benzofurans, indoles, and phthalanes can be also synthesized by the tandem Ullmann–Goldberg cross-coupling and cyclopalladation-reductive elimination reactions, as well as by the related Pd-catalyzed processes involving hetero-Michael additions and cyclization in a one-pot process [50].

The tandem Heck/Suzuki reaction can be applied for the enantioselective intramolecular cyclization/cross-coupling of olefin-tethered aryl halides with various organoboronic acids [51]. Under the optimized reaction conditions, a number of dihydrobenzofuran derivatives were synthesized (Figure 10) with high yields (up to 99%) and e.r.s (up to 97:3). Moreover, by the same approach, different indolines, chromanes, and indanes were synthesized with yields of up to 99% and high e.r.s, depending on the substituents.

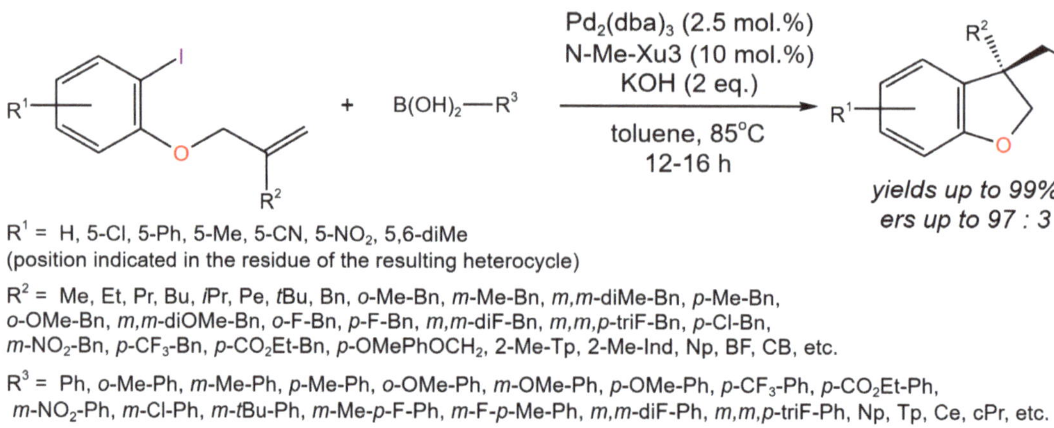

Figure 10. Scheme of the tandem Heck/Suzuki process for the synthesis of chiral compounds by the example of dihydrobenzofurans bearing a quaternary stereocenter [51].

Yokoya et al. [52] suggested fabricating the benzo[*de*]chromene ring followed by its oxidation in order to produce benzo[*de*]chromene-7,8-dione derivatives, which are known to have significant biological and pharmacological effects (Figure 11). Pd- and Cu-catalyzed Sonogashira coupling and cyclization was applied to produce benzo[*de*]chromene-7,8-dione derivatives with yields of up to 69%. It was found that the coupling process was catalyzed by Cu and Pd complexes, according to the established mechanism, while the further cyclization was catalyzed by CuI alone.

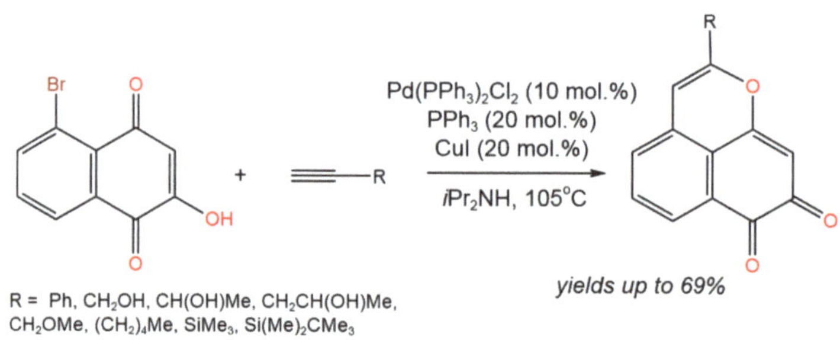

Figure 11. Tandem Sonogashira coupling and intramolecular cyclization with terminal alkynes for the synthesis of benzo[*de*]chromene-7,8-dione derivatives [52].

For the one-pot synthesis of alkenyl-substituted boron dipyrromethene (BODIPY) from Cl-BODIPY and alkyne, the coupling-reduction tandem process (Figure 12) was reported [53]. In comparison to traditional approaches, this synthesis enabled higher yields (up to 80%), a wider range of substrates, and a faster reaction rate by combining the Sonogashira coupling and the reduction process, avoiding the use of a reductant.

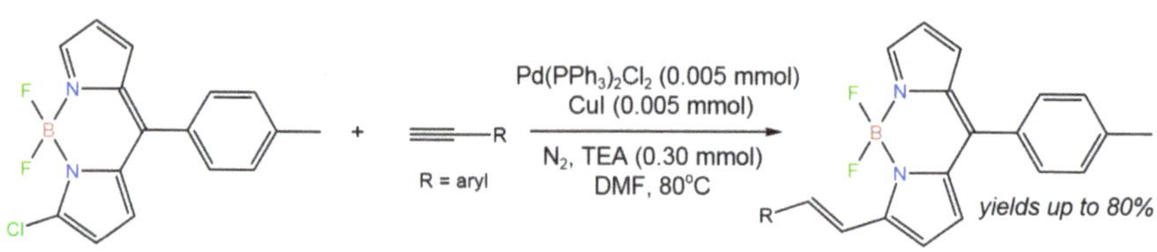

Figure 12. Synthesis of alkenyl-substituted BODIPYs [53].

One of the reactions attributed to the tandem Pd-catalyzed catalytic processes is the synthesis of imidazole derivatives [54–56]. The tandem synthesis of imidazole-fused polyheterocycles (Figure 13) from 2-vinyl imidazoles and aryl halides was reported by Li et al. [54] and involved intermolecular Heck arylation of 2-vinyl imidazoles, followed by an intramolecular aerobic oxidative C–H amination reaction.

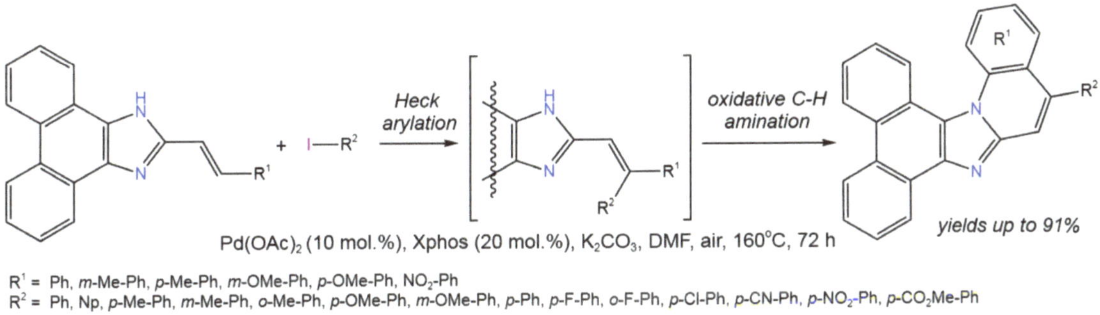

Figure 13. Pd-catalyzed annulation of phenanthroimidazoles with aryl iodides for the synthesis ring-fused phenanthroimidazoles [54].

In this tandem reaction, the Pd(0)↔Pd(II) catalytic cycle was used to simultaneously break two or three C–H bonds, one C–X bond, and one N–H bond. Different catalysts were used ($Pd(OAc)_2$, $PdCl_2$, $Pd(PPh_3)_4$, $Pd_2(dba)_3$), among which the palladium acetate allowed the highest yields of the target products (up to 91%). Potassium carbonate was chosen as an optimum base (other bases were Cs_2CO_3, NaO*t*Bu, NaOH, and Et_3N). DMF was shown to be the most preferable solvent as compared to DMA, DMSO, toluene, and 1,4-dioxane [54].

A tandem reaction involving boronic acids and 2-(bromobenzylsulfenyl)-1-propargyl benzimidazoles catalyzed by Pd was developed [55] (Figure 14). This process involved three different reactions: (i) Suzuki debrominative cross-coupling; (ii) Cu-free desulfenylative coupling at the 2-position of the benzimidazole; and (iii) intermolecular regioselective and stereoselective hydrothiolation of the triple C–C bond of the *N*-propargyl benzimidazole.

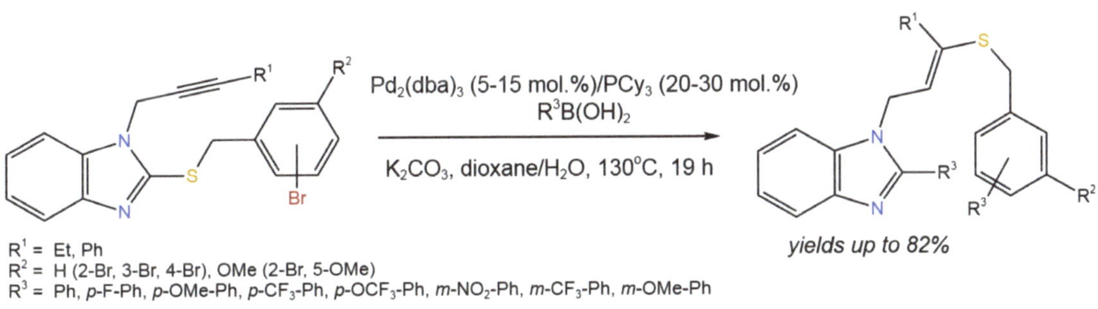

Figure 14. Tandem Pd-catalyzed reaction between boronic acids and 2-(bromobenzylsulfenyl)-1-propargyl benzimidazoles for the synthesis of substituted benzimidazoles bearing a stereodefined alkenyl sulfide [55].

The whole procedure resulted in an alkenyl benzyl sulfide and doubled the amount of boronic acid partner that was incorporated into the final structure. The developed tandem process proceeded with moderate up to high yields (82%), depending on the substituents and ligands. It is noteworthy that the heterocycles other than benzimidazole (i.e., imidazole and indole) were studied, and $Pd(PPh_3)_4$ (15 mol.%) was found to provide higher yields of target products as compared to $Pd_2(dba)_3/PCy_3$ [55].

Luo et al. [56] developed the cross-coupling reaction of isocyanides with α-diazoacetates catalyzed by palladium to form ketenimines, which were subsequently underwent the DABCO-catalyzed aza-Mannich type reaction (Figure 15) under mild reaction conditions (80 °C, CH_3CN—solvent). Different 2-mercaptoimidazoles or 1*H*-benzo[*d*]imidazol-2-oles were transformed to 1,3-bis(β-aminoacrylate)-substituted 2-mercaptoimidazoles and 2-benzimidazolinones with the yields up to 68%.

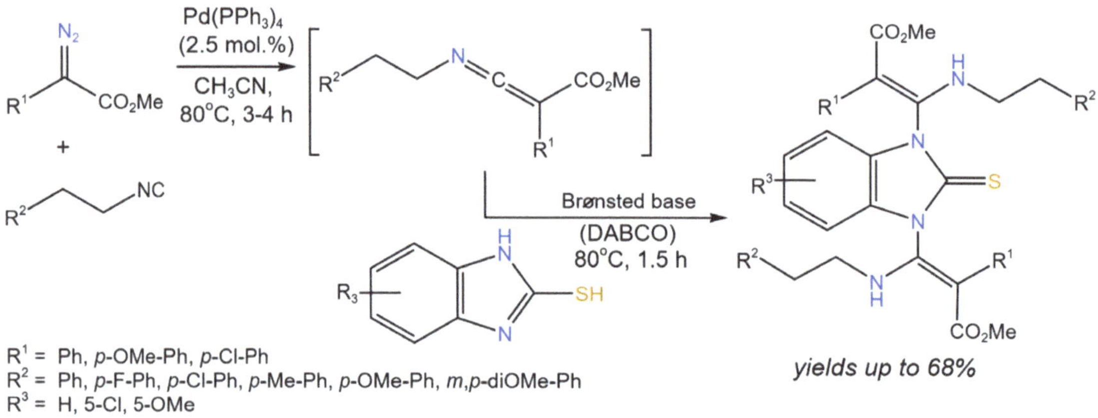

Figure 15. Scheme of the synthesis of 1,3-bis(β-aminoacrylate)-substituted 2-mercaptoimidazole derivatives synthesized in one pot using sequential Pd and Brønsted base catalysis [56].

The proposed reaction mechanism involved two cycles: (i) Pd-catalyzed, and (ii) DABCO-catalyzed. The starting isocyanide compound was coordinated with the palladium catalyst. Then, the diazo compound reacted with the formation of the Pd-carbene intermediate and the release of nitrogen. After that, the resulting intermediate produced ketenimine with regeneration of the catalyst. Ketenimine was trapped by one of the NH groups in DABCO-activated 2-mercaptoimidazole. The intermediate of this step reacted with ketenimine to form the target product [56]. Choi et al. [57] reported the tandem Diels–Alder/cross-coupling (either Stille or Suzuki) process using 2-bromo-1,3-butadienes as substrates, which underwent intermolecular cycloaddition with a variety of activated dienophiles in the presence of Lewis acids (e.g., $BF_3 \cdot OEt_2$, $TiCl_4$, $SnCl_4$). The reaction proceeded with high yields (>70%) and good endo diastereoselectivity with respect to vinyl bromide cycloadducts, which further underwent cross-couplings in the presence of Pd catalysts ($Pd_2(dba)_3$, $Pd(Ph_3P)_4$).

2.2. Isomerization Tandem Processes

Isomerization tandem reactions catalyzed by Pd are possible due to the simultaneous existence of two oxidation states of this metal. Thus, there is double terminology for these reactions: some authors attribute them to orthogonal catalysis [58], othersto tandem catalysis [59]. Arroniz et al. [58] demonstrated the tandem orthogonal process of the conversion of alkynyl epoxides to furans (Figure 14). The developed procedure allowed for high yields (up to 93%) of furan derivatives.

It was found that the reaction proceeded through the formation of allenyl ketone and β,γ-alkynyl ketone as intermediates. The prposed reaction mechanism involved the oxidative addition of Pd(0) into the alkynylepoxide, resulting in the formation of allenylpalladium(II) species, which underwent reductive elimination to enol (shown in brackets in Figure 16). The latter tautomerized to alkynyl ketone and to allenyl ketone. The furanylpalladium complex was formed by the Pd(II)-activated cyclization of allenyl ketone. After the protodepalladation, the target furan's derivatives were produced under mildly acidic conditions. Moreover, a possible alternative mechanism for the conversion of the alkyne to the furan was suggested [58].

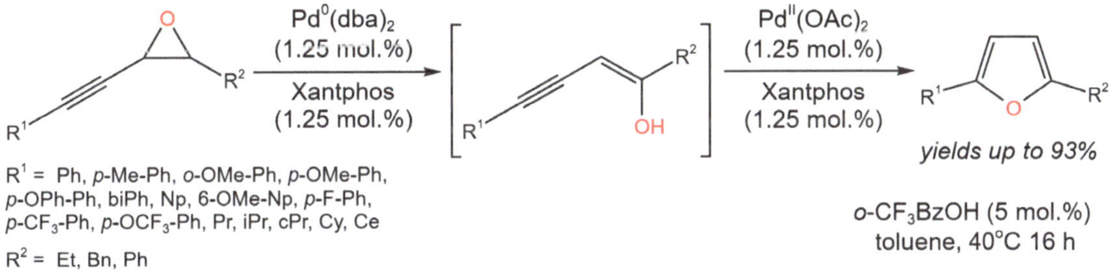

Figure 16. Dual catalytic isomerization of alkynyl epoxides for the synthesis of furans [58].

The synthesis of furans and pyrroles was carried out via the Pd-catalyzed tandem addition/ring-opening/cyclization reaction of 2-(3-aryloxiran-2-yl)acetonitriles with arylboronic acids in aqueous medium (Figure 17) [59].

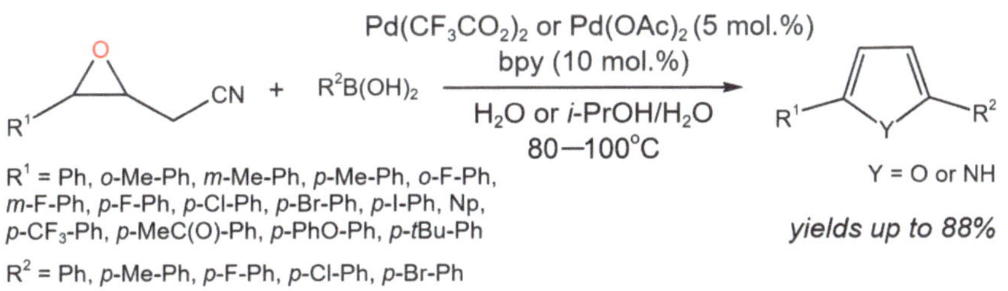

Figure 17. Tandem reaction of epoxyacetonitriles with arylboronic acids for the synthesis of furans and pyrroles [59].

Yu et al. [59] noticed that despite the advances in the Pd-catalyzed ring-opening of epoxides with nonpolar multiple bonds (such as vinyl and alkynyl), the reactions of boronic acids with epoxides with polar groups are less studied due to the lower reactivity.

As a result of the study, an efficient approach for the synthesis of furans and pyrroles with high selectivity was developed, allowing, depending on the catalyst and additives (*p*-toluenesulfonic acid, trifluoromethanesulfonic acid, D-camphorsulfonic acid, methanesulfonic acid, etc.), yields of up to 73% of furan derivatives ($R^1 = R^2$ = Ph) in aqueous medium at 80 °C under air. It is noteworthy that the replacement of air to the oxygen allowed for a noticeable increase in the yield of the target product (up to 86%). In the case of the synthesis of 2,5-diarylpyrroles at 100 °C under nitrogen using a*i*-PrOH/H_2O mixture as a solvent, up to 84% yield was achieved, depending on the substituents. Moreover, it was found that the steric effects of arylboronic acids primarily controlled the chemo-selectivity of the developed tandem process: when 2-(3-aryloxiran-2-yl)acetonitriles were reacting with the less sterically hindering arylboronic acids, the only products were furans due to the hydrolysis of the ketimine intermediate, arising from the carbopalladation of the cyano group, whereas when reacting with the more sterically bulky arylboronic acids, pyrroles were selectively produced [59].

Another one example of isomerization tandem processes is the tandem isomerization/hydrothiolation of allylbenzenes (Figure 18) [60]. In the case of $n = 0$, using $Pd(OAc)_2$ (3 mol.%) as a catalyst in the presence of 3.9 mol.% of the ligand and 12 mol.% of CF_3SO_3H, up to a 78% yield of the target product was able tobe achieved at 40 °C for 15 h, depending on the substituents. When $n = 1–3$ and R = H or 4-OMe, the one-pot process was carried out: (i) isomerization stage in the presence of $PdCl_2(PhCN)_2$ (5 mol.%) in toluene at 60 °C for 4 h; (ii) hydrothiolation stage in the presence of the ligand (10 mol.%), acid (22 mol.%), and thiol at 50 °C for 22 h, which allowed an up to 43% yield of the target product.

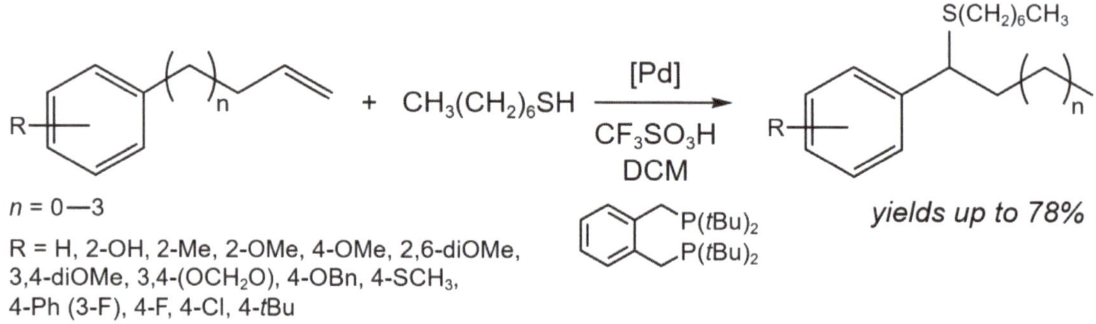

Figure 18. Tandem isomerization/hydrothiolation of allylarenes for the synthesis of benzylic thioethers [60].

It was demonstrated that the highly regioselective catalytic system composed of a Pd(II) precursor, bidentate phosphine ligand, and strong Brønsted acid may convert a wide range of 3-arylpropenes and thiols to branched benzylic thioethers. The reaction was facilitated by an in situ-generated Pd hydride [60].

2.3. Carbonylation Tandem Processes

Over 80% of all the APIs are known to have heterocyclic aromatic rings, with predominating N-containing aromatic heterocycles. Many APIs have the carbolinone ring system as a substructure [61]. Thus, carbonylation reactions are in high demand by the modern chemical industry. These reactions can be carried out in tandem with processes such as cross-coupling, dealkylation, and dehydrogenation.

A tetracyclic isoindoloindole skeleton with three new C–C/C–N bonds that was simultaneously generated, for instance, could be constructed in a one-step process using tandem Pd-catalyzed carbonylation and C–C cross-coupling via C–H activation (Figure 19) [62]. In more detail, the production of *6H*-isoindolo[2,1-*a*]indol-6-ones from commercially accessible substrates involved the carbonylation of aryl dibromides with indoles and the C–H activation of in situ-produced *N*-(2′-bromoaroyl)-indole. C–H activation likely occurred via conventional Pd(0)/Pd(II) transformations. It is noteworthy that glyoxylic acid monohydrate was used as an eco-friendly CO surrogate in the aminocarbonylation stage of the proposed tandem process. As a result, a number of *6H*-isoindolo[2,1-*a*]indol-6-ones were produced with yields of 40–64%.

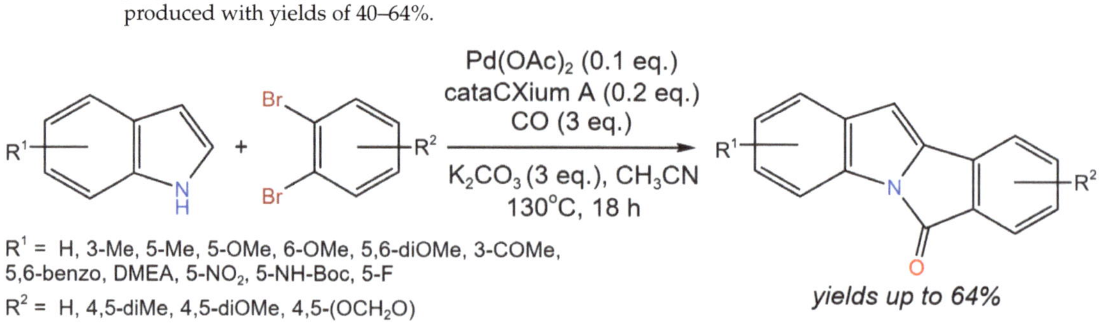

Figure 19. Tandem Pd-catalyzed carbonylation and C-C cross-coupling [62].

Li et al. [63] reported the enantioselective dearomative carbonylation of *N*-arylacyl indoles (Figure 20). The following reaction conditions were varied: Pd source ($Pd_2(dba)_3 \cdot CHCl_3$, $Pd(dba)_2$, $Pd_2(dba)_3$, $Pd(OAc)_2$, $Pd(TFA)_2$), chiral ligands, base additives (Et_3N, DIPEA, DABCO, K_2CO_3), and solvents (MTBE, MeCN, THF, Et_2O, PhOMe, toluene). Under the optimal conditions ($Pd_2(dba)_3$, Feringa ligand (*S,S,S*)-L1, Et_3N, MTBE) under the atmosphere of CO, the target products were formed yields of up to 89%, up to 97% ee, and dr > 20:1.

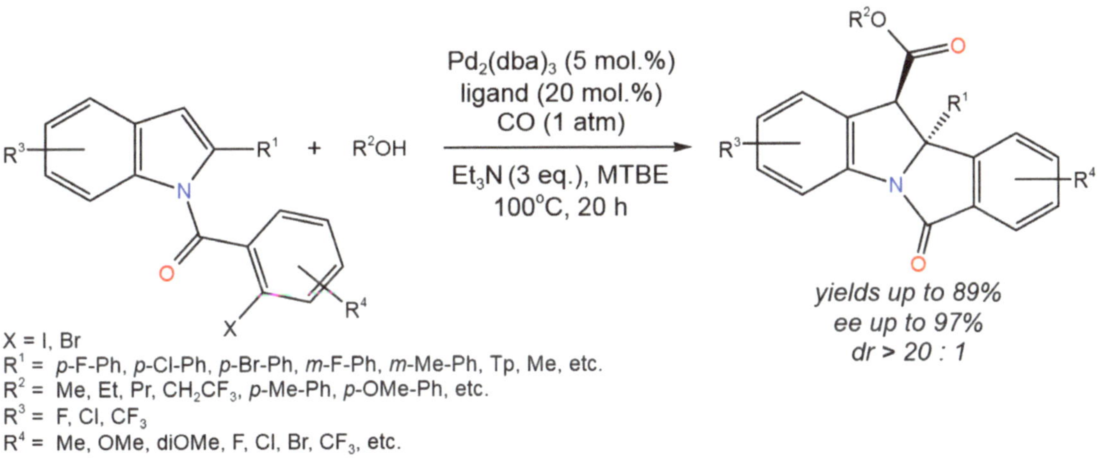

Figure 20. The scheme of thetandem Heck/carbonylation process for the asymmetric dearomatization of *N*-arylacyl indoles [63].

Wang et al. [64] developed the Pd-catalyzed multi-step tandem carbonylation/N-dealkylation/carbonylation reaction with alkyl as the leaving group and tertiary anilines as the nitrogen nucleophile, which allowed for the effective production of isatoic anhydride derivatives (Figure 21). Moderate to good (up to 80%) yields were obtained, depending on the substituents. Moreover, the developed approach was successfully used to synthesize Evodiamine—a biologically active alkaloid—with a 70% yield. The reaction mechanism involved two cycles. In the first cycle, the starting compound interacted with Pd(II) and CO. After the CO insertion and reductive elimination, the intermediate was formed, which then interacted with Cu(II) and O_2 and underwent C-N cleavage. Then, the resulting intermediate entered the second cycle and underwent another one CO insertion with the participation of Pd(II). After the reductive elimination and formation of the target product, the Pd(0) was oxidized by Cu(II) and O_2 to regenerate Pd(II).

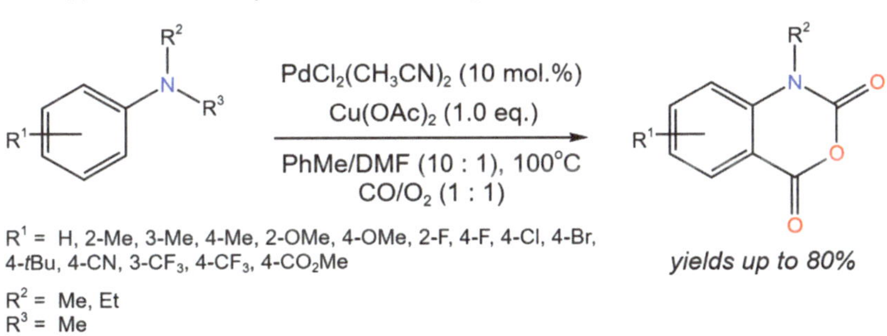

Figure 21. Tandem Pd-catalyzed carbonylation/N-dealkylation/carbonylation reaction of *N,N*-dialkyl anilines for the synthesis of isatoic anhydride derivatives [64].

For the synthesis of a variety of carbolinones under mild reaction conditions, Han et al. [61] proposed an effective tandem oxidative C–H aminocarbonylation and a dehydrogenation reaction co-catalyzed by Pd and Cu (Figure 22).

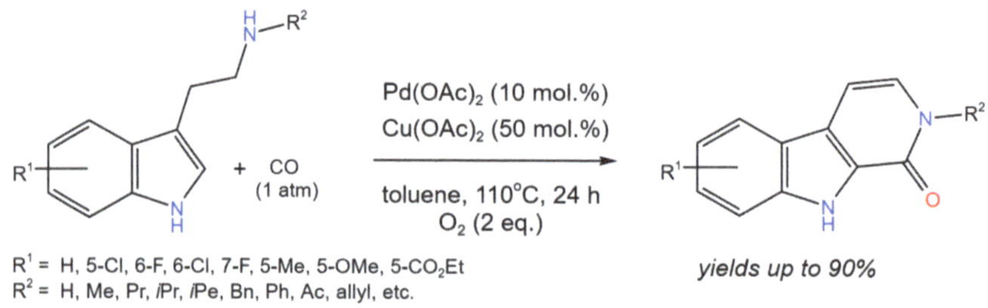

Figure 22. The scheme of the synthesis of β-carbolinones via the Pd/Cu co-catalyzed tandem C–H aminocarbonylation and dehydrogenation [61].

Carbolinones, tetrahydro-β-carbolinones, and tetrahydro-γ-carbolinoneswere selectively obtained with the yields of up to 90% by simply changing the reaction conditions. Strychnocarpine, a natural alkaloid, and its analogs were also produced.

The reaction proceeded through the well-known Pd(0)/Pd(II) transformations. The role of Cu(II) was in the oxidation of Pd(0) to regenerate Pd(II). In this process, Cu(II) was reduced to Cu(I), which was reoxidized with O_2.

A new P,O-hybrid ligand (L1) was developed by Zhao et al. [65] for the Pd-catalyzed bis-hydroaminocarbonylation of alkynes, allowing for the effective production of *N*-aryl substituted succinimides with the isolated yield of 57–90% (Figure 23) in the presence of methanesulfonic acid (MSA) as an additive. The catalyst L1-$Pd(CH_3CN)_2Cl_2$ system was reused for five runs without the precipitation of Pd-black.

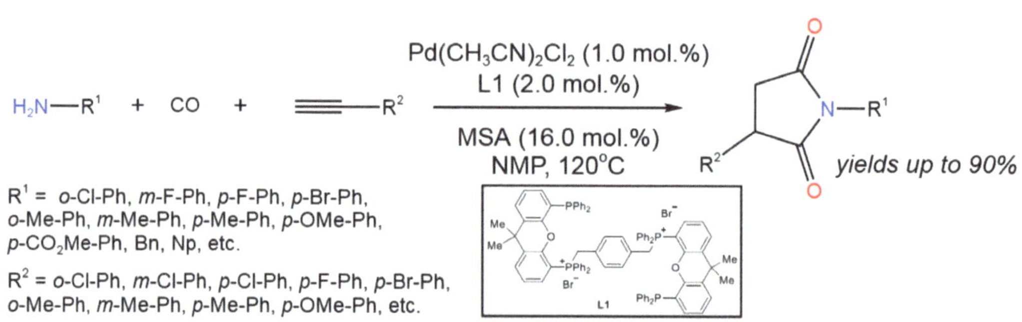

Figure 23. Pd-catalyzed bis-hydroaminocarbonylation of alkynes for the synthesis of *N*-aryl substituted succinimides [65].

One of the widespread reactions for the synthesis of cyclic compounds is the Diels–Alder reaction, which can be also combined with the carbonylation process [66]. Such an approach can provide a variety of functionalized carbocycles with complex architectures. Most of the methods of the synthesis of lactone-containing bridged polycyclic compounds are based on a stepwise protocol, restricting the overall scope and practicality of these transformations.

Therefore, the Pd-catalyzed tandem carbonylative Diels–Alder process was developed by Wang et al. [66] (Figure 24).

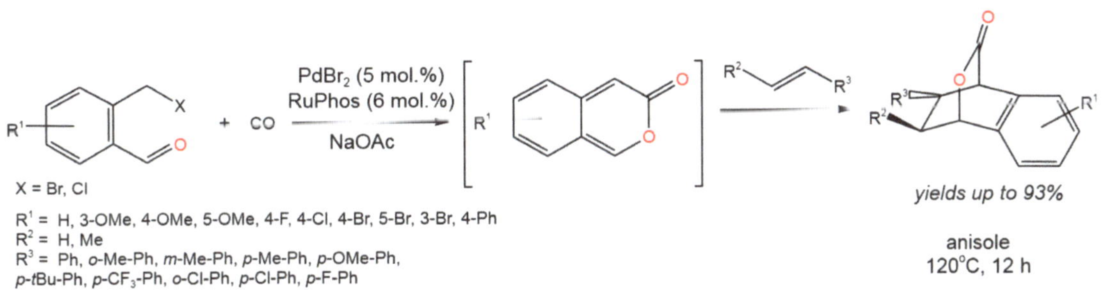

Figure 24. Scheme of the reaction of aldehyde-tethered benzylhalides and alkenes via the Pd-catalyzed tandem carbonylative lactonization and Diels–Alder cycloaddition [66].

It was proposed that the use of the aldehyde functionality as the reactive directing group resulted in good chemo- and stereoselectivity. Moderate to high (93%) yields were achieved, depending of the palladium source, base, and nature of substituents in the initial compounds [66]. The reaction was catalyzed by RuPhos-stabilized Pd(0), which was formed by the reduction of the Pd(II) precursor with CO or the ligand (RuPhos). Then, Pd(0) reacted with the substrate (benzaldehyde compound), producing an intermediate, which further underwent the insertion of CO. The subsequent reductive elimination resulted in the regeneration of Pd(0) and the formation of an intermediate, which underwent deprotonation due to interaction with NaOAc. Finally, the resulting benzopyran-2-one (shown in brackets in Figure 24) underwent [4 + 2]-cycloaddition with an alkene.

2.4. Cyclization Tandem Processes

Cyclization is another type of reaction widely used in fine synthesis [67–76]. Such reactions can proceed with the in situ-generated active species as a part of tandem catalytic transformations. For example, in order to produce EGFR (epidermal growth factor receptor) inhibitors, Ansari et al. [72,73] established a four-component procedure for the synthesis of pyrazolo[1,5-*c*]quinazolines with a small polar substitution of the pyrazole ring (Figure 25). In this tandem process, azomethine imine was generated in situ. Then, acetonitrile, having the electron-withdrawing group (EWG) at α-position (malononitrile, α-cyanocaboxyalate, orβ-ketonitrile), participated in the reaction.

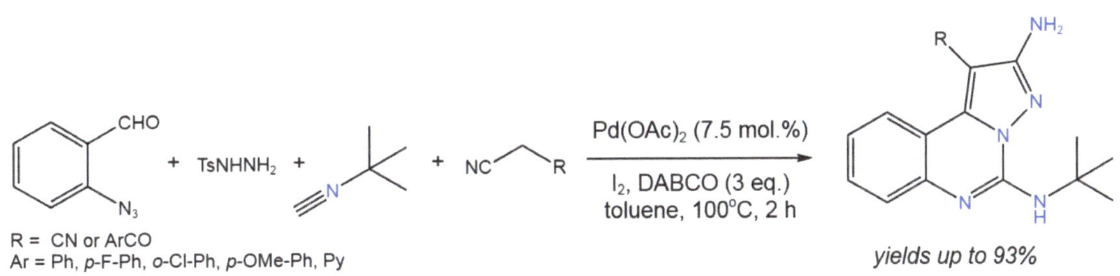

Figure 25. Four-component one-pot synthesis of pyrazolo[1,5-*c*]quinazoline [72].

In order to create 6-fluoroalkyl-phenanthridines in the absence of an oxidant, Bao et al. [77] developed a Pd-catalyzed tandem cyclization of fluorinated imidoyl chlorides using 2-bromophenylboronic acid (Figure 26). The developed procedure allowed for high yields (up to 97%) of target compounds using fluorinated imidoyl chlorides as fluorine-containing synthons. The authors [77] proposed the generally accepted Pd(0)/Pd(II) mechanism with Pd(0) as a starting active form, which interacts with fluorinated imidoyl chloride to form Pd(II). However, it was mentioned that the other pathway Pd(II)/Pd(IV) cannot be excluded.

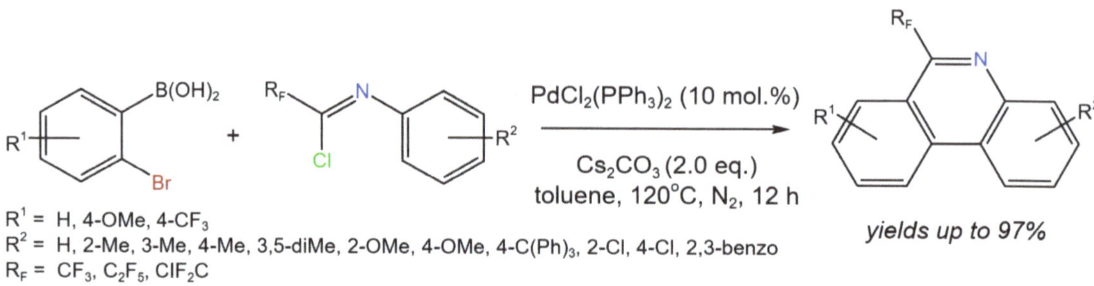

Figure 26. Synthesis of 6-fluoroalkyl-phenanthridines [77].

Another example of cyclization reactions proceeding with the participation of organic cyanides is the Pd-catalyzed tandem reaction of cyanomethyl benzoates with arylboronic acids developed by Dai et al. [78]. According to this approach, the selective synthesis of oxazoles and isocoumarins, through the intermediate imine–Pd complex obtained by carbopalladation of the nitrile, was found to depend strongly on substitution at the 2-position of starting cyano-compounds. Thus, 2,4-diaryloxazoles were selectively produced from cyanomethyl benzoates (3-benzoyl-4-aryl-isocoumarinsfrom 2-benzoyl-substituted cyanomethyl benzoates) with yields of up to 93% (Figure 27). To achieve high yields, different catalysts ($Pd(acac)_2$, $Pd(TFA)_2$, etc.), ligands (2,2′-bipyridine and its derivatives), additives (TfOH, TFA), and reaction conditions, such as solvent nature (THF, dioxane) and atmosphere (air, N_2), were applied.

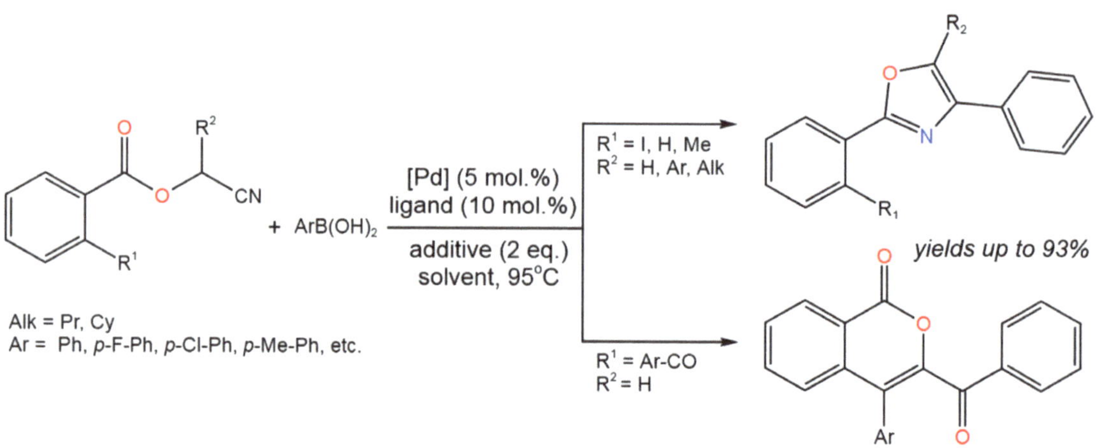

Figure 27. Scheme of the tandem reaction of cyanomethyl benzoates with arylboronic acids for the synthesis of oxazoles and isocoumarins [78].

By using the Pd-catalyzed tandem reaction of 2-aminostyryl nitriles with arylboronic acids, Xu et al. [79] developed a new protocol for the synthesis of 2-arylquinolines (Figure 28). The proposed procedure is an alternative synthetic route as compared to the common condensation reaction of (*E*)-2-aminostyryl ketones. It is noteworthy that the reaction was catalyzed by Pd(II), which interacted with arylbotonic acid, the resulting aryl-palladium complex, then reacted with the nitrile.

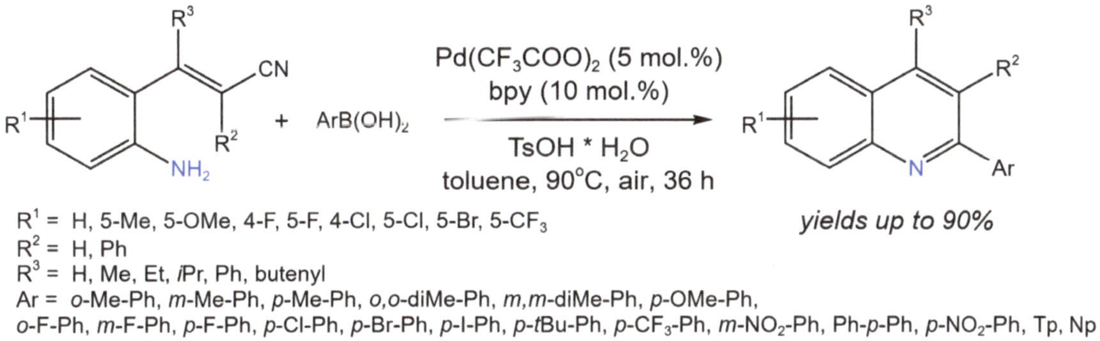

Figure 28. Pd-catalyzed tandem reaction of 2-aminostyryl nitriles with $ArB(OH)_2$ [79].

In a similar manner, Yao etal. [80] implemented the tandem Pd-catalyzed cascade carbopalladation/cyclization/aromatization of arylboronic acids with 5-oxohexanenitrile and its derivatives to synthesize 2-methyl-6-arylpyridines. Different catalysts ($Pd(OAc)_2$, $Pd(CH_3CN_2)Cl_2$, $Pd(CF_3CO_2)_2$, $PdCl_2$, $Pd(PPh_3)_2Cl_2$, $Pd_2(dba)_3$), ligands, and solvents (THF, DMSO, DMF, 1,4-dioxane, toluene, acetone, MeOH, EtOH, nPrOH, *i*PrOH, nBuOH, 1-PeOH, H_2O) were studied. As a result, the optimum conditions were found ($Pd_2(dba)_3$ (5 mol.%), 2,2′-bipyridine ligand (10 mol.%), MeOH), which allowed for up to 94% yields of the target product for 12 h, depending on the substituents in the initial substrate molecules, with CF_3COOH as an additive at 90 °C in air.

Ye et al. [81] established a palladium-catalyzed approach for the tandem reaction of 2-(arylamino)benzonitrile with arylboronic acids in water to produce 9-arylacridine derivatives posessing estrogenic biological activity (Figure 29).

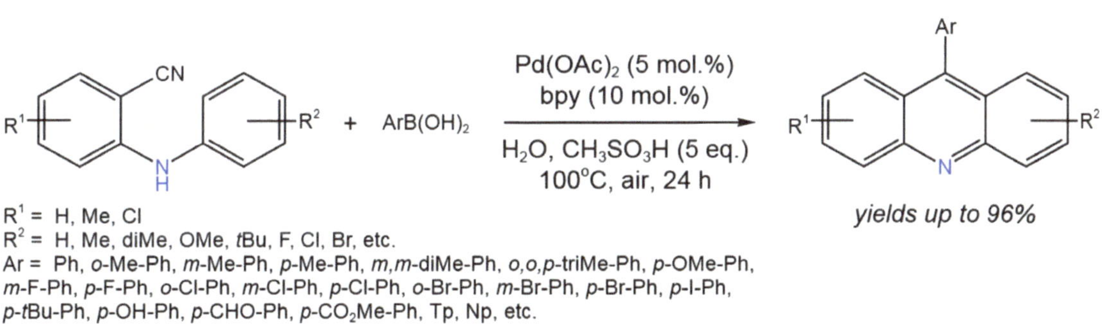

Figure 29. Tandem reaction of 2-(arylamino)benzonitrile with arylboronic acids for the synthesis of 9-arylacridine derivatives [81].

High yields of the desired products (up to 96%) were achieved via the proposed tandem process, involving the nucleophilic addition of aryl Pd species to the nitrile to produce an aryl ketone intermediate, which then underwent an intramolecular Friedel–Crafts acylation and dehydration to produce acridines.

A method for the production of benzoxazinones or quinazolinones was reported by Lang et al. (Figure 30) [82].

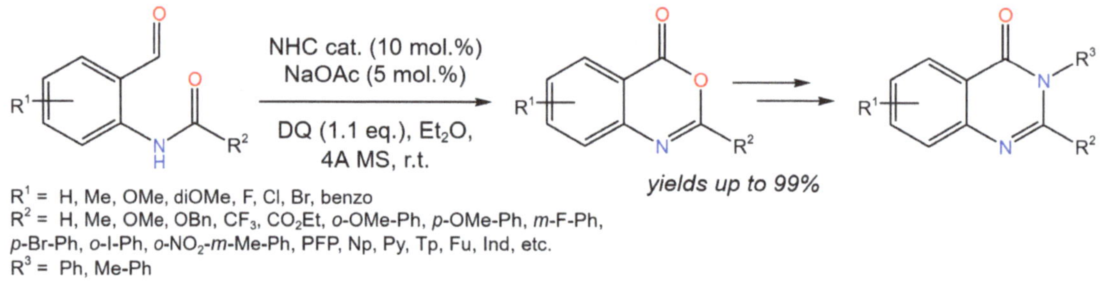

Figure 30. Carbene-catalyzed tandem isomerization/cyclisation for the synthesis of benzoxazinones or quinazolinones [82].

It was mentioned that other protocols are carried out under harsh conditions, restricting the functional group tolerance. According to the proposed strategy, benzoxazinone skeletons were synthesized via utilizing an oxidative carbene-catalyzed tandem isomerization/cyclisation process, which, under mild conditions, allowed for the production of a number of valuable benzoxazinones or quinazolinones from a wide range of substrates.

By using a one-pot intramolecular C–N coupling cyclization reaction, Patel et al. [83] created promising catalytic process to incorporate four-membered ring systems between the benzimidazole and 4-phenyl quinoline cores (Figure 31). This protocol was proposed to be an attractive route for the synthesis of different heterocycles, considering its high efficiency (yields up to 94%), wide substrate scope, and mild reaction conditions [83].

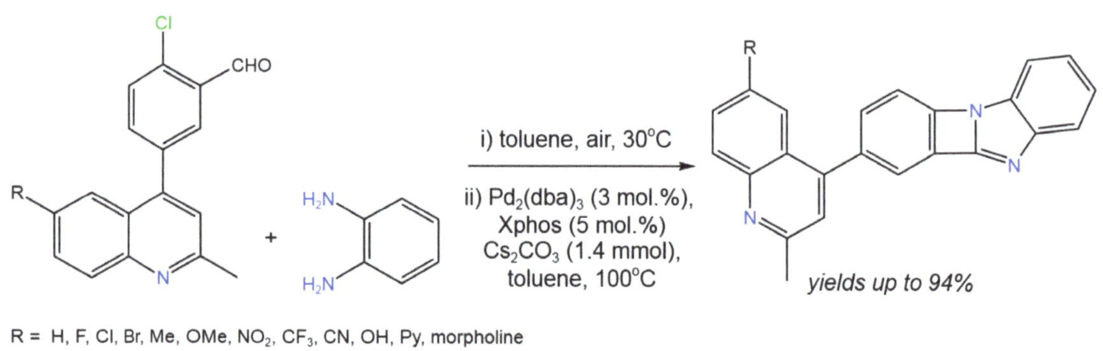

Figure 31. Scheme of the Pd-catalyzed intramolecular C–N coupling cyclization reaction [83].

The reaction likely begun with the condensation of *o*-phenylenediamine with 2-chloro-5-(*N*-substituted-2-methylquinoline-4-yl) benzaldehyde. The resulting intermediate then underwent C–N coupling via conventional Pd(II)/Pd(0) chemistry, starting with the oxidative addition to Pd(0) [83].

Another example of intramolecular cyclization is the work of Jin et al. [84], who developed new method for the synthesis of a cyclopenta-fused acenaphtho[1,2-*b*]indole (ANI) scaffold (Figure 32).

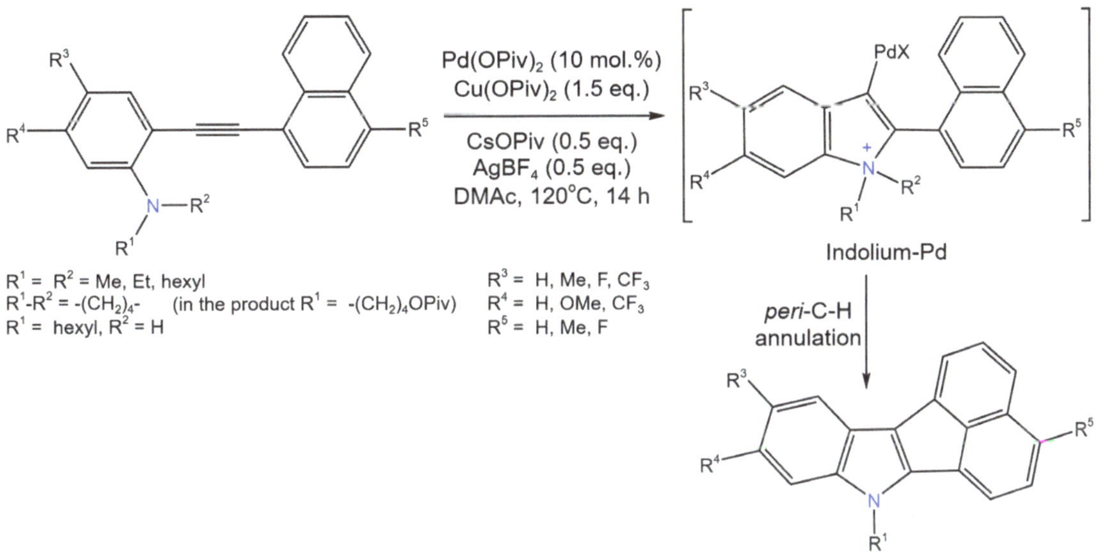

Figure 32. Scheme of Pd-catalyzed intramolecular C–N coupling cyclization reaction [84].

This Pd-catalyzed cascade process involved indolization, *peri*-C–H annulation, and N-dealkylation. Numerous *o*-alkynylanilines containing different substituents and tethered with various polyaromatic hydrocarbons having a *peri*-C–H bond were studied, allowing an up to 79% yield of the target product. It is noteworthy that, in contrast to the secondary aniline, the substrate with a *N,N*-dihexyl substituent resulted in an ANI with 100% selectivity [84].

In a similar way, Siciliano et al. [85] carried out the cyclisation of *o*-alkynyl-anilines using a ligand-free $Pd(OAc)_2$ catalyst in 3 wt.% TPGS-750-M/H_2O medium. Since the amino group was unprotected, the 2-substituted indoles were obtained with a yield of up to 76%, depending on the conditions (catalyst loading (2–10 mol.%), temperature (80–100 °C), type of heating (MW or oil bath), additive (AcOH, $C_{11}H_{23}COOH$, DPBA), and reaction duration (from 10 min to 16 h). Moreover, the tandem process Sonogashira/indole cyclisation was implemented according to the Cu-free mechanism, with the Pd(II) as the active species, which allowed for an up to 40% yield of the target product [85]. It is noteworthy that 2-substituted indoles can be also obtained via tandem Sonogashira coupling, followed by reductive cyclization with 1-halo-2-nitrobenzenes and terminal alkynes as starting compounds [69].

The Pd-catalyzed cascade addition/cyclization/aromatization of heteroarenes (e.g., thiophenes, furans, pyrroles, and indoles) with 2-(cyanomethoxy)chalcones was shown to be a promising approach for the synthesis of various benzofuro [2,3-*c*]pyridines (Figure 33) [86]. The following reaction conditions were varied: catalyst ($PdCl_2$, $Pd(CF_3CO_2)_2$, $Pd(acac)_2$, $Pd(OAc)_2$), solvent (DMA, NMA, THF, DMF, toluene), temperature (60–120 °C), oxidant (Ag_2CO_3, $Cu(Oac)_2$, $AgNO_3$, Ag_2O, $AgSbF_6$, $AgCF_3CO_2$), and additive (D-CSA, CF_3CO_2H, CF_3SO_3H, *p*-TsOH, CH_3CO_2H). As a result, the optimum parameters were chosen (catalyst $Pd(OAc)_2$; solvent NMA; additive CH_3CO_2H; temperature $\geq$ 80 °C; and, optionally, $AgCF_3CO_2$), which provided the yields of target products of up to 86%.

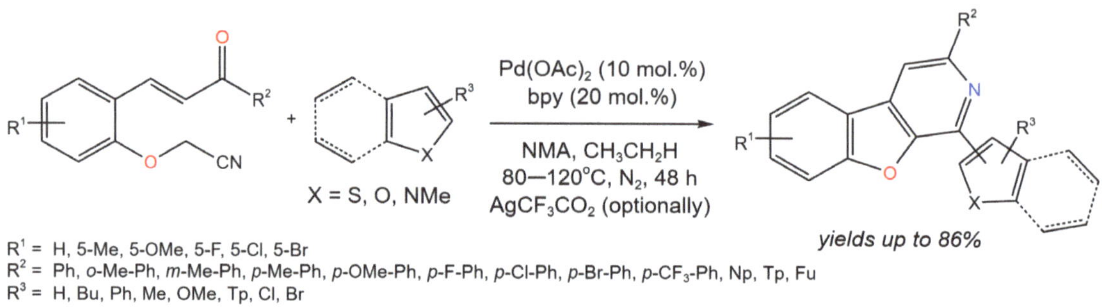

Figure 33. Pd-catalyzed cascade reaction of 2-(cyanomethoxy)chalcones with heteroarenes [86].

A novel procedure for the tandem intramolecular addition of active methylene compounds to internal alkynes, followed by coupling with aryl and heteroaryl bromides, was described by Błocka et al. [87] (Figure 34).

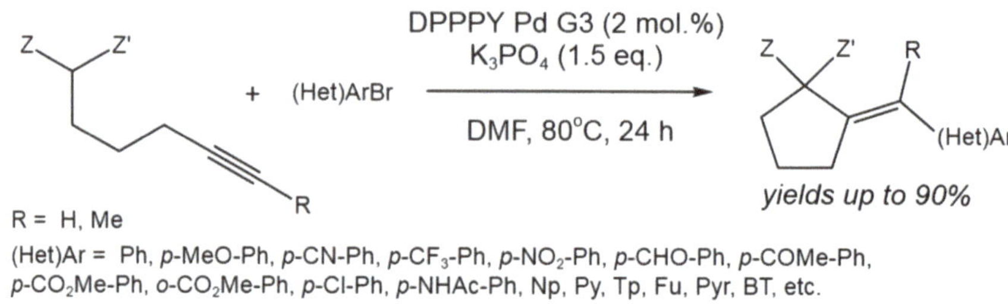

Figure 34. Tandem Pd-catalyzed carbocyclizationcoupling of methylene compounds bearing an internal alkyne group with aryl and heteroaryl bromides [87].

As a catalyst, third-generation (G3) Buchwald palladium precatalysts in combination with diphenyl-2-pyridylphosphine (DPPPY) ligand was used. This method allowed for the obtaining of a variety of vinylidenecyclopentanes with yields of up to 90%, providing high regio- and stereoselectivity and tolerance to the wide range of substituents. Thus, the the *5-exo-dig* intramolecular addition proceeded with high efficiency.

The mechanism involving oxidative addition, cyclization, and reductive elimination was also confirmed by the computational study. The rate- and configuration-determining step was found to be the 5-*exo-dig* intramolecular nucleophilic addition of the enol intermediate to the alkyne activated via coordination with Pd(II) [87].

2.5. Other Tandem Processes

There are many other tandem reactions homogeneously catalyzed by palladium. Below, we provide several examples, among which are alkylation, alkenylation arylation, cycloaddition, and carboannulation.

For example, Hu et al. [88] developed a one-pot approach for the synthesis of a triphenylene core under ligand-free conditionsusing commercially available (hetero)aromatic carboxylic acids and cyclic diaryliodonium salts (Figure 35). The obtained triphenylenes can be used as the electrontransport materials. The proposed reaction mechanism involved an acid-directed *ortho*-C–H arylation/intramolecular decarboxylative annulation sequence mediated by Pd(II)/Pd(IV) transformations.

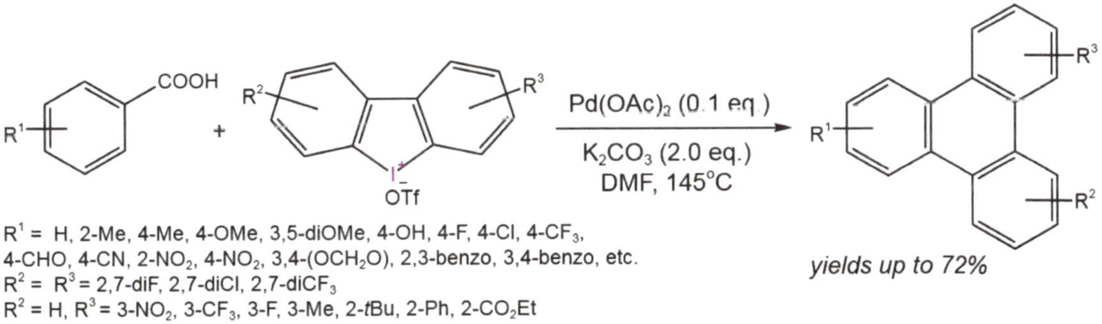

Figure 35. Scheme of tandem C–H arylation/decarboxylative annulation for the synthesis of functionalized triphenylenes [88].

The novel procedure of arylation-cyclization was developed for the synthesis of 3-aryl-2-quinolone (Figure 36) and 4-aryl-2-quinolone derivatives from simple *E*/*Z* geometric isomer precursors using oxygen as a co-oxidant [89].

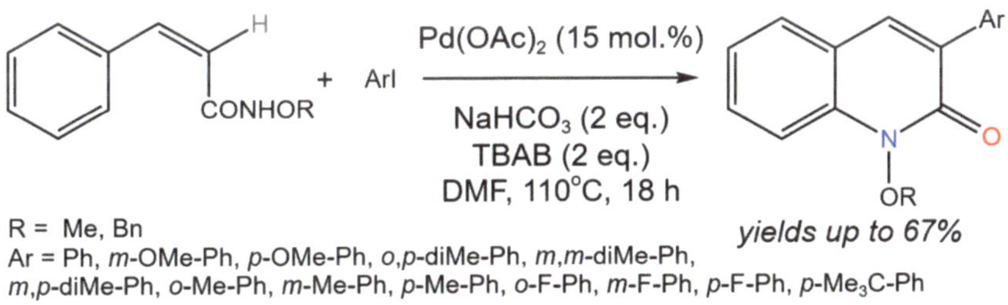

Figure 36. Scheme of tandem arylation-cyclization for the synthesis of 3-aryl-2-quinolones [89].

Ghosh and Chattopadhyay [74] identified the following Pd-catalyzed tandem procedures, which can be applied for the synthesis of 4-aryl-2-quinolones:

(i) intramolecular hydroarylation of yanamides;
(ii) hydroarylation-heterocyclization of 2-aminophenyl propiolate;
(iii) Heck-heterocyclization;
(iv) oxidative Heck-heterocyclization;
(v) arylation of *ortho*-halo-cinnamides, followed by Buchwald-type intramolecular amidation.

In the developed approach, the unusual complementary regioselectivity was found in the tandem C–H-arylation C–H-amidation of isomeric substrates. The origin of selectivity was proposed at the C–H-arylation step [89].

Domańskiet al. [90] described the tandem Pd-catalyzed three-component reaction that allowed for regio- and stereoselective perfluoroalkylative borylation of a variety of terminal and internal alkynes in the presence of perfluoroalkyl iodide and $(Bpin)_2$. Thus, the first example of anti-addition across C–C multiple bonds of groups originating from two separate electrophiles to functionalize alkynes by reductive dicarbofunctionalization was shown. Iodoperfluoroalkylation, borylation, and coupling are the three fundamental processes responsible for this sequential transformation, and mechanistic studies have shown that their rates significantly differ (Figure 37).

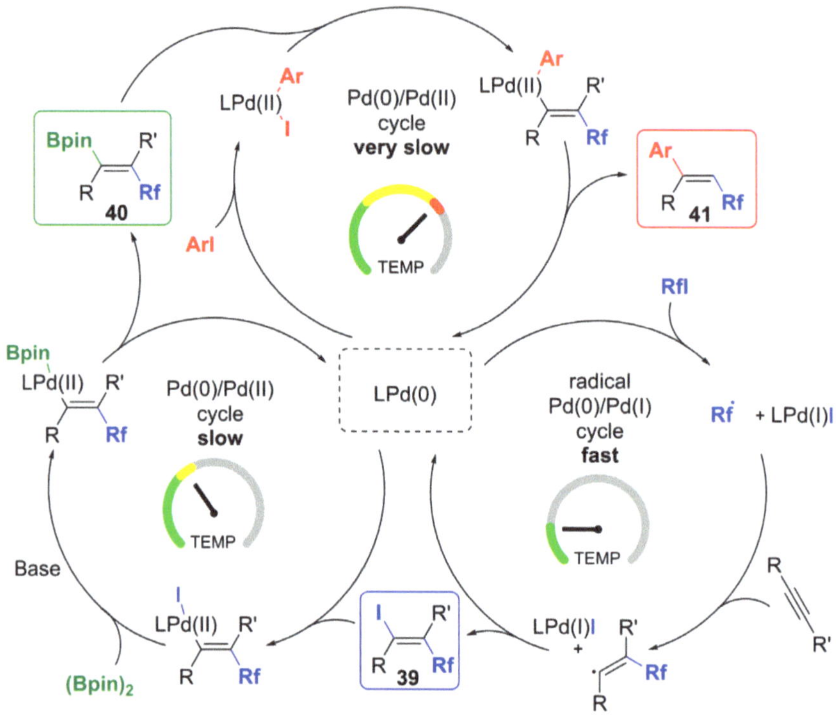

Figure 37. Proposed reaction mechanism of perfluoroalkylative borylation. Reproduced with permission from [90]. Copyright 2019 American Chemical Society.

According to the developed tandem process, from the same reaction mixture and under the same reaction conditions, the fluoroalkyl-substituted vinyl iodides, vinyl boronates, or olefins can be produced by changing the temperature program [90].

The Pd-catalyzed direct tandem C–O/C–H activation method for the C–C bond formation was described by Fernández et al. [91]. To accomplish base-free direct C–H alkenylation, the novel approach combined coordinated metalation–deprotonation of functionalized heterocycles with a C–O oxidative addition at enol pivalates (Figure 38).

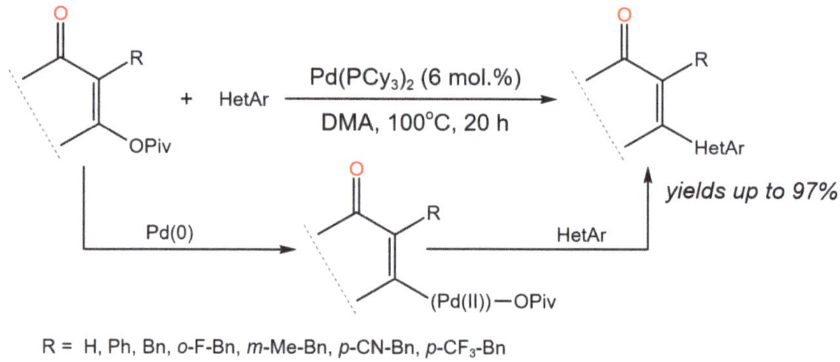

Figure 38. General scheme of Pd-catalyzed tandem C–O/C–H activation [91].

According to mechanistic studies, the C–O oxidative addition to Pd(0) was shown to be reversible, and the product of the Pd(II) C–O oxidative addition directly resulted in C–H activation [91].

Zhang et al. [92] developed a prospective route for the synthesis of poly-substituted quinolines via a three-component tandem reaction. 2-Aminobenzonitriles, arylboronic acids, and ketones were used as the substrates. Pd-catalyzed aryl addition to the cyano group was followed by hydrolysis and Friedländer-type cyclization to produce the quinoline compounds, with the yields of up to 96%. The described above cyclization tandem process developed by Xu et al. [68] resulted in lower yields, while comparing the same structures of target products.

Through the three-component tandem arylation and allylic etherification of 2,3-allenol with aryl iodides and alcohols, an effective approach for the synthesis of arylated allylic ethers was implemented [93]. By this method, functionalized 1-arylvinylated 1,2-diol derivatives were produced with yields of up to 83% and complete selectivities.

Tang et al. [94] reported a Pd-catalyzed tandem reaction of 3-allyloxybenzocyclobutenols that involved proximal C–C bond cleavage, C–O bond cleavage, and allylic alkylation of the C–H bond. Thus, a novel approach to *meta*-β-keto phenols bearing an allylic group with 100% atom economy was developed. It is noteworthy that, by adjusting substituents on various positions of benzocyclobutenols, sequential proximal or distal C–C bond cleavage/deallylation could be implemented.

Li et al. [95] developed a tandem process for the selective assembly of tri- or tetrasubstituted vinylsilanes. It was shown that using this method, *ortho*-vinyl bromobenzenes can be formed in situ from 1-bromo-2-iodobenzenes and *N*-tosylhydrazones and disilylated to produce two C–Si bonds and two C–C bonds (Figure 39).

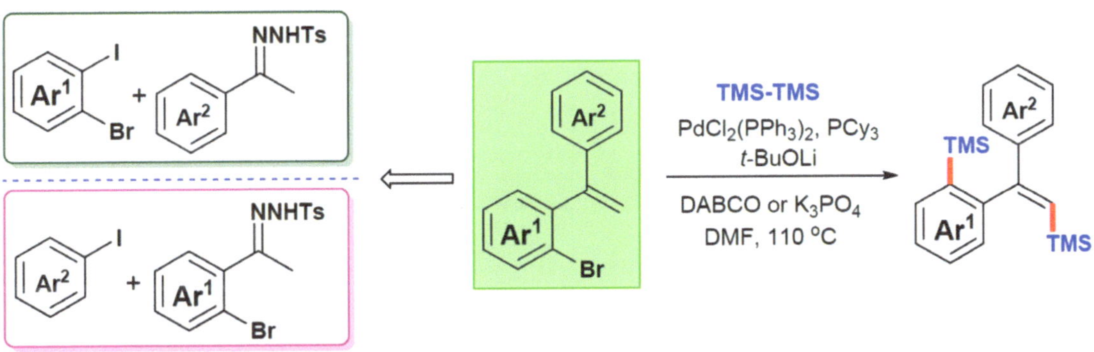

Figure 39. General scheme of the disilylation of C(aryl),C(vinyl)-palladacycles. Reproduced with permission from [95], RCS, 2020.

It was determined that the *ortho*-vinyl bromobenzenes produced from 1-bromo-2-iodobenzenes and *N*-tosylhydrazones are the crucial intermediates needed to manufacture vinylsilanes by disilylating the C(aryl),C(vinyl)-palladacycle, which is formed by a direct vinylic C–H bond activation [95].

By the reaction with aldehydic *N*-tosylhydrazones, the treansformation of 1-aryl- and 2-aryl-1,2-dihydro-3*H*-indazol-3-ones into 1,2-di(hetero)aryl- and 2,3-di(hetero)aryl-2,3-dihydroquinazolin-4(1*H*)-ones was carried out (Figure 40) [96]. The method involved a cascade process that included base-mediated Pd-carbenoid production via the decomposition of *N*-tosylhydrazones, a nucleophilic attack of indazolone on the Pd-carbenoid complex, and intramolecular ring expansion by N–N bond cleavage.

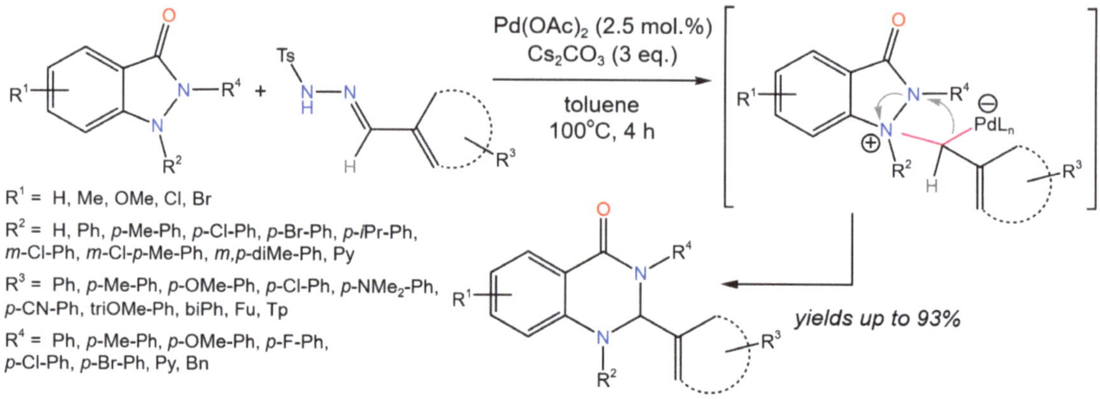

Figure 40. General scheme of Pd-catalyzed tandem C–O/C–H activation for the synthesis of 1,2-di(hetero)aryl- and 2,3-di(hetero)aryl-2,3-dihydroquinazolin-4(1*H*)-ones [96].

An interesting approach to the synthesis of [4,5]-spirocycles in a cascade Pd-catalyzed reaction was reported by Azizollahi et al. [97]. It was found that, when varying the type of ligand, the reaction pathway can be switched from carbopalladation to β-C-elimination: monodentate phosphine ligand (IMes·HCl, PCy3) affords [4,5]-spirocycles, while bidentate phosphines (Xantphos, DPPF, DPE-Phos) results in the formation of heteroaromatics (Figure 41).

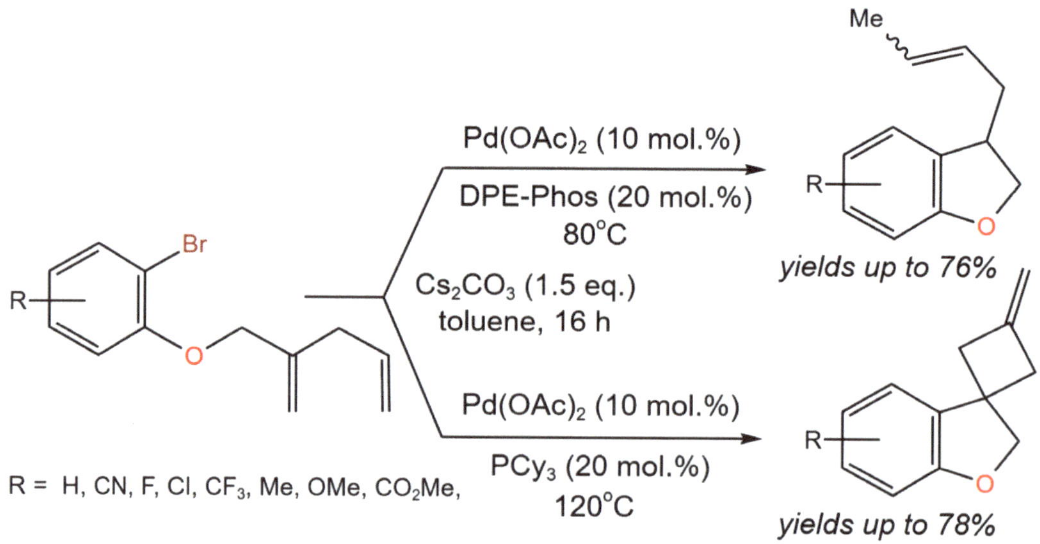

Figure 41. Ligand-controlled cascade synthesis of [4,5]-spirocycles from skipped dienes [97].

Another example of switchable cascade reactions was the oxidation of *N*-allyl-2-aminophenols [98] in the presence of hypervalent iodines and Pd catalyst ($Pd(OAc)_2$). In the absence of palladium, the dearomatization of the substrate and intramolecular Diels-Alder reaction occurred with the formation of tricyclic systems (yields of up to 72%), while the Pd-catalyzed process resulted in the methylacyloxylated dihydro-1,4-benzoxazines (yields of up to 86%).

Activated tetrasubstituted alkenes made from phthalides or butyrolactone were combined with vinylethylene carbonates in an unprecedented [5 + 2] cycloaddition/ring-contraction tandem process by Xiong et al. [99] under Pd(0) catalysis (Figure 42). The proposed approach was practical and mechanistically novel. In contrast to the traditional spirolactonization method, benzo [5,5] spiroketal lactones and [5,5]-spiroketal lactones bearing two vicinal tetrasubstituted centers can be obtained by this tandem [5 + 2] cycloaddition/ring-contraction mechanism with moderate to high yields.

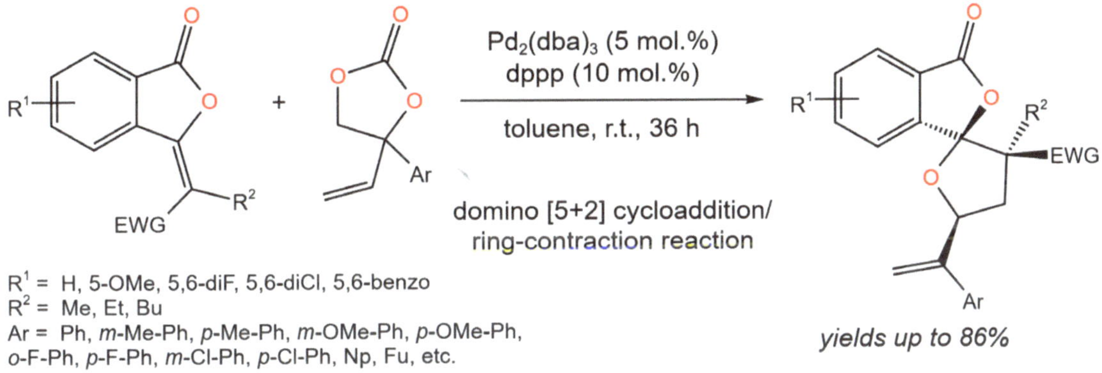

Figure 42. Pd-catalyzed [5 + 2] cycloaddition/ring-contraction tandem process [99].

It was proposed that the reaction was started from the formation of a zwitterionic π-allyl palladium intermediate via the interaction of vinylethylene carbonate with Pd(0), accompanied by the decarboxylation. Then, the intermediate reacted with activated phthalide or butyrolactone through [5 + 2] cycloaddition. Thus, a complex of Pd(0) and a spiro-oxepene phthalide was formed, which further facilitated the ring-opening reaction, resulting in a new zwitterionic π-allyl palladium intermediate. Finally, the branched allylic etherification occurred with the formation of a target product [99].

The first example of a transition-metal-catalyzed multicomponent carboannulation reaction of [60]fullerene was reported by Liu et al. [100]. In a Pd-catalyzed three-component tandem coupling–carboannulation process, [60]fullerene, 2-(2,3-allenyl)-malonates, and (hetero)aryl iodides were converted into a variety of polysubstituted [60]fullerene-fused cyclopentanes with a wide range of substrates and excellent functional group compatibility (Figure 43).

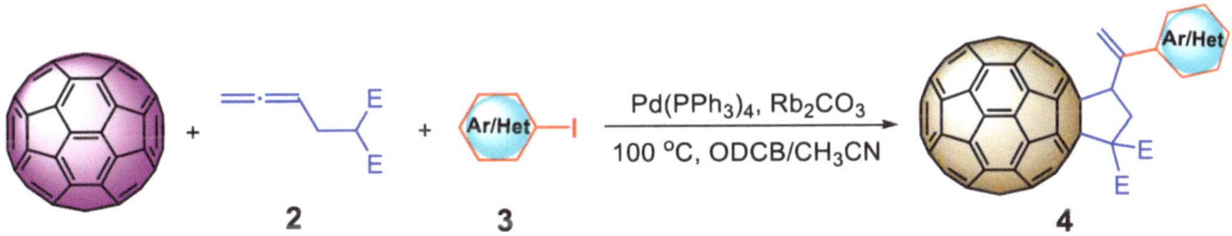

Figure 43. Scheme of Pd-catalyzed tandem coupling–carboannulation for the regioselective synthesis of polysubstituted C_{60}-fused cyclopentanes. Reproduced with permission from [100]. Copyright 2020 American Chemical Society.

Different reaction conditions were explored: catalysts ($Pd(PPh_3)_4$, $Pd(dba)_2$, $Pd_2(dba)_3\cdot$ $CHCl_3$), bases (Li_2CO_3, Cs_2CO_3, K_2CO_3, K_3PO_4, Rb_2CO_3, DMAP, and DABCO), and cosolvents (CH_3CN, THF, 1,4-dioxane, DMF, and DMSO). Thus, the optimal conditions were found ($Pd(PPh_3)_4$, Rb_2CO_3, 100 °C, solvent 1,2-dichlorobenzene (ODCB), cosolvent CH_3CN), which allowed for an up to 62% yield of the reaction product [100].

Zhang et al. [101] developed the Pd-catalyzed tandem synthesis of 2-trifluoromethylthio-(seleno)-substituted benzofurans, benzothiophenes, and indoles (Figure 44) in acceptable to good yields (up to 93%). The reaction likely proceeded via the intramolecular cross-coupling followed by trifluoromethylthiolation.

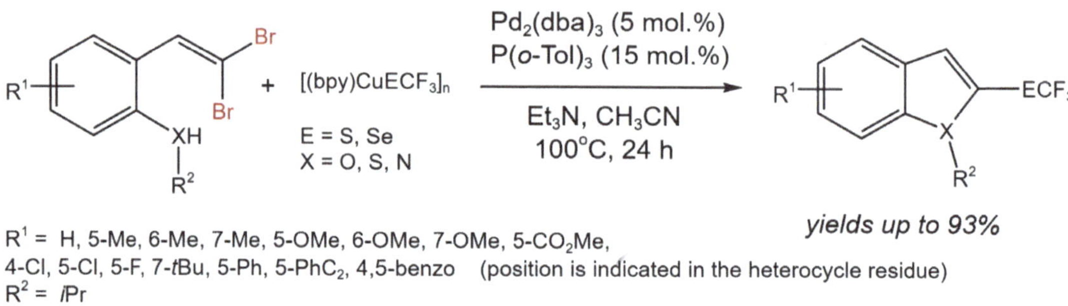

Figure 44. Pd-catalyzedtandem synthesis of 2-trifluoromethylthio-(seleno)-substituted benzofurans, benzothiophenes, and indoles [101].

Yang et al. [102] reported the asymmetric tandem C–C bond activation/Cacchi reaction between cyclobutanones and *o*-ethynylanilines (Figure 45). The chiral σ-alkylpalladium intermediates were formed via the enantioselective C(sp^3)–C(sp^2) bond activation of cyclobutanones and then promoted the cyclization of *o*-ethynylanilines, leading to one-carbon-tethered chiral indanone-substituted indoles. An all-carbon quaternary stereocenter was simultaneously formed, along with the two C–C bonds and one C–N bond. Under the optimum conditions (catalyst: $[Pd(allyl)Cl]_2$ (0.05 eq.), TADDOL-derived phosphoramidite: L1 (0.1 eq.), base: K_2CO_3 (2.5 eq.), solvent: 1,4-dioxane, temperature: 90 °C), the indanone-substituted indoles with both central and axial stereogenic elements were synthesized with good yields (up to 85%) and excellent enantioselectivity (up to 99:1 e.r.).

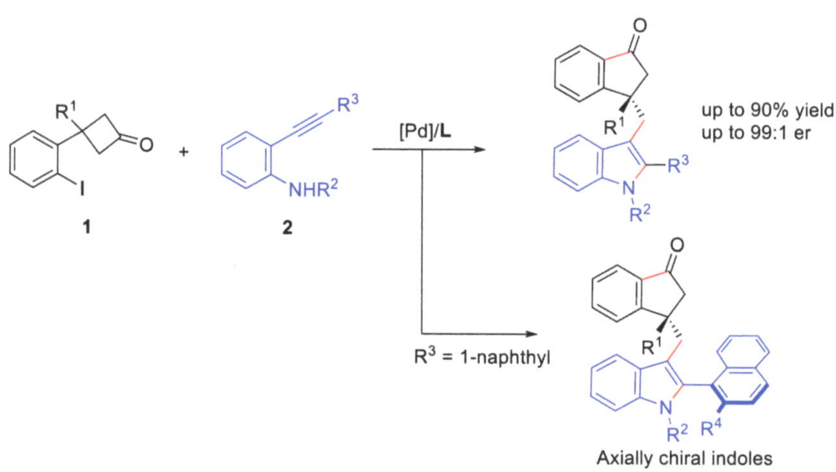

Figure 45. Pd-catalyzed enantioselective tandem ring-opening/Cacchi reaction. Reproduced with permission from [102]. Copyright 2021 American Chemical Society.

The borrowing hydrogen (BH) reaction (hydrogen autotransfer) is another example of Pd-catalyzed tandem processes that can be mentioned [103] as a powerful strategy that combines transfer hydrogenation (avoiding the direct use of molecular hydrogen) with one or more intermediate reactions to synthesize more complex molecules without the need for time-consuming separation or isolation processes. The strategy of the BH process relies on three steps, namely, (i) dehydrogenation, (ii) intermediate reaction, and (iii) hydrogenation, among which the intermediate reaction works in tandem with the metal-catalyzed hydrogenation/dehydrogenation step.

3. Heterogenized PdComplexes in Tandem Processes

Homogeneous Pd complexes can be anchored to the different organic or inorganic supports. Such heterogenized complexes can be also used in tandem catalytic processes.

As for the examples of heterogenized Pdcomplexes reported for the last five years in terms ofapplication in tandem reactions, the works of Fan et al. [104] and Esfandiary et al. [105] can be mentioned. In both cases, Pd catalyzed the chosen tandem process homogeneously, while the support facilitated the separation and reuse of the catalysts.

Fan et al. [104] synthesized a polyethyleneimine (PEI)-capped microcrystalline cellulose(MCC)-supported polyamidoamine (PAMAM) dendrimer (MCC-PAMAM$_{G2.5}$-PEI) for the loading of Pd^{2+} and Cu^{2+} ions. In this catalytic system, PAMAM$_{G2.5}$ served as the carrier to immobilize Pd^{2+} ions, and PEI provided the coordinating sites to selectively chelate Cu^{2+} species. The Pd/Cu@MCC-PAMAM$_{G2.5}$-PEI was tested in a Sonogashiracyclization tandem reaction (Figure 46), revealing excellent catalytic activity, recyclability, and robustness. The synergetic effect between Pd and Cu was observed.

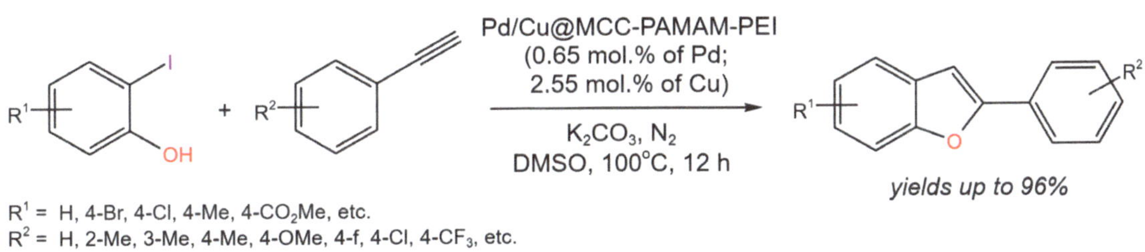

Figure 46. Scheme of Pd/Cu@MCC-PAMAM$_{G2.5}$-PEI-catalyzed Sonogashira cyclization tandem reaction [104].

A novel magnetic γ-Fe_2O_3@Cu-LDH@Cysteine-Pd dual nanocatalyst system was designed by Esfandiary et al. [105]. The surface of magnetic NPs was modified by the substitution of Cu(II) metal cations with layered double hydroxide (LDH) cations. Then, the Pd(II) complex was immobilized on the interface defect structure of LDH (Figure 47).

Using the synthesized catalyst, the A^3 coupling followed by C–N coupling and intramolecular cyclization was carried out for the synthesis of quinolines (yields of up to 94%) from 2-bromobenzaldehyde as a low-cost simple starting material (Figure 48). The suggested nanocatalyst easily activated terminal alkynes in the absence of base using recyclable choline azide as a green solvent and reagent. The catalyst could be reused up to four times without significant loss of activity [105].

Meng et al. [106] created a bifunctional catalytic system based on a water-soluble thermoresponsive polymer (poly(ethane-*co*-*N*-isopropylacrylamide) and hollow-shell-structured mesoporous silica as a support for Ru-(MesRuArDPEN) and Pdcomplexes ($Pd_2(dba)_3$ + PMe_3). This catalytic system (polymer@Pd@Ru) was applied in the cascade Suzuki cross-coupling/asymmetric transfer hydrogenation reactions of various iodoacetophenones and aryl boronic acids, allowing high yields and ee (up to 97%). By regulating the polymer layers on the outer silica shell, two catalysts selectively controlled the catalytic behavior for the implementation of switchable enantioselective cascades.

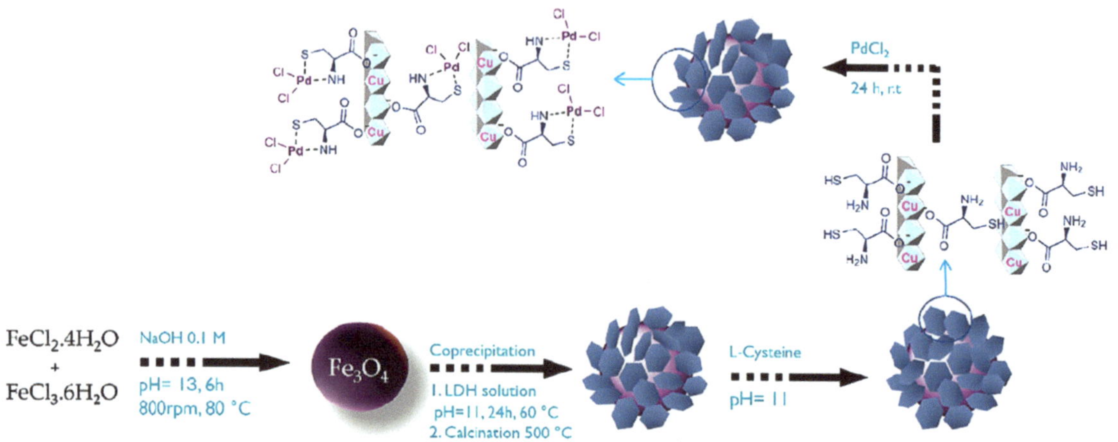

Figure 47. Schematic procedure of the synthesis of γ-Fe_2O_3@Cu-LDH@Cysteine-Pd. Reproduced with permission from [105], Elsevier, 2020.

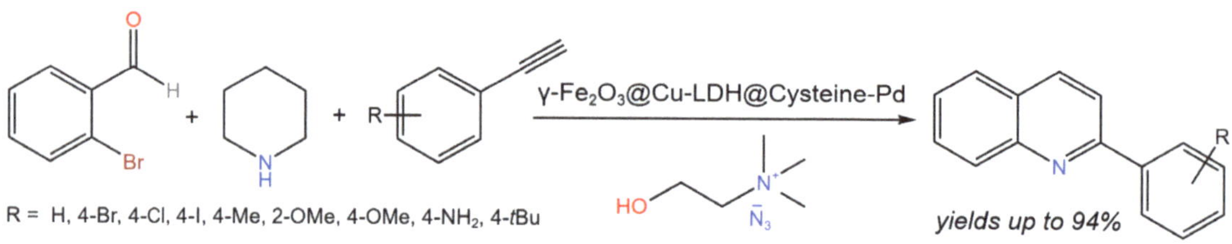

Figure 48. Scheme of the synthesis of 2-substituted quinolines in the presence of γ-Fe_2O_3@Cu-LDH@Cysteine-Pd [105].

Rajabi et al. [107] developed a new environmentally friendly universal protocol for the one-pot conversion of aliphatic, aromatic, and unsaturated aldehydes into esters under ambient conditions in the presence of oxygen using a palladium nanocatalyst supported on mesoporous organosilica (SBA-15). Palladium bearing a functionalized cytosine (Cyt) on the surface of mesoporous silica (Pd-Cyt@SBA-15) allowed for a selective large-scale conversion of a broad range of short-chain alcohols with aldehydes, including aromatic and aliphatic aldehydes to the corresponding ester products with yields of up to 98% and TON (98,000). The catalyst Pd-Cyt@SBA-15 revealed high stability and reusability over ten reaction runs.

4. Heterogeneous Pd-Catalyzed Tandem Processes

Aiming at the implementation of tandem catalysis, multiple reactions can be carried out in one pot by using bifunctional heterogeneous systems for the simultaneous activation of substrates and reagents in various Lewis/Brønsted acid, acid/base, metal/base, metal/acid, or metal/metal-catalyzed processes. The bifunctional catalyst can be designed in such a way that the two different catalytic functions (e.g., acidic and basic site) act in a collaborative way in the transition state, or each one catalyzes a different reaction in a multistep process [19].

Besides the common approaches to the desing of the multifunctional catalyst, Cho et al. [15] defined the following additional directions in the desing of tandem processes:

(i) size and shape selectivity, with active sites only accessible to substrates and intermediates with specific sizes and shapes;
(ii) surface and solvent engineering that exploits differences in the hydrophobicity, hydrophilicity, and other properties of the catalizate's components;

(iii) metal site engineering through controlled size, exposed facets, composition, and their spatial distribution;
(iv) reactor and process engineering (i.e., chemical looping, reactive separations, andmultiple sequential catalyst beds in flow reactors) [15].

For example, Sheetal et al. [108] developed the direct one-pot carbonylation of iodobenzene and NaN_3 as a N-atom source under Pd@PS catalyzed conditions utilizing $(CO_2H)_2$ as an environmentally benign CO surrogate in a DMF solvent system. The proposed approach was based on the use of two vials with different solvents: (i) an outer vial with $(CO_2H)_2$ and DMF, and (ii) an inner vial containing iodobenzene and NaN_3, along with a catalyst in xylene solvent conditions (Figure 49).

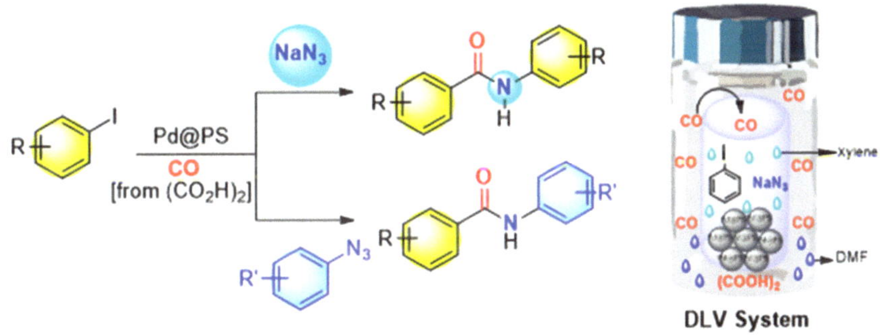

Figure 49. Tandem approach to the synthesis of *N*-aryl benzamides through the bifunctional transformation of aryl iodides in a double-layer vial (DLV) system. Reproduced with permission from [108], Elsevier, 2021.

Thus, the use of aryl iodides as bifunctional reactants under base, ligand, and additive-free conditions allowed for the carrying out of the simultaneous C–C and C–N bond formations to obtain the desired products (*N*-phenyl benzamide derivatives) with good to moderate yields (up to 70%) at 140 °C for 24 h [108].

Heterogeneous catalysts with acid sites in the composition of supports are widely used in catalysis, especially in biomass processing. Zeolites are a common example of such supports since they have high thermal and hydrothermal stabilities, homogeneous porosity, shape-selective properties, and tunable acidities [109]. Amoo et al. [109] in their review mentioned that the design of metal-zeolite composite catalysts is prospective for syngas conversion. For example, alkali surfaces are known to favor the adsorption and subsequent conversion of CO to olefins over Fe-based catalysts due to the preffered formation of Fe–C over a high pH surface. The acid sites of zeolites might alter the intrinsic pH of the Fe-zeolite composite catalyst, and thus the interaction between these two active components located in proximity will benefit in tandem catalysis (such as oligomerization, isomerization, alkylation, and hydrocracking) [109].

Mesoporous silica materials, which have ordered pore size, a high specific surface area, a large pore volume, and the ability to synthesize a wide range of morphologies and shapes, can be utilized as an alternative to zeolites [110–116]. Mesoporous silica, such as MCM-41 and SBA-15, used as supports for metal NPs and widely applied in catalysis, including tandem processes. Recently, propylamine, diethylamine, and pyrrolidine were used by Hernández-Soto et al. [115] to create the single basic sites in organic–inorganic hybrid bifunctional organosiliceous catalysts with pendant amine groups in addition to Pd NPs.In the developed catalysts, heterogenized amine groups and palladium NPs were found to be homogeneously distributed and stabilized in the mesochannels of the MCM-41. The steric effects around the amine groups in mesochannels were proposed to have a strong influence on the catalytic activity. Moreover, longer amines exhibited higher interaction with the silica surface, decreasing the catalytic activity [115].

In the presence of Pd/MCM-41 bearing the propylamine groups, furfural and methyl isobutyl ketone underwent a tandem aldolcondensation/crotonization reaction in a single reactor, followed by hydrogenation (Figure 50). At 100 °C in the presence of toluene, an almost complete conversion of furfural and 82% yield of 1-(furan-2-yl)-5-methylhexan-3-ol was achieved. It was shown that in a dual fixed-bed reactor, the catalyst robustness may be improved, providing 20% furfural conversion for 12 h on stream with the preferential production of 1-(furan-2-yl)- 5-methylhexan-3-one [115]. It is noteworthy that the reaction products can be used as a renewable biosolvent and biofuel precursors.

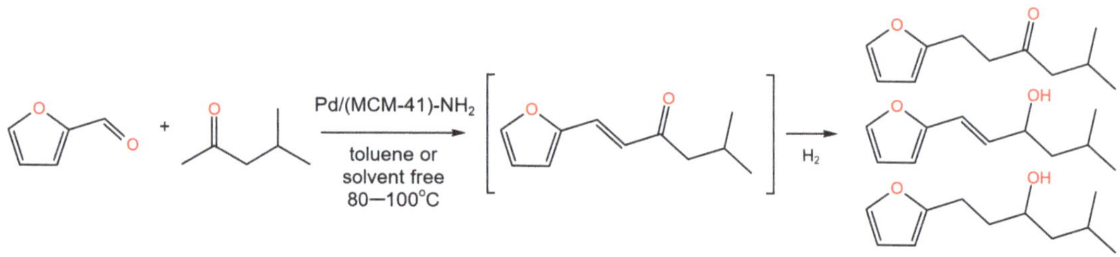

Figure 50. Scheme of tandem aldol condensation/crotonization of furfuraland methyl isobutyl ketone, followed by hydrogenation [115].

Maties et al. [116] synthesized Pd-containing Al-SBA-15 materials for the valorization of trans-ferulic acid into stilbenes via tandem decarboxylation/Mizoroki–Heck coupling (Figure 51). Under mild reaction conditions (100 °C, 3–6 h), quantitative product yields were obtained with over 90% selectivity to target stilbene products. The sizes of Pd NPs were found to be an important factor with a significant impact on the catalytic activity. The catalyst deactivation via sintering of Pd NPs was also observed.

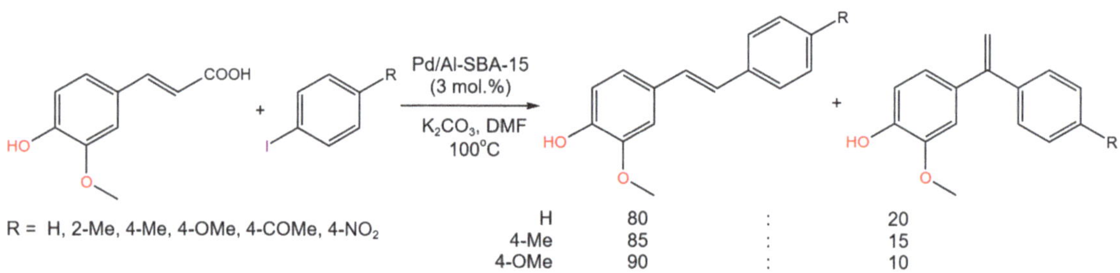

Figure 51. Stilbene synthesis by decarboxylative C–C coupling of trans-ferulic acid and iodoanisole over Pd/Al-SBA-15 catalyst [116].

Tungsten oxide (WO_x) is one of the most wellknown and widely used oxides with surface acid–base properties, being often combined with other oxydes (ZrO_2 [117–119], SiO_2 [120–122], Al_2O_3 [123,124]), and zeolites [121,125]. Chu et al. [126] reported the synthesis of a multifunctional Pd-Cu-WO_x/SiO_2 catalyst for the one-pot conversion of cellulose to ethanol. This catalyst allowed an ethanol yield of 42.5% at 300 °C under 4 MPa H_2 in aqueous medium. It was shown that the cellulose conversion to ethanol followed the following consecutive steps: (i) cellulose hydrolyzed to glucose over acid sites; (ii) glucose then converted to glycolaldehyde over W species; (iii) glycolaldehyde hydrogenated to ethylene glycol over Pd; (iv) ethylene glycol hydrolyzed to ethanol over Cu. It was shown that the three metal components, Pd, Cu, and WO_x, were in appropriate balance, allowing for the achievement of an ethanol formation rate of 0.163 g/(g_{cat}·h).

The versatility of MOFs as highly porous acidic supports for metal NPscan be used for one-pot tandem processes [16,127–129].

MIL-101(Cr) (Figure 52) contining Pd NPs (about 3 nm) at 0.2–1.0 wt.% was used to catalyse one-pot tandem reductive amination of 4′-fluoroacetophenone with benzylamine (Figure 53) under 10 bar of H_2 [16].

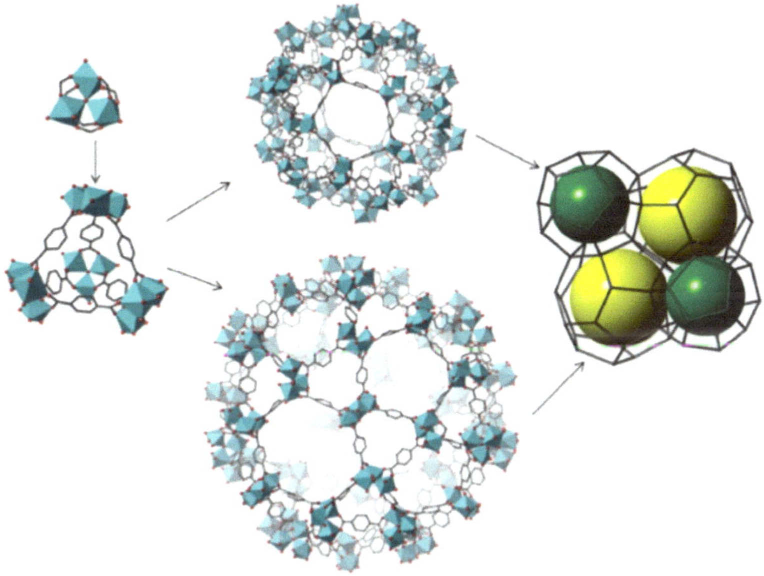

Figure 52. Scheme of MIL-101(Cr), showing the $Cr_3O(O_2C–)_6$ cluster, the super-tetrahedral building units, and the smaller (green spheres) and larger cages (yellow spheres). Reproduced with permission from [16], Springer Nature, 2018.

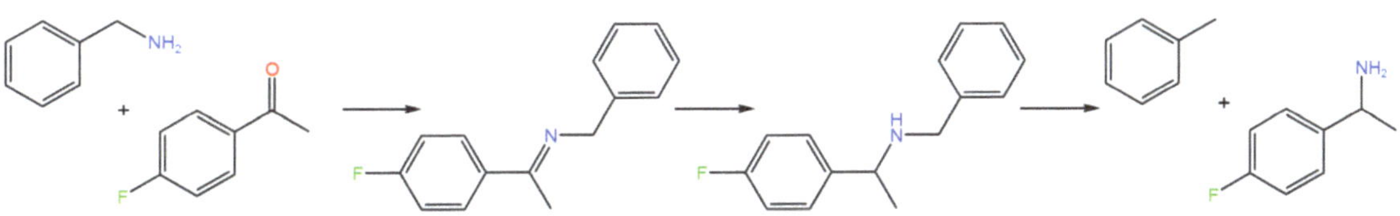

Figure 53. Synthesis of 4′-fluoro-α-methyl-*N*-(phenylmethyl)benzenemethanamine via the reductive amination reaction [16].

The MOF's Lewis acidity and the Pd NPs' capacity for catalytic hydrogenation werecombined in the bifunctional catalyst. The selectivity in the reductive amination reaction was found to be significantly influenced by altering the Pd loading within the MOF to tune the ratio of active sites. Higher metal loadings led to a significant amount of undesired product (4′-fluoro-α-methylbenzenemethanamine). Higher selectivity was achieved by decreasing the total number of Pd sites compared to Lewis acidic sites; the 0.4 wt.% Pd MIL-101 allowed for nearly 90% of the target amine (4′-fluoro-methyl-*N*-(phenylmethyl)benzenemethanamine) after 7 h. Moreover, the synthesized catalysts were reusable and maintained crystallinity and small highly dispersed NPs after reaction.

The bifunctional catalyst Pd@MIL-101-SO_3H was developed by Liu et al. [127] and was applied in the one-pot oxidation–acetalization reaction (Figure 54), the products of which are widely utilized as fuel additives, perfumes, and pharmaceuticals, and in polymer chemistry.

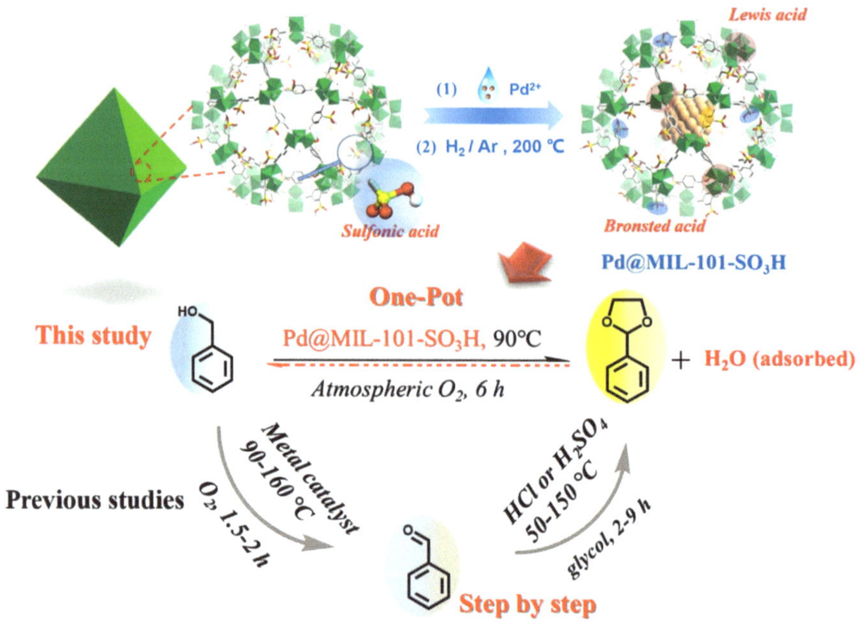

Figure 54. Scheme of the synthesis of Pd NPs immobilized in MIL-101-SO_3H and one-pot transformation of benzaldehyde glycol acetal. Reproduced with permission from [127], Elsevier, 2019.

High yields (>99%) were ascribed to the combined actions of Pd NPs, responsible for oxidation, and Lewis and Brønsted acid sites, responsible for acetalization, which wereencapsulated in sulfonated MIL-101(Cr). The high capacity of MIL-101(Cr) for water adsorption aided the shift in equilibrium by impeding the reversible process.Thus, the nonpolar solvents were demonstrated to be the optimum ones.The reaction mechanism is presented in Figure 55. Moreover, the Pd@MIL-101-SO_3H could be reused for at least eight times without loss of catalytic activity [127].

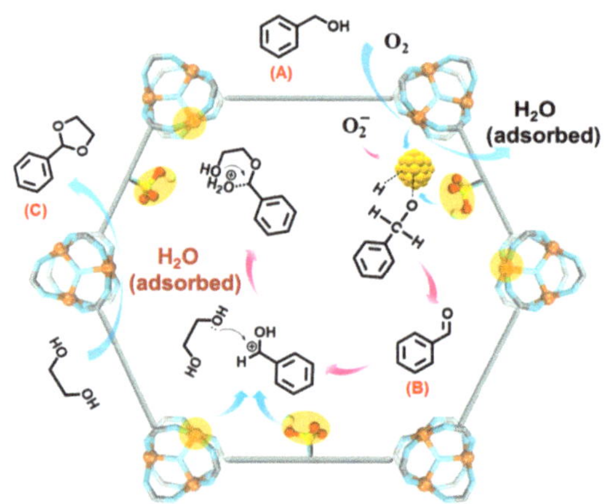

Figure 55. Proposed mechanism of the one-pot oxidation–acetalization. Reproduced with permission from [127], Elsevier, 2019.

Tandem heterogenous catalysis of bimetallic Cu-Pd on an aminofunctionalized Zr-based metal–organic framework (UiO-66(NH_2)) incorporated into sulfonated graphene oxide (Cu-Pd/UiO-66(NH_2)@SGO) was investigated by Insyani et al. [128] for the one-pot conversion of disaccharides and polysaccharides into 2,5-dimethylfuran (2,5-DMF) (Figure 56).

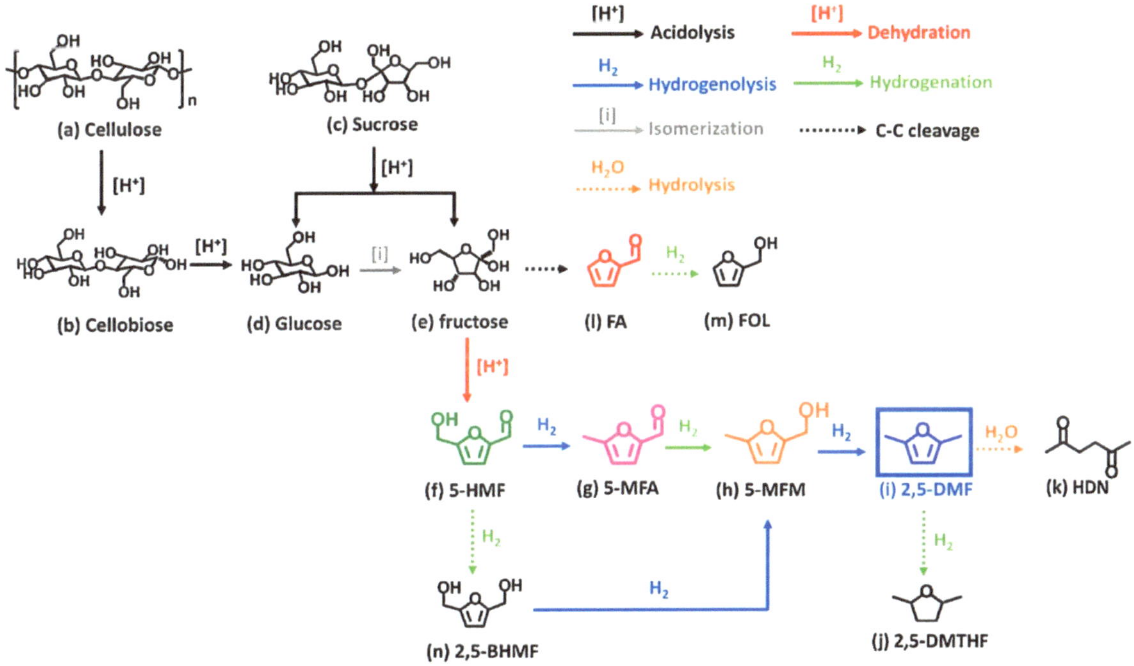

Figure 56. Reaction pathway for di- and polysaccharide conversion into 2,5-DMF: (**a**) cellulose, (**b**) cellobiose, (**c**) sucrose, (**d**) glucose, (**e**) fructose, (**f**) 5-hydroxymethylfurfural (5-HMF), (**g**) 5-methylfurfural (5-MFA), (**h**) 5-methyl-2-furanmethanol (5-MFM), (**i**) 2,5-dimethylfuran (2,5-DMF), (**j**) 2,5-dimethyltetrahydrofuran (2,5-DMTHF), (**k**) 2,5-hexanedione (HDN), (**l**) furfural (FA), (**m**) furfuryl alcohol (FOL), and (**n**) 2,5-bis-hydroxymethylfurfural (2,5-BHMF). Reproduced with permission from [128], Elsevier, 2019.

Sequential glycosidic bond cleavage, isomerization, and dehydration of sucrose led to a high yield (75.8%) of 5-(hydroxymethyl)furfural (5-HMF) by altering the strength and ratio of the Lewis and Brønsted acid sites by varying the ratios of UiO-66(NH_2) to SGO. Bimetallic Cu-Pd, in contrast to monometallic Cu and Pd, favored sequential C–OH hydrogenolysis and C=O hydrogenation of the intermediates, resulting in the production of 2,5-DMF with a high yield of 73.4% during the one-pot conversion of sucrose at 200 °C and 1 MPa H_2 for 3 h. 2,5-DMF was produced during the conversion of starch with a53.6% yield. In the presence of 0.01 M HCl, cellulose was converted into 2,5-DMF with ayield of 29.8% [128].

Another example of tandem catalytic processes is the one-pot three-step deacetalization–Knoevenagel–hydrogenation (D–K–H) reaction (Figure 57), requiring the synergetic catalysis and the closelocation of the acid, base, and metal sites.D–K–H can be successfully carried out using a trifunctional integrated catalyst (Pd@HPW@HP-UiO-66-NH_2): phosphotungstic acid (HPW) immobilized on the hierarchically porous UiO-66-NH_2 acting as a support for Pd NPs [129].

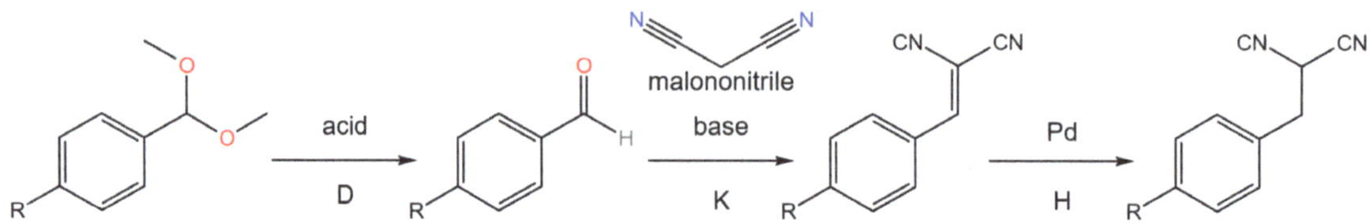

Figure 57. Proposed mechanism of the one-pot oxidation–acetalization [129].

While controlling the ratio of immobilized HPW and amino groups, acidic and basic properties can be easily tuned, and thus the catalytic performance of Pd NPs and its stability can be controlled.The as-prepared catalyst revealed good catalytic activity in the one-pot D–K–H tandem processes, allowing up to 99% substrate conversion and 97% yield of the target product (toluene, 80 °C, 12 + 24 h). Moreover, due to the strong interaction between the MOF (HP-UiO-66-NH_2) and the guest molecules, the catalyst could be reused at least five times without noticeable loss of its activity [129].

By using a straightforward two-step post-synthetic modification, Lee et al. [130] synthesized an acid–base bifunctional zeolitic imidazolate framework catalyst (ZIF-8-A61-SO_3H) with amine and sulfonic acid groups. First, amine-functionalized ZIF-8 with amine contents of 61% (ZIF-8-A61) was obtained by the ligand exchange of 2-mIM with 3-amino-1,2,4-triazole (Atz). Then, the sulfonic acid functionalization by the ring-opening reaction of 1,3-propanesultone with –NH_2 groups in ZIF-8-A61 was carried out. Different amine-functionalized ZIF-8-A materials (15%, 34%, and 61% ofamine content) were prepared by controlling the synthesis time. The catalysts were used for one-pot deacetalization–Knoevenagel (D-K) condensation tandem reaction (the reaction is similar to the one presented in the Figure 51). The ZIF-8-A61-SO_3H catalyst allowed for 100% conversion of the reactant and 98% selectivity of the final Knoevenagel product at mild conditions (0.1 g of catalyst, 1,4-dioxane/H_2O, 80 °C, 4 h) [130]. The developed ZIF-8-A61-SO_3H [130] seemedto possess higher efficiencyin in the D-K process as compared to HPW@HP-UiO-66-NH_2 [129]. Thus, the immobilization of Pd NPs in ZIF-8-A61-SO_3H may result in the further improvement of the D–K–H tandem reaction.

As an another example of a bifunctional catalyst, a double-shelled hollow polymer microsphere wasdeveloped [131]. Pd NPs were found mostly in the outer shell of the catalytic material and stabilized by pyrrolidone groups. The inner shell P(EGDMA-co-AA) contained the acid sites. Theresulting bifunctional catalytic system (void@PAA/PNVP@Pd) was applied for the deacetalization–hydrogenation tandem process (Figure 58), demonstrating the viability of the plan to load distinct catalytic sites onto the walls of double-shelled hollow polymer microspheres.Deacetalization of benzaldehyde dimethyl acetal was followed by hydrogenation with the formation of benzyl alcohol. As a result, the conversion of benzaldehyde dimethyl acetal reached 99% with a 96% yield of the benzyl alcohol [131].

Bifunctionalcatalytic system of Pd/C and water-tolerant Lewis acid (i.e., Sm(OTf)$_3$, La(OTf)$_3$, Cu(OTf)$_2$) were applied for the synthesis of fuel precursors from biomass liquefaction with high efficiency in both water and ethanol [132]. The maximum yield of bio-oil (49.71 wt.%)was achieved for 30 min at supercritical ethanol (300 °C) in the presence of Pd/C and La(OTf)$_3$.

Raza et al. [133] synthesized highly dispersed Pd NPs immobilized over covalent triazine polymer (CTP) functionalized with sulfonic acid groups (CTP-SO_3H/Pd). The sulfonic acid groups were shown to be responsible for the uniform dispersion of palladium NPs over the CTP. The obtained bifunctional catalyst was used in the one-pot hydrogenation–esterification (OHE) reaction and revealed promising catalytic activity with a 94% yield of the target product at 95% conversion (Figure 59).

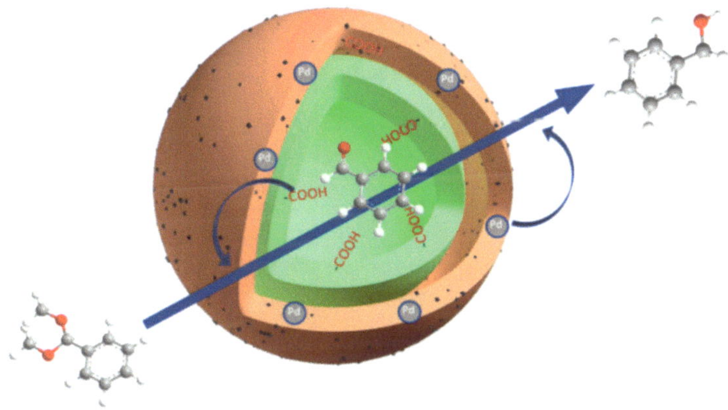

Figure 58. The void@PAA/PNVP@Pd hollow multishell microspheres as a nanoreactor for the tandem reaction. Reproduced with permission from [131], Elsevier, 2023.

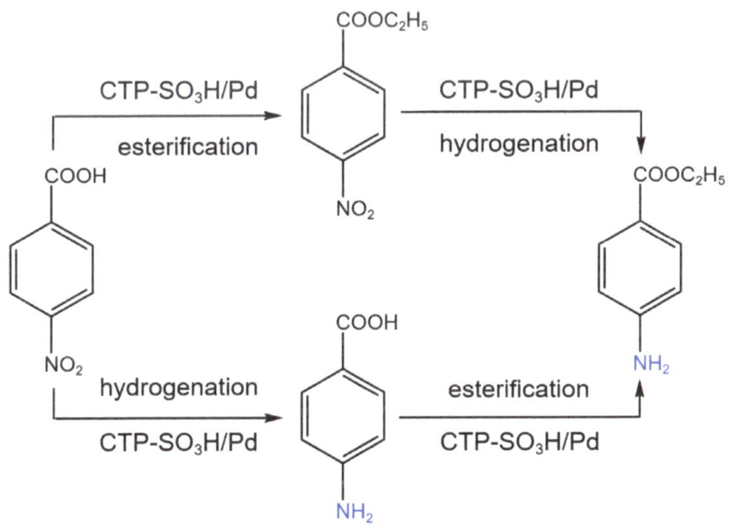

Figure 59. Scheme of the OHE reaction [133].

After the reaction, the catalyst was separated by simple filtration. It was shown that the CTP-SO_3H/Pd catalyst could be reused for at least five times with a slight loss of catalytic activity, indicating its potential usage in OHE reactions. The cooperative effect of functional acidic and metal sites was proposed [133].

Covalent organic frameworks (COFs) are widely used as a support for catalytic applications [134–137], including tandem one-pot processes [138]. The COFs possess a variety of chemical structures, which may be precisely tuned to control the surrounding coordination environment and electronic interaction between metal NPs and supports, in contrast to other materials used for NPs immobilization.The uniform porous structure in COFs allows for numerous metal active sites, which enhances the catalytic activity [139]. A simple hydrothermal synthesis was recently used by Wang et al. [139] to produce COF. Noble metals (Pt, Pd, and Rh) were introduced to the COF matrix for further application in the reductive amination of benzaldehyde.The optimum Pd/COF catalyst revealed a 91% yield of a secondary aminesat 1.2:1 M ratio of aldehyde and ammonia under mild reaction conditions (2 MPa of H_2 and 90 °C, 15 h) [139].

5. Chemoenzymatic Processes

At present, the enzymatic catalysis is considered a powerful tool for synthetic chemists, providing access to a wide range of compounds. Advances in the field of immobilization, molecular biology, and bioinformatics have paved the way for biotransformation in various environments, improving the stability and activity of biocatalysts, opening up new enzymatic pathways. The capacity of enzymes is attracting increasingly more attention, since they can be successfully combined with other types of catalysts, as shown by recent achievements in their joint action with metals [140–142]. This section is devoted to the consideration of the combination of palladium and enzymatic catalysts for chemoenzymatic processes. This combination gives the advantages of conducting tandem processes in a single reactor, which simplifies the complex routes, avoiding the separation of unstable intermediates to obtain final products with a higher yield.

In the recent review by Gonzalez-Granda et al. [140], the combination of enzymes and transition metals in catalysis for asymmetric synthesis was considered. Reactions such as Suzuki cross-coupling [143–145], Wacker–Tsuji oxidation [146–148], and Buchwald–Hartwig cross-coupling [149–151] were discussed.

Metal-enzymatic parallel and sequential transformations were described, involving Pd, Ru, Au, Ir, and Fe, which catalyze numerous organic transformations (C-C coupling, isomerization, hydrogenation, etc.). The combination of catalytically active metals with enzymes in the chemoenzymatic processes allows for the obtaining of chiral products due to the action of stereospecific enzymes, including alcohol oxidases, aldolases, alcohol dehydrogenases, amine dehydrogenases, amino acid dehydrogenases, aminotransaminases, arylmalonate decarboxylase, enreductase, iminoreductase, nitrile hydratase, or phenylalanine monialiases, among others. It was noticed [140] that incompatibilities between chemical and enzymatic steps can be often found, such as cross-prohibitions due to the presence of cofactors, reagents, or intermediates, or the preference for different solvents or temperatures, and thus these protocols must be accomplished sequentially.

A stereoselective one-pot tandem reduction of 3-methyl-2-cyclohexenone to 3-(1*S*,3*S*)-methylcyclohexanol (Figure 60) was carried out by Coccia et al. [152] using Pd and Pt NPs as the metal precatalyst and a NAD^+-dependent thermostable alcohol dehydrogenase isolated from *Thermus* sp. ATN1 (TADH).

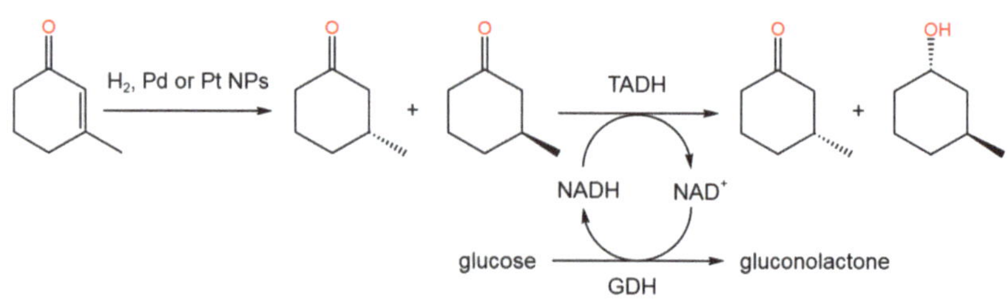

Figure 60. Scheme of the tandem selective reduction of the C=C bond with metal (Pd or Pt) NPs and the asymmetric reduction of the keto group by TADH(GDH—glucose dehydrogenase) [152].

The Pd and Pt NPs possess a high surface-to-volume ratio, simple preparation, and good "solubility" in water. Moreover, the metal NPs can work without significant pH or temperature limitations and can be used in different reactions such as oxidation and reduction. TADH revealed a broad substrate scope including aldehydes, aliphatic ketones, cyclic ketones, and double-ring systems [152].

The assumptions were made on the interactions between the chemo- and the biocatalyst. The sizes of NPs were demonstrated to be a crucial parameter for mutual inhibition: the larger the NPs, the higher the enzyme inhibition, and vice versa: the smaller the NPs, the lower the TADH denaturation. In general, the chemocatalysts possessed high deactivation sensitivity, which was highly dependent on the amount of enzyme utilized, i.e., the inhibition of the biocatalyst could be greatly decreased by reducing the NPs/TADH ratio. In order to avoid the direct binding of NPs to TADH, the use of large Pd NPs protected with a silica shell is promising: the yield of 3-(1*S*,3*S*)-methylcyclohexanol was increased up to 36% [152].

Combining metal ions or NPs with biocatalysts in a single system is a promising strategy for implementation in the one-pot chemoenzymatic cascade reactions [153], especially for asymmetric syntheses. For example, Li et al. [154] developed a one-pot chemoenzymatic cascade reaction to asymmetric synthesize (*R*)-1-(4-biphenyl) ethanol (Figure 61) while using a highly active and selective enzyme–metal–single-atom catalyst. In the facilitation of the production of chiral biaryl alcohols, the Pd single-atom-anchored lipase (Pd1/CALB-P) may effectively drive one-pot cascade reactions in aqueous solution at 30 °C. The production rate was 30 times higher than that catalyzed by the commercial Pd/C and CALBP.

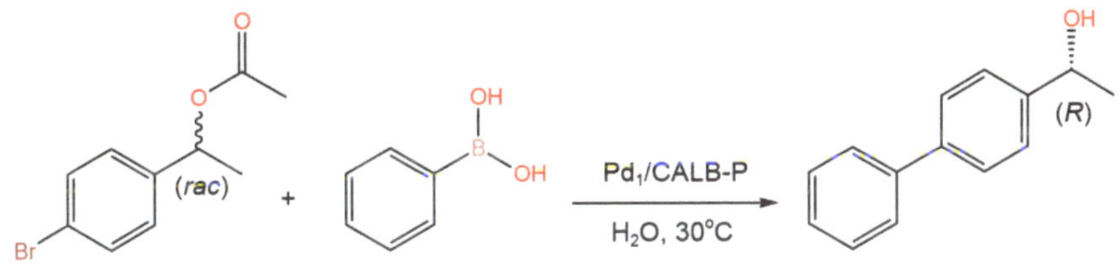

Figure 61. Synthesis of (*R*)-1-(4-biphenyl)ethanol by a one-pot chemoenzymatic reaction [154].

In an other recent work of Li et al. [155], a bimetallic PdCu/CALB CLEAs hybrid catalyst was synthesized by the in situ reduction of PdCu nanoclusters immobilized on cross-linked lipase aggregates (CALB CLEAs).

Cross-linked enzyme aggregates (CLEAs) were prepared from *Candida antarctic* lipase B (CALB) by their precipitation and subsequent cross-linking with the glutaraldehyde. The bimetallic PdCu nanoclusters were formed in situ on CALB CLEAs (Figure 62) by the reduction of Pd^{2+} and Cu^{2+} ions in an aqueous solution containing 20% (v/v) methanol, which served as an additional reducing agent. The obtained PdCu nanoclusters were highly dispersed (sizes of 1.5 ± 0.2 nm). The strong synergistic effect between Pd and Cu in PdCu/CALB CLEAs allowed for high activities in the Sonogashira cross-coupling reaction and one-pot chemoenzymatic reaction, resulting in the synthesis of (*R*)-*N*-[1-(4-(phenylethynyl)phenyl)ethyl]acetamide (Figure 63) [155].

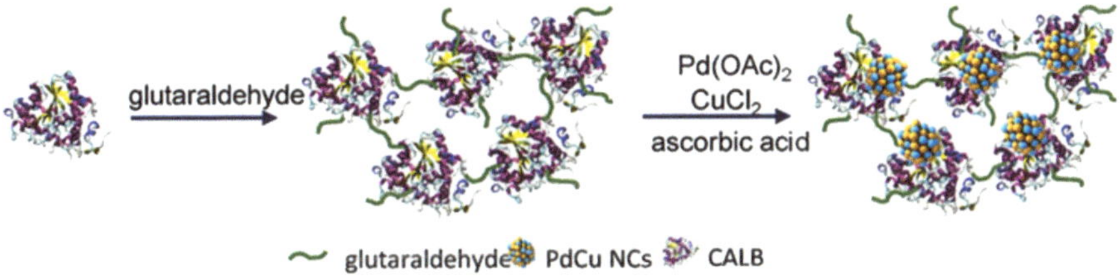

Figure 62. Schematic illustration of the strategy used to synthesize PdCu/CALB CLEAs via in situ reduction. Reproduced with permission from [155], Elsevier, 2021.

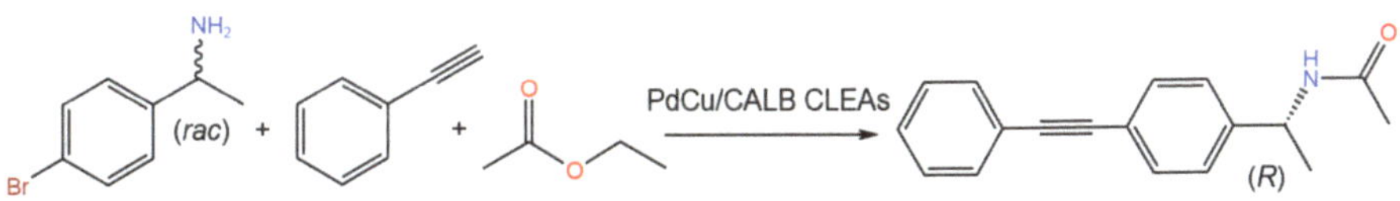

Figure 63. Synthesis of (*R*)-*N*-[1-(4-(phenylethynyl)phenyl)ethyl]acetamide by one-pot chemoenzymatic reaction [155].

In this reaction (Figure 63), CALB catalyzed the acylation of (*R*)-enantiomer of the amine, while PdCu nanoclusters were responsible for the Sonogashira cross-coupling and the racemization of (*S*)-enantiomer of the amine. It was proposed that Cu-alkynyl and Pd-aryl particles were formed simultaneously by surface coordination. The coordinated Pd and Cu particles were then subjected to the transmetalation stage, which facilitated the cross-coupling reaction. Catalytic activity increased with the increase in the content of PdO [155]. It is noteworthy that the higher activity of PdO as compared to metallic palladium was described earlier for the reaction of Suzuki cross-coupling by Collins et al. [156].

Deiana et al. [157] developed the bioinspired multicatalytic system based on an artificial plant cell wall (APCW) containing the lipase and Pd NPs for the transformation of racemic amine into the corresponding enantiomerically pure amide with a yield of up to 99% (Figure 64), which involves a synergistic interaction between the racemization reaction catalyzed by Pd(0) and the enantioselective amidation catalyzed by CALB.

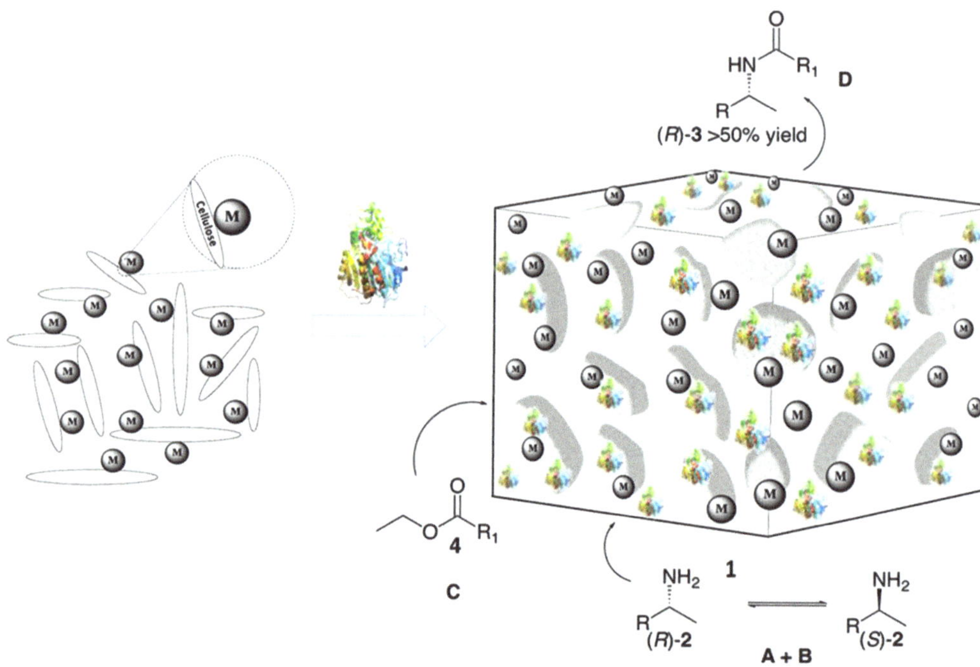

Figure 64. Simplified scheme of self-assembly of components to create an APCW, which simultaneously catalyzes racemization and amidation of racemic amines. M = metal. Reproduced with permission from [157], RSC, 2021.

The main component of the developed multicatalytic system is MCC or nanofibrillated cellulose, the surface of which is modified with aminopropylsilane, which allows for the retention of Pd NPs. The best approach for the self-assembly of the resulting hybrid catalyst was a non-covalent modification of CALB with surfactant polyethylene glycol hexadecyl ether (Brij) in a phosphate buffer. TEM data confirmed the formation of Pd NPs with a narrow size distribution (from 1.6 nm up to 2.8 nm). The hybrid heterogeneous multicatalytic system APCW9 was more efficient and chemoselective as compared to the mixture of Pd NPs on MCC and CALB on MCC [157].

By the example of different aryl and heteroaryl scaffolds, Craven et al. [158] showed that flavin adenine dinucleotide (FAD)-dependent halogenases (Fl-Hal) can be used in combination with Pd-catalyzed cyanation to affect the C–H functionalization cascades that deliver nitriles in a highly regioselective manner. As a non-toxic cyanation reagent,

$K_4[Fe(CN)_6]$ was used. It was shown that this approach can be extended by including nitrile hydratase (NHase) or nitrilase (NITR) to create integrated cascades of three catalysts for the regioselective installation of amide and carboxylic acid groups (Figure 65).

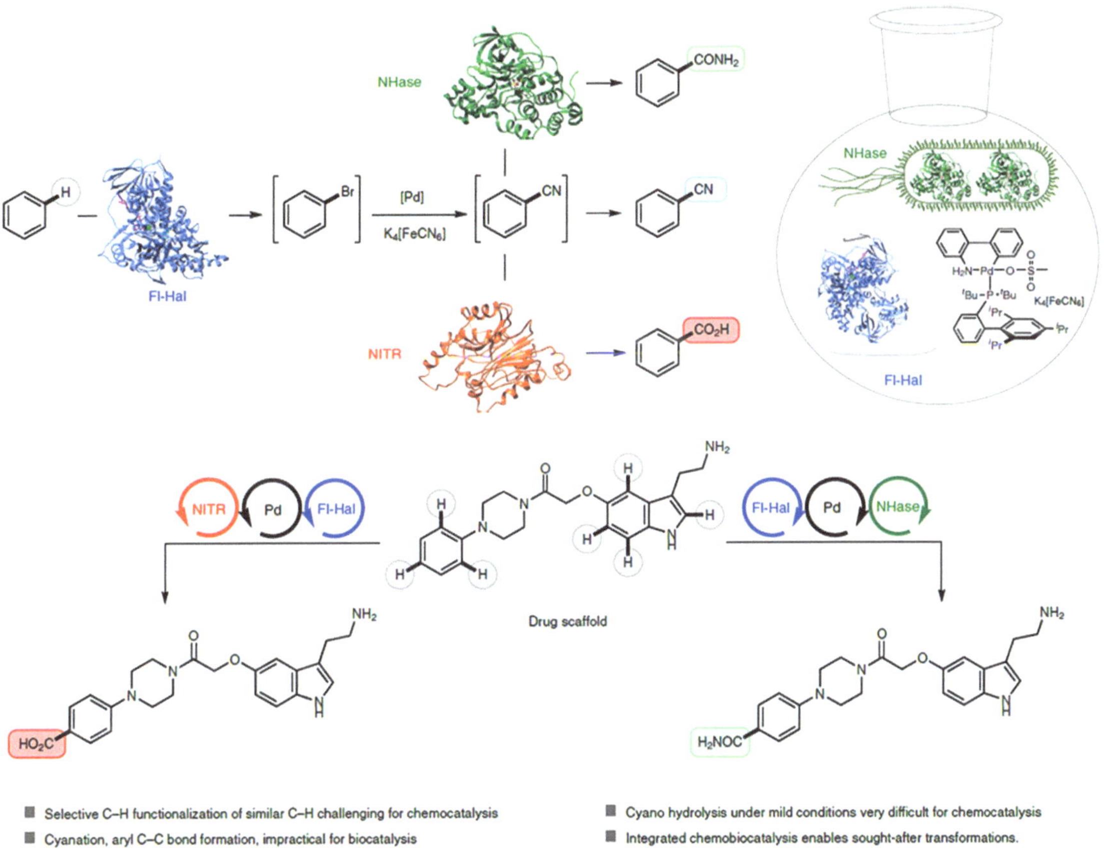

Figure 65. Overview of site-selective C–H functionalization cascades. Reproduced with permission from [158], Springer Nature, 2021.

It was emphasized [158] that the recent advances in the discovery, engineering, and synthesis of Fl-Hal biocatalysts can significantly broaden the application of this strategy. New Fl-Hals with improved catalytic characteristics, modified substrate specificities, and altered regioselectivity are continuously emerging from genome mining and directed-evolution programs. The inclusion of the Fl-Hals will boost the potential of the proposed programmable integrated cascades, enabling the integration of rich functionality into a variety of candidate scaffolds and facilitating the manufacturing of target molecules.

PdCu hydrogel nanozymes with a hierarchically porous structure were synthesized by Huang et al. [159] to immobilize horseradish peroxidase (HRP) (PdCu@HRP). The 3D porous nanowire networks of resulting hydrogels with high porosity served as biocompatible supports for immobilizing HRP. The affinity of Cu and Pd to proteins simplified the procedures of the enzyme immobilization without the addition of other cross-linking agents. The immobilization of HRP on PdCu hydrogels enhanced the thermal and chemical stabilities of HRP, realizing the reuse of the enzyme. Moreover, PdCu@HRP exhibited synergistically enhanced HRP activity as compared to native HRP and PdCu hydrogel nanozymes. The improved catalytic activity was likely due to specific interactions between

PdCu hydrogel nanozymes and enzymes as well as the enrichment of substrates around enzymes by electrostatic adsorption of hydrogels. Using catalytic cascade reactions, colorimetric biosensing of the carcinoembryonic antigen (CEA) was carried out while applying the PdCu@HRP and the glucose oxidase encapsulated in ZIF-8.The obtained biosensor allowed for the quantitative probing of the CEA concentration in a wide range from 5 to 1000 pg/mL with a LOD of 1.4 pg/mL and nearly 6.1-fold increase in the detection sensitivity as compared to the conventional HRP-based enzyme-linked immunosorbent assay [159].

It is important to highlight the work of Ming et al. [160], who developed a novel tandem Pd-Ru/Uricase@RBC nanoreactor, including Pd-Ru nanosheets, uricase, and red blood cell (RBC) membrane for hyperuricemia treatment (Figure 66). A new highly active 2D nanozyme Pd-Ru was covalently bounded with the uricase and immobilized on the surface of RBCs. It is noteworthy that the RBC membrane coating is a new biomimetic technique that gives nanomaterials a natural surface, and it can significantly increase the time that they circulate in the bloodstream [160].

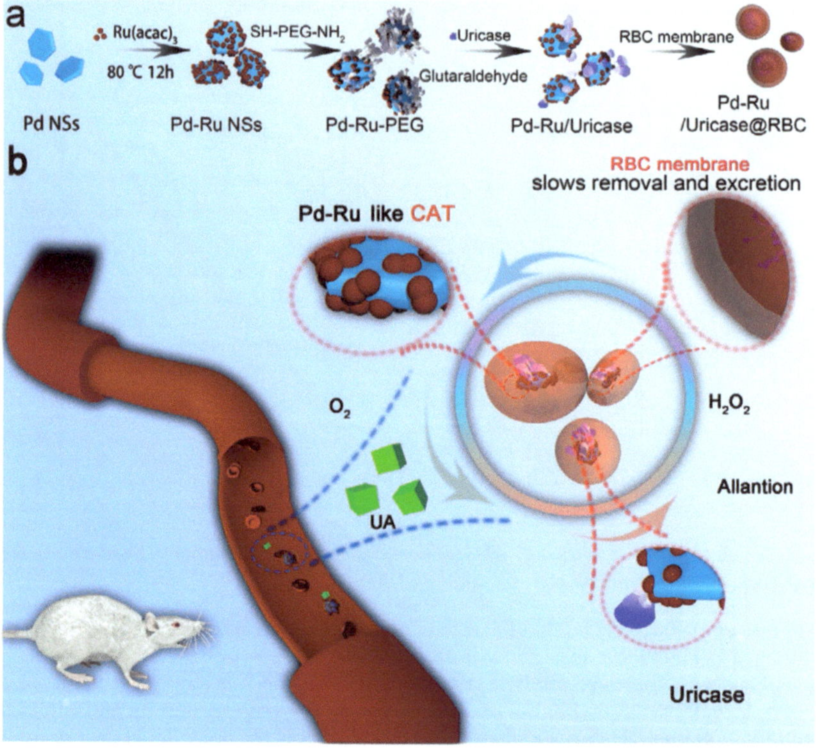

Figure 66. Schematic illustration of the construction procedure of Pd-Ru/Uricase@RBC (**a**) and its treatment of hyperuricemia (**b**). Reproduced with permission from [160], John Wiley and Sons, 2021.

The prepared Pd-Ru/Uricase@RBC demonstrated high catalase-like activity (Pd-Ru nanosheets decomposed H_2O_2 to rapidly generate O_2) and stability against various extreme pH values, temperatures, and forms of proteolytic degradation during biological transport. Moreover, the Pd-Ru nanozyme and uricase being in close proximity to each other allowed for the achievement of the efficient cascade reactions: degradation of the uric acid by uricase to allantoin and H_2O_2, as well as the removal of H_2O_2 by Pd-Ru nanosheets. The generated O_2 even facilitated the catalytic degradation of the uric acid [160]. Thus, the cascade reactions based on the nanozymes were shown to be an effective strategy for the treatment of diseases due to its high efficiency and low level of side effects.

Zhang et al. [161] developed bifunctional biocatalysts based on mesoporous silica NPs (MSN), the surface hydrophobicity of which was created via alkylation. Pd NPs and the enzyme CALB were separately loaded into compartmentalized locations (Figure 67).

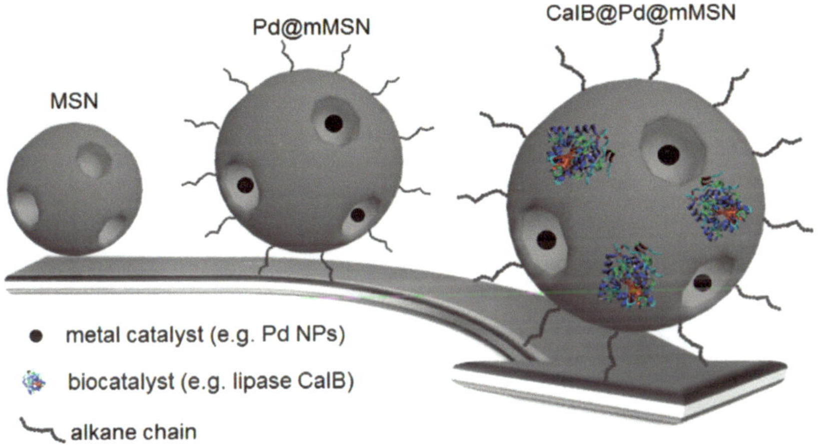

Figure 67. Schematic illustration of the construction of CalB@Pd@mMSN. Reproduced with permission from [161], Copyright 2018 American Chemical Society.

Pd(0) NPs were loaded into the MSN by the in situ reduction of Pd acetatewith $NaBH_4$. Then, the surface of Pd@MSN was modified with long-chain alkanes (obtaining Pd@mMSN). Subsequently, CALB was immobilized via hydrophobic interactions (bifunctional biocatalyst was denoted as CalB@Pd@mMSN) [161].

The CalB@Pd@mMSN was tested in a one-pot cascade reaction, in which Pd NPs first reduced the benzaldehyde, and then the immobilized CalB converted the benzyl alcohol into benzyl hexanoate (Figure 68). The developed catalyst was highly active and reusable: after four times of reuse, the reaction yield (4 h) remained higher than 80%.

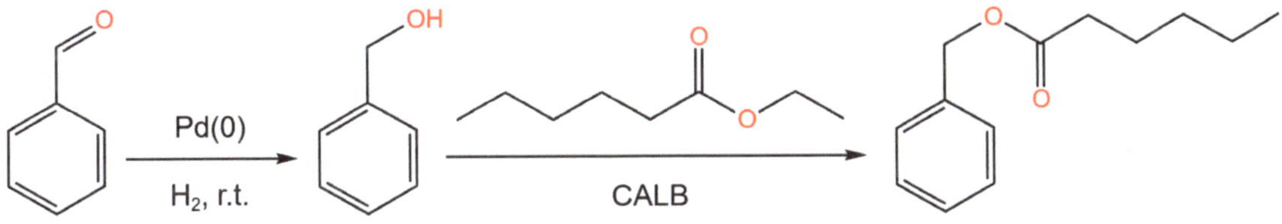

Figure 68. Scheme of the one-pot cascade reaction of benzaldehyde with ethyl hexanoate [161].

The developed bifunctional biocatalyst provides a universal platform with a large surface area for the transfer of catalysts to various organic solvents. Thus, it can be developed as a tool for adding other chemical and biocatalysts (such as glucose oxidase and other proteins) to the reaction medium of interest [161].

As was shown in the review by Metzger et al. [162], the combination of inorganic supports and catalysts into tandem systems is an actively developing area of tandem catalysis, since the durability of the systems allows for their application in industrial processes. Compartmentalization was shown to be a method for overcoming the difficulties associated with mixing several incompatible catalysts into a one-pot system. Compartmentalization can decrease side reactions, inhomogeneity, and catalyst deactivation that are often experienced when incompatible catalysts are in close proximity. Thus, the application of MSNs and MOFs as supports in tandem chemoenzymatic processes is promising.

Chemoenzymatic tandem reactions for the synthesis of pharmacologically active compounds in continuous flow was developed by Lackner et al. [163]. While using two sequential packed bed reactors hosting encapsulated phenolic acid decarboxylase from *B. subtilis* (*Bs*PAD) facilitating enzymatic decarboxylation as well as a heterogeneous Pd catalyst for Heck coupling, stilbene derivativeswere synthesized (Figure 69).

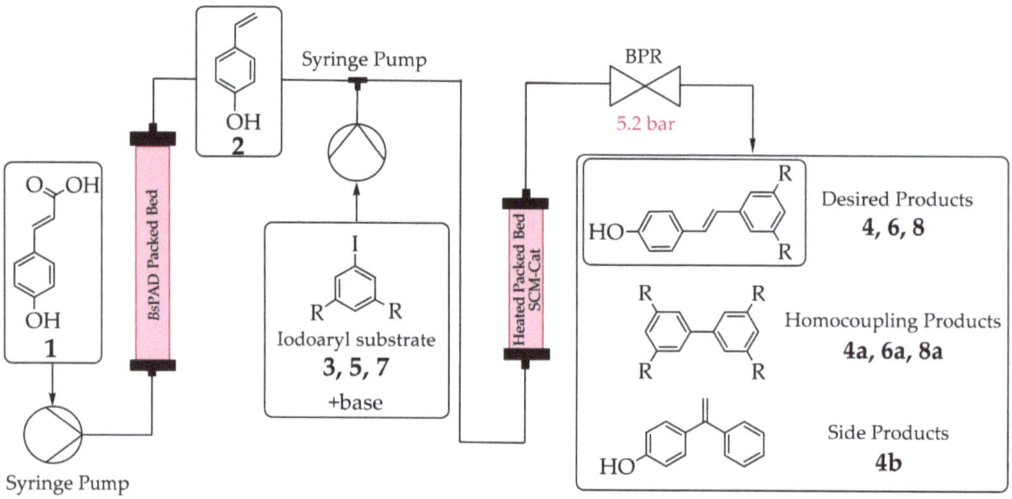

Residue [1]	Product	Iodoaryl substrate	Homoproduct
R = H	4-hydroxystilbene **4**	3	**4a**
R = OH	resveratrol **6**	5	**6a**
R = OMe	pterostilbene **8**	7	**8a**

[1] All compounds in the same row have the same residue

Figure 69. Scheme of the chemoenzymatic cascade for the synthesis of stilbenes in continuous flow [163].

Varying the reaction conditions of the cross-coupling step (pH, carbonate concentration, temperature) allowed for the obtaining of the 4-hydroxystilbene and other products (resveratrol and pterostilbene) with yields of 32–54%. By changing the substrate being subjected to enzymatic decarboxylation and using more complex iodoaryl coupling partners, this chemoenzymatic cascade is likely to be further expanded to the production of other valuable stilbenes [163].

6. Conclusions

Tandem processes is a hot area of catlytic reasearch aimed at performing the traditional stepwise reactions in one-pot mode. Palladium, as one of the most abundant metalcatalysts, is oftenly used in tandem processes.

In many homogeneouslycatalyzed reactions, the tandem processes can be implemented due to the known ability of palladium to constantly change its oxydation state during the reaction course. This ability of Pd to catalyze the reactions while being in different oxidation states generated from a single precatalyst is effectively used in those tandem processes, which include, for example, the C–C coupling (Heck, Suzuki, Sonogashira, etc.), isomerization, cyclization, and carbonylation, as some of the reactions in tandem. Thus, some authors refer such processes to the orthogonal catalysis, considering Pd(II) and Pd(0) as two different catalysts. Since in homogeneouslycatalyzed tandem processes, the catalysts are rather ordinar (Pd salts or complexes), the current trends are towards the development of new routes to the one-pot tandem syntheses of certain structures, which would allow for higher efficiency at milder reaction conditions.

In heterogeneous catalysis for tandem processes, two or more active sites are required with their rational desing to act independently or synergistically. Many of such processes are not truly tandem and rather can be attributed to orthogonal catalysis, especially those related to the biomass processing in the presence of supported Pd-containg catalytic systems bearing acid sites (Figure 70). For the last years, materials such as zeolites, oxides (especially mesoporous), MOFs, and COFs gained attention as supports for the development of heterogeneous Pd-containing materials for tandem processes.

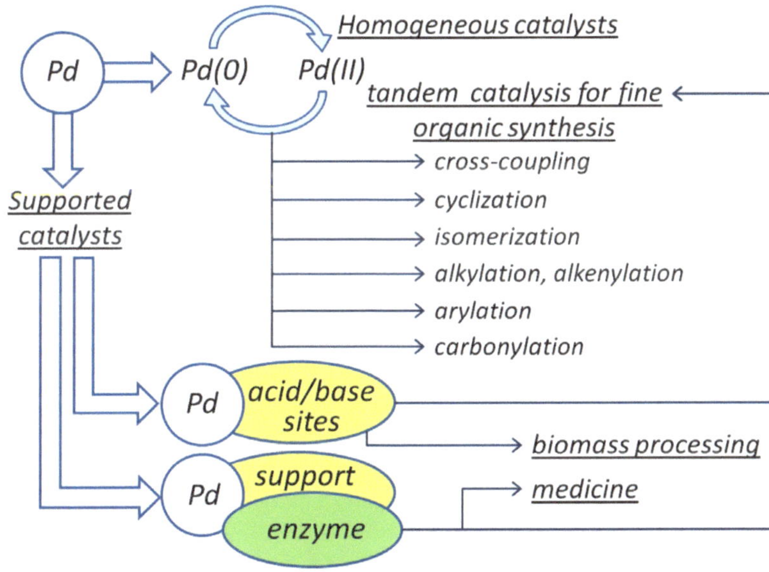

Figure 70. General scheme of the application of Pd-catalyzed tandem processes.

The third type of prospective catalysts for tandem processes is the systems combining the action of metals and enzymes (chemoenzymatic catalysis). Such mulifunctional metal-enzyme catalysts are highly perspective for asymmetric catalysis due to their known extremely high substrate sensitivity and the selectivity of enzymes as compared to inorganic catalysts.

In spite of some existing difficulties due to the different reaction conditions required for the optimum reactivity of inorganic and organic parts as well as the possibility of enzyme inactivation with metal NPs, chemoenzymatic tandem processes are highly demanded by modern organic synthesis and biotechnology.

Author Contributions: Conceptualization, L.Z.N.; methodology, L.Z.N.; software, L.Z.N.; validation, V.G.M.; formal analysis, L.Z.N.; investigation, L.Z.N.; resources, L.Z.N.; data curation, V.G.M.; writing—original draft preparation, L.Z.N. and V.G.M.; writing—review and editing, L.Z.N. and V.G.M.; visualization, L.Z.N.; supervision, V.G.M.; project administration, V.G.M.; funding acquisition, V.G.M. All authors have read and agreed to the published version of the manuscript.

Funding: This research was funded by the Russian Science Foundation, grant number 23-79-00009.

Data Availability Statement: Not applicable.

Conflicts of Interest: The authors declare no conflict of interest.

References

1. Campos, J.F.; Berteina-Raboin, S. Tandem Catalysis: Synthesis of Nitrogen-Containing Heterocycles. *Catalysts* **2020**, *10*, 631. [CrossRef]
2. Camp, J.E. Auto-Tandem Catalysis: Activation of Multiple, Mechanistically Distinct Process by a Single Catalyst. *Eur. J. Org. Chem.* **2017**, *2017*, 425–433. [CrossRef]
3. Fogg, D.E.; dos Santos, E.N. Tandem Catalysis: A Taxonomy and Illustrative Review. *Coord. Chem. Rev.* **2004**, *248*, 2365–2379. [CrossRef]
4. Das, S.; Hong, D.; Chen, Z.; She, Z.; Hersh, W.H.; Subramaniam, G.; Chen, Y. Auto-Tandem Palladium Catalysis: From Isoxazoleto 2-Azafluorenone. *Org. Lett.* **2015**, *17*, 5578–5581. [CrossRef]
5. Chen, P.; Chen, Z.-C.; Li, Y.; Ouyang, Q.; Du, W.; Chen, Y.-C. Auto-Tandem Cooperative Catalysis Using Phosphine/Palladium: Reactionof Morita–Baylis–Hillman Carbonates and Allylic Alcohols. *Angew. Chem.* **2019**, *131*, 4076–4080. [CrossRef]
6. Alexander, J.R.; Shchepetkina, V.I.; Stankevich, K.S.; Benedict, R.J.; Bernhard, S.P.; Dreiling, R.J.; Cook, M.J. Pd-Catalyzed Rearrangement of *N*-Alloc-*N*-allyl Ynamides via Auto-Tandem Catalysis: Evidence for Reversible C–N Activation and Pd(0)-Accelerated Ketenimine Aza-Claisen Rearrangement. *Org. Lett.* **2021**, *23*, 559–564. [CrossRef]
7. Jeske, K.; Rösler, T.; Belleflamme, M.; Rodenas, T.; Fischer, N.; Claeys, M.; Leitner, W.; Vorholt, A.J.; Prieto, G. Direct Conversion of Syngas to Higher Alcohols via Tandem Integration of Fischer–Tropsch Synthesis and Reductive Hydroformylation. *Angew. Chem. Int. Ed.* **2022**, *61*, e202201004. [CrossRef]
8. Araújo, M.; MuñozCapdevila, I.; Díaz-Oltra, S.; Escuder, B. Tandem Catalysis of an Aldol-'Click' Reaction System within a Molecular Hydrogel. *Molecules* **2016**, *21*, 744. [CrossRef] [PubMed]
9. Wei, W.; Thakur, V.K.; Chew, Y.M.J.; Li, S. Towards Next Generation "Smart" Tandem Catalysts with Sandwiched Mussel-inspired Layer Switch. *Mater. Today Chem.* **2020**, *17*, 100286. [CrossRef]
10. Li, N.; Huang, B.; Dong, X.; Luo, J.; Wang, Y.; Wang, H.; Miao, D.; Pan, Y.; Jiao, F.; Xiao, J.; et al. Bifunctional Zeolites-silver Catalyst Enabled Tandem Oxidation of Formaldehyde at Low Temperatures. *Nat. Commun.* **2022**, *13*, 2209. [CrossRef]
11. Gumus, I.; Ruzgar, A.; Karatas, Y.; Gülcan, M. Highly Efficient and Selective One-pot Tandem Imine Synthesis via Amine-alcohol Cross-coupling Reaction Catalysed by Chromium-based MIL-101 Supported Au Nanoparticles. *Mol. Catal.* **2021**, *501*, 111363. [CrossRef]
12. Liu, J.; Cui, J.; Chen, L.; Chen, J.; Zheng, H.; Oyama, S.T. Advantages of Tandem versus Simultaneous Operation: The Case of Isomerization/Hydrogenation of Terpinolene Epoxideto Terpinen-4-ol usinga Ni/TiO_2-SiO_2 Bifunctional Catalyst. *Chem. Eng. Sci.* **2022**, *259*, 117828. [CrossRef]
13. Borguet, Y.; Sauvage, X.; Zaragoza, G.; Demonceau, A.; Delaude, L. Tandem Catalysis of Ring-closing Metathesis/AtomTransfer Radical Reactions with Homobimetallic Ruthenium-Arene Complexes. *Beilstein J. Org. Chem.* **2010**, *6*, 1167–1173. [CrossRef] [PubMed]
14. Wang, L.; Yang, Y.; Shi, Y.; Liu, W.; Tian, Z.; Zhang, X.; Zheng, L.; Hong, S.; Wei, M. Single-atom Catalysts with Metal-acid Synergistic Effect Toward Hydrodeoxygenation Tandem Reactions. *Chem. Catal.* **2023**, *3*, 100483. [CrossRef]
15. Cho, H.J.; Xu, B. Enabling Selective Tandem Reactions via Catalyst Architecture Engineering. *Trends Chem.* **2020**, *2*, 929–941. [CrossRef]
16. Anderson, A.E.; Baddeley, C.J.; Wright, P.A. Tuning Pd-nanoparticle@MIL-101(Cr) Catalysts for Tandem Reductive Amination. *Catal. Lett.* **2018**, *148*, 154–163. [CrossRef] [PubMed]
17. Pellissier, H. Recent Developments in Enantioselective Multicatalyzed Tandem Reactions. *Adv. Synth. Catal.* **2020**, *362*, 2289–2325. [CrossRef]
18. Reen, G.K.; Kumar, A.; Sharma, P. Recent Advances on the Transition-metal-catalyzed Synthesis of Imidazopyridines: An Updated Coverage. *Beilstein J. Org. Chem.* **2019**, *15*, 1612–1704. [CrossRef]
19. Climent, M.J.; Corma, A.; Iborra, S.; Sabater, M.J. Heterogeneous Catalysis for Tandem Reactions. *ACS Catal.* **2014**, *4*, 870–891. [CrossRef]
20. She, W.; Wang, J.; Li, X.; Li, J.; Mao, G.; Li, W.; Li, G. Bimetallic CuZn-MOFs derived Cu-ZnO/C catalyst for reductive amination of nitroarenes with aromatic aldehydes tandem reaction. *Appl. Surf. Sci.* **2021**, *569*, 151033. [CrossRef]
21. Zhang, S.; Ma, H.; Sun, Y.; Liu, X.; Zhang, M.; Luo, Y.; Gao, J.; Xu, J. Selective Tandem Hydrogenation and Rearrangement of Furfural to Cyclopentanone over CuNi Bimetallic Catalyst in Water. *Chin. J. Catal.* **2021**, *42*, 2216–2224. [CrossRef]
22. George, J.; Kim, H.Y.; Oh, K. Cooperative Pd/Cu Catalysis to Spiro[indoline-2,3′-pyrrolidin]-2′-ones: Tandem Benzylation of α-Isocyano Lactams, Amine Addition, and N-Arylation. *Org. Lett.* **2019**, *21*, 5747–5752. [CrossRef] [PubMed]
23. Chen, H.; Li, H.; Chen, S.; Sheng, L.; Zhang, Z.; Wu, W.; Fan, M.; Wang, L.; Yang, B. Atomic Pd dispersion in triangular Cu nanosheets with dominant (111) plane as a tandem catalyst for highly efficient and selective electrodehalogenation. *Appl. Catal. B Environ.* **2023**, *328*, 122480. [CrossRef]
24. Dong, Y.; Li, W.-H.; Dong, Y.-B. Dual-Metal *N*-Heterocyclic Carbene Complex (M= Au and Pd)-Functionalized UiO-67MOF for Alkyne Hydration-Suzuki Coupling Tandem Reaction. *J. Org. Chem.* **2021**, *86*, 1818–1826. [CrossRef]
25. Nishad, R.C.; Kumar, S.; Rit, A. Hetero- and Homobimetallic Complexes Bridged by a Bis(NHC) Ligand: Synthesis via Selective Sequential Metalation and Catalytic Applications in Tandem Organic Transformations. *Organometallics* **2021**, *40*, 915–926. [CrossRef]

26. Dehury, N.; Tripathy, S.K.; Sahoo, A.; Maity, N.; Patra, S. Facile Tandem Suzuki Coupling/Transfer Hydrogenation Reaction with a bis-Heteroscorpionate Pd–Ru Complex. *Dalton Trans.* **2014**, *43*, 16597–16600. [CrossRef] [PubMed]
27. Pezük, L.G.; Şen, B.; Hahn, F.E.; Türkmen, H. Heterobimetallic Complexes Bridged by Imidazol{[4,5-*f*][1,10]-phenanthrolin}-2-ylidene: Synthesis and Catalytic Activity in Tandem Reactions. *Organometallics* **2019**, *38*, 593–601. [CrossRef]
28. Mandegani, Z.; Nahaei, A.; Nikravesh, M.; Nabavizadeh, S.M.; Shahsavari, H.R.; Abu-Omar, M.M. Synthesis and Characterization of Rh^{III} M^{II} (M = Pt, Pd) Heterobimetallic Complexes Based on a Bisphosphine Ligand: Tandem Reactions Using Ethanol. *Organometallics* **2020**, *39*, 3879–3891. [CrossRef]
29. Liu, X.; Liu, H.; Wang, Y.; Yang, W.; Yu, Y. Nitrogen-richg-C_3N_4@AgPd Mott-Schottky heterojunction boosts photocatalytic hydrogen production from water and tandem reduction of NO^{3-} and NO^{2-}. *J. Colloid Interface Sci.* **2021**, *581*, 619–626. [CrossRef]
30. Zhao, M.; Zhao, L.; Cao, J.; Jiang, W.; Xie, J.-X.; Zhu, C.; Wang, S.-Y.; Wei, Y.-L.; Zhao, X.-Y.; Bai, H.-C. Water-involved tandem conversion of arylethers to alcohols over metal phosphide catalyst. *Chem. Eng. J.* **2022**, *435*, 134911. [CrossRef]
31. Jia, H.L.; Yang, Y.Y.; Chow, T.H.; Zhang, H.; Liu, X.Y.; Wang, J.F.; Zhang, C.-Y. Symmetry-Broken Au-Cu Heterostructures and their Tandem Catalysis Processin Electrochemical CO_2 Reduction. *Adv. Funct. Mater.* **2021**, *31*, 2101255. [CrossRef]
32. Li, X.-Q.; Duan, G.-Y.; Wang, R.; Han, L.-J.; Wang, Y.-F.; Xu, B.-H. Poly(ionicliquid)-based Bimetallic Tandem Catalysts for Highly Efficient Carbon Dioxide Electroreduction. *Appl. Catal. B Environ.* **2022**, *313*, 121459. [CrossRef]
33. Balanta, A.; Godard, C.; Claver, C. Pd Nanoparticles for C–C Coupling Reactions. *Chem. Soc. Rev.* **2011**, *40*, 4973–4985. [CrossRef] [PubMed]
34. Biffis, A.; Centomo, P.; DelZotto, A.; Zecca, M. Pd Metal Catalysts for Cross-Couplings and Related Reactions in the 21st Century: A Critical Review. *Chem. Rev.* **2018**, *118*, 2249–2295. [CrossRef] [PubMed]
35. Dobrounig, P.; Trobe, M.; Breinbauer, R. Sequential and Iterative Pd-catalyzed Cross-coupling Reactions in Organic Synthesis. *Monatsh. Chem.* **2017**, *148*, 3–35. [CrossRef] [PubMed]
36. Ueno, M.; Miyoshi, N.; Hanada, K.; Kobayashi, S. Three-Component, One-Pot Tandem Sonogashira/Suzuki-Miyaura Coupling Reactions for the Synthesis of a Library of Ceramide-Transport Protein Inhibitors Designed In Silico. *Asian J. Org. Chem.* **2020**, *9*, 267–273. [CrossRef]
37. Zhou, B.; Wang, H.; Cao, Z.-Y.; Zhu, J.-W.; Liang, R.-X.; Hong, X.; Jia, Y.-X. Dearomative 1,4-Difunctionalization of Naphthalenes via Palladium-Catalyzed Tandem Heck/Suzuki Coupling Reaction. *Nat. Commun.* **2020**, *11*, 4380. [CrossRef]
38. Matsude, A.; Hirano, K.; Miura, M. Highly Stereoselective Synthesis of 1,2-Disubstituted Indanes by Pd-Catalyzed Heck/Suzuki Sequence of Diarylmethyl Carbonates. *Org. Lett.* **2020**, *22*, 3190–3194. [CrossRef]
39. Pagliaro, M.; Pandarus, V.; Ciriminna, R.; Béland, F.; Carà, P.D. Heterogeneous versus Homogeneous Palladium Catalysts for Cross-Coupling Reactions. *ChemCatChem* **2012**, *4*, 432–445. [CrossRef]
40. Eremin, D.B.; Ananikov, V.P. Understanding Active Species in Catalytic Transformations: From Molecular Catalysis to Nanoparticles, Leaching, "Cocktails" of Catalysts and Dynamic Systems. *Coord. Chem. Rev.* **2017**, *346*, 2–19. [CrossRef]
41. Lamb, C.J.C.; Nderitu, B.G.; McMurdo, G.; Tobin, J.M.; Vilela, F.; Lee, A.-L. Auto-Tandem Catalysis: Pd^{II}-Catalysed Dehydrogenation/Oxidative Heck Reaction of Cyclopentane-1,3-diones. *Chem. Eur. J.* **2017**, *23*, 18282–18288. [CrossRef] [PubMed]
42. Wang, Z.; Li, P.; Fu, H.; Dai, Q.; Hu, C. Palladium-Catalyzed Synthesis of Indolines from Aroyloxycarbamates through a Tandem Decarboxylative Amination/Heck/Annulation Reaction. *Adv. Synth. Catal.* **2019**, *361*, 192–200. [CrossRef]
43. Song, J.; Chi, X.; Meng, L.; Zhao, P.; Sun, F.; Zhang, D.; Jiao, L.; Liu, Q.; Dong, Y.; Liu, H. Pd-Catalyzed Tandem Coupling Reaction of 2-*gem*-Dibromovinylanilines and *N*-Tosylhydrazones to Construct 2-(1-phenylvinyl)-indoles. *Adv. Synth. Catal.* **2019**, *361*, 3599–3604. [CrossRef]
44. Revathi, L.; Ravindar, L.; Balakrishna, M.; Qin, H.-L. SO_2F_2 mediated dehydrative cross-coupling of alcohols with electron-deficient olefins in DMSO using Pd-catalyst: One-pot transformation of alcohols to 1,3-diene. *Org. Chem. Front.* **2019**, *6*, 796–800. [CrossRef]
45. Manikandan, T.S.; Ramesh, R.; Semeril, D. The Tandem C–H/N–H Activation of *N*-Methyl Arylamide Catalyzed by Dinuclear Pd(II)Benzhydrazone Complex: A Concise Access to Phenanthridinone. *Organometallics* **2019**, *38*, 319–328. [CrossRef]
46. Yadav, S.; Dash, C. One-pot Tandem Heck Alkynylation/cyclization Reactions Catalyzed by Bis(Pyrrolyl)pyridine Based Palladium Pincer Complexes. *Tetrahedron* **2020**, *76*, 131350. [CrossRef]
47. Karu, R.; Gedu, S. Microwave Assisted Domino Heck Cyclization and Alkynylation: Synthesis of Alkyne Substituted Dihydrobenzofurans. *Green Chem.* **2018**, *20*, 369–374. [CrossRef]
48. Ho, H.E.; Stephens, T.C.; Payne, T.J.; O'Brien, P.; Taylor, R.J.K.; Unsworth, W.P. Merging π-Acid and Pd Catalysis: Dearomatizing Spirocyclization/Cross-Coupling Cascade Reactions of Alkyne-Tethered Aromatics. *ACS Catal.* **2019**, *9*, 504–510. [CrossRef]
49. Wei, W.-X.; Li, Y.; Wen, Y.-T.; Li, M.; Li, X.-S.; Wang, C.-T.; Liu, H.-C.; Xia, Y.; Zhang, B.-S.; Jiao, R.-Q.; et al. Experimental and Computational Studies of Palladium-Catalyzed Spirocyclization via a Narasaka–Heck/C(sp^3 or sp^2)–H Activation Cascade Reaction. *JACS* **2021**, *143*, 7868–7875. [CrossRef]
50. Khan, F.; Fatima, M.; Shirzaei, M.; Vo, Y.; Amarasiri, M.; Banwell, M.G.; Ma, C.; Ward, J.S.; Gardiner, M.G. Tandem Ullmann–Goldberg Cross-Coupling/Cyclopalladation-Reductive Elimination Reactions and Related Sequences Leading to Polyfunctionalized Benzofurans, Indoles, and Phthalanes. *Org. Lett.* **2019**, *21*, 6342–6346. [CrossRef]
51. Zhang, J.; Zhang, Z.-M.; Xu, B.; Wu, L.; Wu, Y.; Qian, Y.; Zhou, L.; Liu, Y. Enantioselective Dicarbofunctionalization of Unactivated Alkenes by Pd-Catalyzed Tandem Heck/Suzuki-Coupling Reaction. *Angew. Chem. Int. Ed.* **2019**, *58*, 14653–14659. [CrossRef] [PubMed]

52. Yokoya, M.; Ishiguro, T.; Sakairi, Y.; Kimura, S.; Morita, Y.; Yamanaka, M. Simple Strategy for Benzo[*de*]chromene-7,8-dione Synthesis via Tandem Sonogashira Coupling and Intramolecular Cyclization Reactions. *Asian J. Org. Chem.* **2022**, *11*, e202200534. [CrossRef]
53. Teng, K.-X.; Niu, L.-Y.; Li, J.; Jia, L.; Yang, Q.-Z. An Unexpected Coupling-Reduction Tandem Reaction for the Synthesis of Alkenyl-Substituted BODIPYs. *Chem. Commun.* **2019**, *55*, 13761–13764. [CrossRef] [PubMed]
54. Li, X.; Chen, X.; Wang, H.; Chen, C.; Sun, P.; Mo, B.; Peng, J. Palladium-Catalyzed Tandem One-pot Synthesis of π-Expanded Imidazoles through a Sequential Heck and Oxidative Amination Reaction. *Org. Biomol. Chem.* **2019**, *17*, 4014–4023. [CrossRef] [PubMed]
55. Lopes, A.B.; Choury, M.; Wagner, P.; Gulea, M. Tandem Double-Cross-Coupling/Hydrothiolation Reaction of 2-Sulfenyl Benzimidazoles with Boronic Acids. *Org. Lett.* **2019**, *21*, 5943–5947. [CrossRef]
56. Luo, J.; Chen, G.-S.; Chen, S.-J.; Li, Z.-D.; Zhao, Y.-L.; Liu, Y.-L. One-Pot Tandem Protocol for the Synthesis of 1,3-Bis(β-aminoacrylate)-Substituted 2-Mercaptoimidazole Scaffolds. *Adv. Synth. Catal.* **2020**, *362*, 3635–3643. [CrossRef]
57. Choi, H.; Shirley, H.J.; Aitken, H.R.M.; Schulte, T.; Söhnel, T.; Hume, P.A.; Brimble, M.A.; Furkert, D.P. Intermolecular Diels–Alder Cycloaddition/Cross-Coupling Sequences of 2-Bromo-1,3-butadienes. *Org. Lett.* **2020**, *22*, 1022–1027. [CrossRef]
58. Arroniz, C.; Chaubet, G.; Anderson, E.A. Dual Oxidation State Tandem Catalysis in the Palladium-Catalyzed Isomerization of Alkynyl Epoxides to Furans. *ACS Catal.* **2018**, *8*, 8290–8295. [CrossRef]
59. Yu, S.; Dai, L.; Shao, Y.; Li, R.; Chen, Z.; Lv, N.; Chen, J. Palladium-Catalyzed Tandem Reaction of Epoxynitriles with Arylboronic Acids in Aqueous Medium: Divergent Synthesis of Furans and Pyrroles. *J. Org. Chem. Front.* **2020**, *7*, 3439–3445. [CrossRef]
60. Kathe, P.M.; Fleischer, I. Palladium-Catalyzed Tandem Isomerization/Hydrothiolation of Allylarenes. *Org. Lett.* **2019**, *21*, 2213–2217. [CrossRef]
61. Han, H.; Yang, S.-D.; Xia, J.-B. Pd/Cu Cocatalyzed Oxidative Tandem C–H Aminocarbonylation and Dehydrogenation of Tryptamines: Synthesis of Carbolinones. *J. Org. Chem.* **2019**, *84*, 3357–3369. [CrossRef] [PubMed]
62. Čarný, T.; Markovič, M.; Gracza, T.; Koóš, P. One-Step Synthesis of Isoindolo[2,1-*a*]indol-6-ones via Tandem Pd-Catalyzed Aminocarbonylation and C–HActivation. *J. Org. Chem.* **2019**, *84*, 12499–12507. [CrossRef] [PubMed]
63. Li, Q.; Zhang, Y.; Zeng, Y.; Fan, Y.; Lin, A.; Yao, H. Palladium-Catalyzed Asymmetric Dearomative Carbonylation of Indoles. *Org. Lett.* **2022**, *24*, 3033–3037. [CrossRef] [PubMed]
64. Wang, S.; Li, X.; Zang, J.; Liu, M.; Zhang, S.; Jiang, G.; Ji, F. Palladium-Catalyzed Multistep Tandem Carbonylation/N-Dealkylation/Carbonylation Reaction: Access to Isatoic Anhydrides. *J. Org. Chem.* **2020**, *85*, 2672–2679. [CrossRef] [PubMed]
65. Zhao, K.-C.; Zhuang, Y.-Y.; Jing, T.-H.; Shi, G.-H.; Guo, L.; Zhao, X.-L.; Lu, Y.; Liu, Y. Pd-Catalyzed Tandem Bis-Hydroaminocarbonylation of Terminal Alkynes for Synthesis of N-Aryl Substituted Succinimides with Involvement of Ionic P,O-hybrid Ligand. *J. Catal.* **2023**, *417*, 248–259. [CrossRef]
66. Wang, S.; Zhou, Y.; Huang, H. Palladium-Catalyzed Tandem Carbonylative Diels-Alder Reaction for Construction of Bridged Polycyclic Skeletons. *Org. Lett.* **2021**, *23*, 2125–2129. [CrossRef] [PubMed]
67. Cheng, Y.-J.; Zhao, L.-P.; Wang, L.; Tang, Y. Cyclization with Alkyl Substituted Methylene Malonate Enabling Concise Total Synthesis of Four Malagasy Alkaloids. *CCS Chem.* **2023**, *5*, 124–132. [CrossRef]
68. Gu, Z.-Y.; Xia, J.-B. [3+1+1]Cyclization of vinyloxiranes with azides and CO by tandem palladium catalysis: Efficient synthesis of oxazolidinones. *Org. Chem. Front.* **2021**, *8*, 4112–4117. [CrossRef]
69. Lokolkar, M.S.; Mane, P.A.; Dey, S.; Bhanage, B.M. Synthesis of 2-Substituted Indoles by Pd-Catalyzed Reductive Cyclization of 1-Halo-2-nitrobenzene with Alkynes. *Eur. J. Org. Chem.* **2022**, *2022*, e202101505. [CrossRef]
70. Ding, L.; Niu, Y.-N.; Xia, X.-F. Pd-Catalyzed Tandem Isomerization/Cyclization for the Synthesis of Aromatic Oxazaheterocycles and Pyrido[3,4-*b*]indoles. *J. Org. Chem.* **2021**, *86*, 10032–10042. [CrossRef]
71. Fillery, S.M.; Gregson, C.L.; Guérot, C.M. Expeditious Access to Functionalized Tricyclic Pyrrolo-Pyridones via Tandem or Sequential C–N/C–C Bond Formations. *Org. Lett.* **2019**, *21*, 9128–9132. [CrossRef] [PubMed]
72. Ansari, A.J.; Joshi, G.; Yadav, U.P.; Maurya, A.K.; Agnihotri, V.K.; Kalra, S.; Kumar, R.; Singh, S.; Sawant, D.M. Exploration of Pd-Catalysed Four-Component Tandem Reaction for One-Pot Assembly of Pyrazolo[1,5-*c*]quinazolines as Potential EGFR Inhibitors. *Bioorg. Chem.* **2019**, *93*, 103314. [CrossRef] [PubMed]
73. Ansari, A.J.; Joshi, G.; Sharma, P.; Maurya, A.K.; Metre, R.K.; Agnihotri, V.K.; Chandaluri, C.G.; Kumar, R.; Singh, S.; Sawant, D.M. Pd-Catalyzed Four-Component Sequential Reaction Delivers a Modular Fluorophore Platform for Cell Imaging. *J. Org. Chem.* **2019**, *84*, 3817–3825. [CrossRef]
74. Gao, X.; Xia, M.; Yuan, C.; Zhou, L.; Sun, W.; Li, C.; Wu, B.; Zhu, D.; Zhang, C.; Zheng, B.; et al. EnantioselectiveSynthesis of Chiral Medium-Sized Cyclic Compounds via Tandem Cycloaddition/Cope Rearrangement Strategy. *ACS Catal.* **2019**, *9*, 1645–1654. [CrossRef]
75. Ren, Z.-L.; Qiu, J.-Y.; Yuan, L.-L.; Yuan, Y.-F.; Cai, S.; Li, J.; Kong, C.; He, P.; Wang, L. Divergent Conversion of Double Isocyanides with Alkenyl Bromide to Polysubstituted Pyrroles and 4-Imino-4,5-dihydropyrrolo[3,4-*b*]pyrrol-6(1*H*)-one Derivatives by Pd-Catalyzed Tandem Cyclization Reactions. *Org. Lett.* **2022**, *24*, 859–863. [CrossRef] [PubMed]
76. Wu, X.; Tang, Z.; Zhang, C.; Wang, C.; Wu, L.; Qu, J.; Chen, Y. Pd-Catalyzed Regiodivergent Synthesis of Diverse Oxindoles Enabled by the Versatile Heck Reaction of Carbamoyl Chlorides. *Org. Lett.* **2020**, *22*, 3915–3921. [CrossRef] [PubMed]

77. Bao, Y.; Wang, Z.; Chen, C.; Zhu, B.; Wang, Y.; Zhao, J.; Gong, J.; Han, M.; Liu, C. Palladium-Catalyzed Tandem Cyclization of Fluorinated Imidoyl Chlorides with 2-Bromophenylboronic Acid: Synthesis of 6-Fluoroalkyl-Phenanthridines. *Tetrahedron* **2019**, *75*, 1450–1456. [CrossRef]
78. Dai, L.; Yu, S.; Xiong, W.; Chen, Z.; Xu, T.; Shao, Y.; Chen, J. Divergent Palladium-Catalyzed Tandem Reaction of Cyanomethyl Benzoates with Arylboronic Acids: Synthesis of Oxazoles and Isocoumarins. *Adv. Synth. Catal.* **2020**, *362*, 1893–1898. [CrossRef]
79. Xu, T.; Shao, Y.; Dai, L.; Yu, S.; Cheng, T.; Chen, J. Pd-Catalyzed Tandem Reaction of 2-Aminostyryl Nitriles with Arylboronic Acids: Synthesis of 2-Arylquinolines. *J. Org. Chem.* **2019**, *84*, 13604–13614. [CrossRef]
80. Yao, X.; Qi, L.; Li, R.; Zhen, Q.; Liu, J.; Zhao, Z.; Shao, Y.; Hu, M.; Chen, J. Palladium-Catalyzed Cascade Reactions of δ-Ketonitriles with Arylboronic Acids: Synthesis of Pyridines. *ACS Comb. Sci.* **2020**, *22*, 114–119. [CrossRef]
81. Ye, X.; Xu, B.; Sun, J.; Dai, L.; Shao, Y.; Zhang, Y.; Chen, J. Pd-Catalyzed Approach for Assembling 9-Arylacridines via a Cascade Tandem Reaction of 2-(Arylamino)benzonitrile with Arylboronic Acidsin Water. *J. Org. Chem.* **2020**, *85*, 13004–13014. [CrossRef] [PubMed]
82. Lang, M.; Wang, J. Carbene-Catalyzed Tandem Isomerization/Cyclisation Strategy: Efficient Assembly of Benzoxazinones. *Org. Chem. Front.* **2019**, *6*, 1367–1371. [CrossRef]
83. Patel, J.J.; Patel, A.P.; Chikhalia, K.H. An Efficient Pd Catalyzed Intramolecular Cyclization Reaction Followed by Formation of Benzimidazole Derivatives: Synthesis of Novel Quinolin-Fused Benzo[*d*] Azeto[1,2-*a*]benzimidazole Analogues. *Synth. Commun.* **2021**, *51*, 81–93. [CrossRef]
84. Jin, T.; Suzuki, S.; Ho, H.E.; Matsuyama, H.; Kawata, M.; Terada, M. Pd-Catalyzed Indolization/*peri*-C–H Annulation/*N*-Dealkylation Cascade to Cyclopenta-Fused Acenaphtho[1,2-*b*]indole Scaffold. *Org. Lett.* **2021**, *23*, 9431–9435. [CrossRef] [PubMed]
85. Siciliano, S.; Cini, E.; Taddei, M.; Vinciarelli, G. Synthesis of 2-Substitued Indoles via Pd-Catalysed Cyclization in an Aqueous Micellar Medium. *Molecules* **2021**, *26*, 3917. [CrossRef]
86. Xiong, W.; Chen, Z.; Shao, Y.; Li, R.; Hu, K.; Chen, J. The Synthesis of Fluorescent Benzofuro[2,3-c] Pyridines via Palladium-Catalyzed Heteroaromatic C–H Addition and Sequential Tandem Cyclization. *Org. Chem. Front.* **2020**, *7*, 756–762. [CrossRef]
87. Błocka, A.; Chaładaj, W. Tandem Pd-Catalyzed Cyclization/Coupling of Non-Terminal Acetylenic Activated Methylenes with (Hetero)Aryl Bromides. *Molecules* **2022**, *27*, 630. [CrossRef]
88. Hu, T.; Xu, K.; Ye, Z.; Zhu, K.; Wu, Y.; Zhang, F. Two-in-One Strategy for the Pd(II)-Catalyzed Tandem C–H Arylation/Decarboxylative Annulation Involved with Cyclic Diaryliodonium Salts. *Org. Lett.* **2019**, *21*, 7233–7237. [CrossRef]
89. Ghosh, S.; Chattopadhyay, S.K. Unusual Regioselectivity in Palladium-Catalyzed Tandem C–H Arylation and C–H Amidation of *cis*-Cinnamyl Hydroxamates: Facile Synthesis of 3-Aryl-2-quinolones. *Eur. J. Org. Chem.* **2022**, *2022*, e202200391. [CrossRef]
90. Domański, S.; Gatlik, B.; Chaładaj, W. Pd-Catalyzed Boroperfluoroalkylation of Alkynes Opens a Route to One-Pot Reductive Carboperfluoroalkylation of Alkynes with Perfluoroalkyl and Aryl Iodides. *Org. Lett.* **2019**, *21*, 5021–5025. [CrossRef]
91. Fernández, N.P.; Gaube, G.; Woelk, K.J.; Burns, M.; Hruszkewycz, D.P.; Leitch, D.C. Palladium-Catalyzed Direct C–H Alkenylation with Enol Pivalates Proceeds via Reversible C–O Oxidative Addition to Pd(0). *ACS Catal.* **2022**, *12*, 6997–7003. [CrossRef]
92. Zhang, Y.; Chen, L.; Shao, Y.; Zhang, F.; Chen, Z.; Lv, N.; Chen, J.; Li, R. Palladium(ii)-catalyzed three-component tandem reactions: Synthesis of multiplysubstituted quinolines. *Org. Chem. Front.* **2021**, *8*, 254–259. [CrossRef]
93. Li, H.; Li, T.; Hsueh, Y.J.; Wu, X.; Xu, F.; Zhang, Y.J. Tandem arylation and regioselective allylic etherification of 2,3-allenol via Pd/B cooperative catalysis. *Org. Biomol. Chem.* **2019**, *17*, 8075–8078. [CrossRef] [PubMed]
94. Tang, T.-M.; Liu, M.; Wu, H.; Gou, T.; Hu, X.; Wang, B.-Q.; Hu, P.; Song, F.; Huang, G. Pd-Catalyzed tandem C–C/C–O/C–H single bond cleavage of 3-allyloxybenzocyclobutenols. *Org. Chem. Front.* **2021**, *8*, 3867–3875. [CrossRef]
95. Li, W.; Zhang, C.; Lu, H.; Wang, Y.; Deng, G.; Liang, Y.; Yang, Y. Pd-Catalyzed One-Pot Synthesis of Vinylsilanes via a Three-Component Tandem Reaction. *Org. Chem. Front.* **2020**, *7*, 2075–2081. [CrossRef]
96. Mahesha, C.K.; Borade, S.A.; Tank, D.; Bajaj, K.; Bhambri, H.; Mandal, S.K.; Sakhuja, R. Tandem Transformation of Indazolones to Quinazolinones through Pd-Catalyzed Carbene Insertionin to an N−N Bond. *J. Org. Chem.* **2023**, *88*, 1457–1468. [CrossRef] [PubMed]
97. Azizollahi, H.; Mehta, V.P.; García-López, J.-A. Pd-catalyzed cascade reactions involving skipped dienes: From double carbopalladation to remote C–C cleavage. *Chem.Commun.* **2019**, *55*, 10281–10284. [CrossRef]
98. Giofrè, S.; Keller, M.; LoPresti, L.; Beccalli, E.M.; Molteni, L. Switchable Oxidative Reactions of *N*-allyl-2-Aminophenols: Palladium-Catalyzed Alkoxyacyloxylation vs an Intramolecular Diels–Alder Reaction. *Org. Lett.* **2021**, *23*, 7698–7702. [CrossRef]
99. Xiong, Q.; Lu, J.; Shi, L.; Ran, G.-Y. Pd-Catalyzed Tandem [5+2] Cycloaddition/Ring Contraction of Phthalide-Derived Alkenes and Vinylethylene Carbonates for the Construction of Benzo-[5,5]-spiroketal Lactones. *Org. Lett.* **2022**, *24*, 3363–3367. [CrossRef]
100. Liu, Q.; Liu, T.-X.; Ma, J.; Zhang, G. Palladium-Catalyzed Three-Component Tandem Coupling–Carboannulation Reaction Leading to Polysubstituted [60]Fullerene-Fused Cyclopentanes. *Org. Lett.* **2020**, *22*, 284–289. [CrossRef]
101. Zhang, M.; Weng, Z. Palladium-Catalyzed Tandem Synthesis of 2-Trifluoromethylthio(seleno)-Substituted Benzofused Heterocycles. *Org. Lett.* **2019**, *21*, 5838–5842. [CrossRef] [PubMed]
102. Yang, W.-C.; Chen, X.-B.; Song, K.-L.; Wu, B.; Gan, W.-E.; Zheng, Z.-J.; Cao, J.; Xu, L.-W. Pd-Catalyzed Enantioselective Tandem C–C Bond Activation/Cacchi Reaction between Cyclobutanones and *o*-Ethynylanilines. *Org. Lett.* **2021**, *23*, 1309–1314. [CrossRef]
103. Corma, A.; Navas, J.; Sabater, M.J. Advances in One-Pot Synthesis through Borrowing Hydrogen Catalysis. *Chem. Rev.* **2018**, *118*, 1410–1459. [CrossRef] [PubMed]

104. Fan, X.; Lin, D.; Xu, Z.; Li, Y. Pd/Cu bimetallic catalyst immobilized on PEI capped cellulose-polyamidoamine dendrimer: Synthesis, characterization, and application in Sonogashira reactions for the synthesis of alkynes and benzofurans. *Colloids Surf. A Physicochem. Eng. Asp.* **2022**, *648*, 129206. [CrossRef]
105. Esfandiary, N.; Bagheri, S.; Heydari, A. Magnetic γ-Fe_2O_3@Cu-LDH intercalated with Palladium Cysteine: An efficient dual nanocatalyst in tandem C–N coupling and cyclization progress of synthesis quinolines. *Appl. Clay Sci.* **2020**, *198*, 105841. [CrossRef]
106. Meng, J.; Chang, F.; Su, Y.; Liu, R.; Cheng, T.; Liu, G. Switchable Catalysts Used to Control Suzuki Cross-Coupling and Aza–Michael Addition/Asymmetric Transfer Hydrogenation Cascade Reactions. *ACS Catal.* **2019**, *9*, 8693–8701. [CrossRef]
107. Rajabi, F.; Chia, C.H.; Sillanpää, M.; Voskressensky, L.G.; Luque, R. Cytosine Palladium Complex Supported on Ordered Mesoporous Silica as Highly Efficient and Reusable Nanocatalyst for One-Pot Oxidative Esterification of Aldehydes. *Catalysts* **2021**, *11*, 1482. [CrossRef]
108. Sharma, A.K.; Bhattacherjee, D.; Sharma, N.; Giri, K.; Das, P. Supported-Pd catalyzed tandem approach for *N*-arylbenzamides synthesis. *Mol. Catal.* **2021**, *516*, 111948. [CrossRef]
109. Amoo, C.C.; Xing, C.; Tsubaki, N.; Sun, J. Tandem Reactions over Zeolite-Based Catalysts in Syngas Conversion. *ACS Cent. Sci.* **2022**, *8*, 1047–1062. [CrossRef]
110. Martínez-Edo, G.; Balmori, A.; Pontón, I.; MartídelRio, A.; Sánchez-García, D. Functionalized Ordered Mesoporous Silicas (MCM-41): Synthesis and Applications in Catalysis. *Catalysts* **2018**, *8*, 617. [CrossRef]
111. Chaudhary, V.; Sharma, S. An overview of ordered mesoporous material SBA-15: Synthesis, functionalization and application in oxidation reactions. *J. Porous Mater.* **2017**, *24*, 741–749. [CrossRef]
112. Yu, X.; Williams, C.T. Recent advances in the applications of mesoporous silica in heterogeneous catalysis. *Catal. Sci. Technol.* **2022**, *12*, 5765–5794. [CrossRef]
113. Davidson, M.; Ji, Y.; Leong, G.J.; Kovach, N.C.; Trewyn, B.G.; Richards, R.M. Hybrid Mesoporous Silica/Noble-Metal Nanoparticle Materials – Synthesis and Catalytic Applications. *ACS Appl. Nano Mater.* **2018**, *1*, 4386–4400. [CrossRef]
114. Singh, B.; Na, J.; Konarova, M.; Wakihara, T.; Yamauchi, Y.; Salomon, C.; Gawande, M.B. Functional Mesoporous Silica Nanomaterials for Catalysis and Environmental Applications. *Bull. Chem. Soc. Jpn.* **2020**, *93*, 1459–1496. [CrossRef]
115. Hernández-Soto, M.C.; Erigoni, A.; Segarra, C.; Rey, F.; Díaz, U.; Gianotti, E.; Miletto, I.; Pera-Titus, M. Bifunctional hybrid organosiliceous catalysts for aldol condensation–hydrogenation tandem reactions of furfural in continuous-flow reactor. *Appl. Catal. A Gen.* **2022**, *643*, 118710. [CrossRef]
116. Maties, G.; Gonzalez-Arellano, C.; Luque, R.; Montejano-Nares, E.; Ivars-Barceló, F.; Pineda, A. Trans-ferulic acid valorization into stilbene derivatives via tandem decarboxylation/Heck coupling using Pd/Al-SBA-15 materials. *Mater. Today Chem.* **2022**, *25*, 100971. [CrossRef]
117. Scheithauer, M.; Grasselli, R.K.; Knözinger, H. Genesis and Structure of WO_x/ZrO_2 Solid Acid Catalysts. *Langmuir* **1998**, *14*, 3019–3029. [CrossRef]
118. Zhou, W.; Luo, J.; Wang, Y.; Liu, J.; Zhao, Y.; Wang, S.; Ma, X. WO_x domainsize, acid properties and mechanistic aspects of glycerol hydrogenolysis over $Pt/WO_x/ZrO_2$. *Appl. Catal. B Environ.* **2019**, *242*, 410–421. [CrossRef]
119. Rodriguez-Gattorno, G.; Galano, A.; Torres-García, E. Surface acid–basic properties of WO_x–ZrO_2 and catalytic efficiency in oxidative desulfurization. *Appl. Catal. B Environ.* **2009**, *92*, 1–8. [CrossRef]
120. Lwin, S.; Li, Y.; Frenkel, A.I.; Wachs, I.E. Nature of WO_x Sites on SiO_2 and Their Molecular Structure–Reactivity/Selectivity Relationships for Propylene Metathesis. *ACS Catal.* **2016**, *6*, 3061–3071. [CrossRef]
121. Gayapan, K.; Sripinun, S.; Panpranot, J.; Praserthdam, P.; Assabumrungrat, S. Effect of pretreatment atmosphere of WO_x/SiO_2 catalysts on metathesis of ethylene and 2-butene to propylene. *RSC Adv.* **2018**, *8*, 11693–11704. [CrossRef] [PubMed]
122. Watmanee, S.; Suriye, K.; Praserthdam, P.; Panpranot, J. Formation of isolated tungstate sites on hierarchical structured SiO_2– and HY zeolite-supported WO_x catalysts for propene metathesis. *J. Catal.* **2019**, *376*, 150–160. [CrossRef]
123. Janampelli, S.; Sethia, G.; Darbha, S. Selective, bifunctional Cu–WO_x/Al_2O_3 catalyst for hydrodeoxygenation of fatty acids. *Catal. Sci. Technol.* **2020**, *10*, 268–277. [CrossRef]
124. García-Fernández, S.; Gandarias, I.; Requies, J.; Güemez, M.B.; Bennici, S.; Auroux, A.; Arias, P.L. New approaches to the $Pt/WO_x/Al_2O_3$ catalytic system behavior for the selective glycerol hydrogenolysis to 1,3-propanediol. *J. Catal.* **2015**, *323*, 65–75. [CrossRef]
125. Kim, H.; Numan, M.; Jo, C. Catalytic Dehydration of Ethanol over WO_x Nanoparticles Supported on MFI (MobileFive) Zeolite Nanosheets. *Catalysts* **2019**, *9*, 670. [CrossRef]
126. Chu, D.; Luo, Z.; Xin, Y.; Jiang, C.; Gao, S.; Wang, Z.; Zhao, C. One-pot hydrogenolysis of cellulose to bioethanol over Pd-Cu-WO_x/SiO_2 catalysts. *Fuel* **2021**, *292*, 120311. [CrossRef]
127. Liu, Y.; Ma, X.-C.; Chang, G.-G.; Ke, S.-C.; Xia, T.; Hu, Z.-Y.; Yang, X.-Y. Synergistic catalysis of Pd nanoparticles withboth Lewis and Bronsted acid sites encapsulated within a sulfonated metal–organic frameworks toward one-pot tandem reactions. *J. Colloid Interface Sci.* **2019**, *557*, 207–215. [CrossRef] [PubMed]
128. Insyani, R.; Verma, D.; Cahyadi, H.S.; Kim, S.M.; Kim, S.K.; Karanwal, N.; Kim, J. One-pot di- and polysaccharides conversion to highly selective 2,5-dimethylfuran over Cu-Pd/Amino-functionalized Zr-based metal-organic framework (UiO-66(NH_2))@SGO tandem catalyst. *Appl. Catal. B Environ.* **2019**, *243*, 337–354. [CrossRef]

129. Yao, Y.; Huang, K.; Liu, Y.; Luo, T.; Tian, G.; Li, J.; Zhang, S.; Chang, G.; Yang, X. A hierarchically multifunctional integrated catalyst with intimate and synergistic activesites for one-pot tandem catalysis. *Inorg. Chem. Front.* **2021**, *8*, 3463–3472. [CrossRef]
130. Lee, Y.-R.; Do, X.H.; Hwang, S.S.; Baek, K.-Y. Dual-functionalized ZIF-8 as an efficient acid-base bifunctional catalyst for the one-pot tandem reaction. *Catal. Today* **2021**, *359*, 124–132. [CrossRef]
131. Fu, X.; Du, Y.; Liu, F.; Yang, J.; He, R.; Fu, G.; Yang, X. Double-shelled hollow polymer microspheres as acid and metallic colloid bi-functional catalyst for a deactalization-hydrogenation tandem reaction. *Colloids Surf. A Physicochem. Eng. Asp.* **2023**, *659*, 130833. [CrossRef]
132. Hao, N.; Alper, K.; Patel, H.; Tekin, K.; Karagoz, S.; Ragauskas, A.J. One-stept ransformation of biomass to fuel precursors using a bi-functional combination of Pd/C and water tolerant Lewis acid. *Fuel* **2020**, *277*, 118200. [CrossRef]
133. Raza, A.A.; Ravi, S.; Tajudeen, S.S.; Sheriff, A.K.I. Sulfonated covalent triazine polymer loaded with Pd nanoparticles as a bifunctional catalyst for one pot hydrogenation esterification reaction. *J. Solid State Chem.* **2021**, *302*, 122417. [CrossRef]
134. Liu, J.; Wang, N.; Ma, L. Recent Advances in Covalent Organic Frameworks for Catalysis. *Chem. Asian J.* **2020**, *15*, 338–351. [CrossRef]
135. Guo, J.; Jiang, D. Covalent Organic Frameworks for Heterogeneous Catalysis: Principle, Current Status, and Challenges. *ACS Cent. Sci.* **2020**, *6*, 869–879. [CrossRef] [PubMed]
136. Cheng, H.-Y.; Wang, T. Covalent Organic Frameworks in Catalytic Organic Synthesis. *Adv. Synth. Catal.* **2021**, *363*, 144–193. [CrossRef]
137. Alsudairy, Z.; Brown, N.; Campbell, A.; Ambus, A.; Brown, B.; Smith-Petty, K.; Li, X. Covalent organic frameworks in heterogeneous catalysis: Recent advances and future perspective. *Mater. Chem. Front.* **2023**, *7*, 3298–3331. [CrossRef]
138. Gong, K.; Li, C.; Zhang, D.; Lu, H.; Wang, Y.; Li, H.; Zhang, H. Sulfonic acid functionalized covalent organic frameworks as efficient catalyst for the one-pot tandem reactions. *Mol. Catal.* **2022**, *519*, 112139. [CrossRef]
139. Wang, N.; Liu, J.; Li, X.; Ma, L. Selective control in the reductive amination of benzaldehyde towards corresponding amines over COF supported Pt, Pd, and Rh catalysts. *Catal. Commun.* **2023**, *175*, 106620. [CrossRef]
140. Gonzalez-Granda, S.; Escot, L.; Lavandera, I.; Gotor-Fernandez, V. Chemoenzymatic Cascades Combining Biocatalysis and Transition Metal Catalysis for Asymmetric Synthesis. *Angew. Chem.* **2023**, *62*, e202217713. [CrossRef]
141. Bering, L.; Thompson, J.; Micklefield, J. New reaction pathways by integrating chemo- and biocatalysis. *Trends Chem.* **2022**, *4*, 392–408. [CrossRef]
142. Liu, Y.; Gao, S.; Liu, P.; Kong, W.; Liu, J.; Jiang, Y. Integration of Chemo- and Bio-Catalysis to Intensify Bioprocesses. *Phys. Sci. Rev.* **2023**. [CrossRef]
143. Luan, P.; Liu, Y.; Li, Y.; Chen, R.; Huang, C.; Gao, J.; Hollmann, F.; Jiang, Y. Aqueous Chemoenzymatic One-Pot Enantioselective Synthesis of Tertiary α-Aryl Cycloketones via Pd-Catalyzed C–C Formation and Enzymatic C=C Asymmetric Hydrogenation. *Green Chem.* **2021**, *23*, 1960–1964. [CrossRef]
144. Dawood, A.W.H.; Bassut, J.; de Souza, R.O.M.A.; Bornscheuer, U.T. Combination of the Suzuki–Miyaura Cross-Coupling Reaction with Engineered Transaminases. *Chem. Eur. J.* **2018**, *24*, 16009–16013. [CrossRef] [PubMed]
145. Paris, J.; Telzerow, A.; Ríos-Lombardía, N.; Steiner, K.; Schwab, H.; Morís, F.; Gröger, H.; González-Sabín, J. Enantioselective One-Pot Synthesis of Biaryl-Substituted Amines by Combining Palladium and Enzyme Catalysis in Deep Eutectic Solvents. *ACS Sustain. Chem. Eng.* **2019**, *7*, 5486–5493. [CrossRef]
146. Gonzalez-Martinez, D.; Gotor, V.; Gotor-Fernandez, V. Stereoselective Synthesis of 1-Arylpropan-2-amines from Allylbenzenes through a Wacker-Tsuji Oxidation-Biotransamination Sequential Process. *Adv. Synth. Catal.* **2019**, *361*, 2582–2593. [CrossRef]
147. Fernandes, R.A.; Jha, A.K.; Kumar, P. Recent Advances in Wacker Oxidation: From Conventional to Modern Variants and Applications. *Catal. Sci. Technol.* **2020**, *10*, 7448–7470. [CrossRef]
148. Albarran-Velo, J.; Gotor-Fernandez, V.; Lavandera, I. Markovnikov Wacker-Tsuji Oxidation of Allyl(hetero)arenes and Application in a One-Pot Photo-Metal-Biocatalytic Approach to Enantioenriched Amines and Alcohols. *Adv. Synth. Catal.* **2021**, *363*, 4096–4108. [CrossRef]
149. Forero-Cortés, P.A.; Haydl, A.M. The 25th Anniversary of the Buchwald–Hartwig Amination: Development, Applications, and Outlook. *Org. Process Res. Dev.* **2019**, *23*, 1478–1483. [CrossRef]
150. Cosgrove, S.C.; Thompson, M.P.; Ahmed, S.T.; Parmeggiani, F.; Turner, N.J. One-Pot Synthesis of Chiral *N*-Arylamines by Combining Biocatalytic Aminations with Buchwald–Hartwig *N*-Arylation. *Angew. Chem. Int. Ed.* **2020**, *59*, 18156–18160. [CrossRef]
151. Heckmann, C.M.; Paradisi, F. GPhos Ligand Enables Production of Chiral *N*-Arylamines in a Telescoped Transaminase-Buchwald-Hartwig Amination Cascade in the Presence of Excess Amine Donor. *Chem. Eur. J.* **2021**, *27*, 16616–16620. [CrossRef] [PubMed]
152. Coccia, F.; Tonucci, L.; DelBoccio, P.; Caporali, S.; Hollmann, F.; D'Alessandro, N. Stereoselective DoubleReduction of 3-Methyl-2-cyclohexenone, by Use of Palladium and Platinum Nanoparticles, in Tandem with Alcohol Dehydrogenase. *Nanomaterials* **2018**, *8*, 853. [CrossRef] [PubMed]
153. Herman, R.A.; Zhu, X.; Ayepa, E.; You, S.; Wang, J. Advances in the One-Step Approach of Polymeric Materials Using Enzymatic Techniques. *Polymers* **2023**, *15*, 703. [CrossRef] [PubMed]
154. Li, X.; Cao, Y.; Xiong, J.; Li, J.; Xiao, H.; Li, X.; Gou, Q.; Ge, J. Enzyme-Metal-Single-Atom Hybrid Catalysts for One-Pot Chemoenzymatic Reactions. *Chin. J. Catal.* **2023**, *44*, 139–145. [CrossRef]

155. Li, X.; Hu, X.; Qiao, Y.; Lu, T.; Bai, Y.; Xiong, J.; Li, X.; Gou, Q.; Ge, J. Enzyme-bimetallic hybrid catalyst for one-pot chemoenzymatic reactions. *Chem. Eng. J.* **2023**, *452*, 139356. [CrossRef]
156. Collins, G.; Schmidt, M.; O'Dwyer, C.; Holmes, J.D.; McGlacken, G.P. The Origin of Shape Sensitivity in Palladium-Catalyzed Suzuki–Miyaura Cross Coupling Reactions. *Angew. Chem. Int. Ed.* **2014**, *53*, 4142–4145. [CrossRef] [PubMed]
157. Deiana, L.; Rafi, A.A.; Naidu, V.R.; Tai, C.W.; Baeckvall, J.E.; Cordova, A. Artificial Plant Cell Walls as Multi-Catalyst Systems for Enzymatic Cooperative Asymmetric Catalysis in Non-Aqueous Media. *Chem. Commun.* **2021**, *57*, 8814–8817. [CrossRef]
158. Craven, E.J.; Latham, J.; Shepherd, S.A.; Khan, I.; Diaz-Rodriguez, A.; Greaney, M.F.; Micklefield, J. Programmable Late-Stage C−H Bond Functionalization Enabled by Integration of Enzymes with Chemocatalysis. *Nat Catal.* **2021**, *4*, 385–394. [CrossRef]
159. Huang, J.; Jiao, L.; Xu, W.; Fang, Q.; Wang, H.; Cai, X.; Yan, H.; Gu, W.; Zhu, C. Immobilizing Enzymes on Noble Metal Hydrogel Nanozymes with Synergistically Enhanced Peroxidase Activity for Ultrasensitive Immunoassays by Cascade Signal Amplification. *ACS Appl. Mater. Interfaces* **2021**, *13*, 33383–33391. [CrossRef]
160. Ming, J.; Zhu, T.B.; Ye, Z.C.; Wang, J.J.; Chen, X.L.; Zheng, N.F. A Novel Cascade Nanoreactor Integrating Two-Dimensional Pd-Ru Nanozyme, Uricase and Red Blood Cell Membrane for Highly Efficient Hyperuricemia Treatment. *Small* **2021**, *17*, 2103645. [CrossRef]
161. Zhang, N.; Hübner, R.; Wang, Y.; Zhang, E.; Zhou, Y.; Dong, S.; Wu, C. Surface-Functionalized Mesoporous Nanoparticles as Heterogeneous Supports to Transfer Bifunctional Catalysts into Organic Solvents for Tandem Catalysis. *ACS Appl. Nano Mater.* **2018**, *1*, 6378–6386. [CrossRef]
162. Metzger, K.E.; Moyer, M.M.; Trewyn, B.G. Tandem Catalytic Systems Integrating Biocatalysts and Inorganic Catalysts Using Functionalized Porous Materials. *ACS Catal.* **2021**, *11*, 110–122. [CrossRef]
163. Lackner, F.; Hiebler, K.; Grabner, B.; Gruber-Woelfler, H. Optimization of a Catalytic Chemoenzymatic Tandem Reaction for the Synthesis of Natural Stilbenes in Continuous Flow. *Catalysts* **2020**, *10*, 1404. [CrossRef]

catalysts

Review

Plant-Biomass-Derived Carbon Materials as Catalyst Support, A Brief Review

Antonina A. Stepacheva *, Mariia E. Markova, Yury V. Lugovoy, Yury Yu. Kosivtsov, Valentina G. Matveeva and Mikhail G. Sulman

Department of Biotechnology, Chemistry and Standardization, Tver State Technical University, A. Nikitin Str., 22, 170026 Tver, Russia
* Correspondence: a.a.stepacheva@mail.ru; Tel.: +7-(4822)789348

Abstract: Carbon materials are widely used in catalysis as effective catalyst supports. Carbon supports can be produced from coal, organic precursors, biomass, and polymer wastes. Biomass is one of the promising sources used to produce carbon-based materials with a high surface area and a hierarchical structure. In this review, we briefly discuss the methods of biomass-derived carbon supported catalyst preparation and their application in biodiesel production, organic synthesis reactions, and electrocatalysis.

Keywords: plant biomass; carbon; biochar; catalysts

1. Introduction

Carbon materials, such as graphite, graphene, activated carbon, carbon nanotubes (CNT), nanosheets (CNS), and nanofibers (CNF), are widely used in catalysis. Such materials are considered to be promising catalytic supports because of their high specific surface area, resistance to acidic or basic media, amphoteric character, high-temperature stability, tailored pore size distribution, ability for modification, and cheapness [1–3]. Carbon supports can be produced from coal, organic precursors, biomass, and polymer waste through different techniques, which are mainly based on thermochemical degradation/decomposition methods. In recent years, biomass-derived carbonaceous materials have attracted great attention due to their cheapness, simple preparation, and wide range of biomass resources.

To produce carbon materials from biomass, different thermochemical methods can be used. Among them, four main techniques are highlighted (Figure 1). Combustion of biomass is widely used to obtain ashes that are characterized by the contents of alkali and alkali-earth metal compounds with a low content of carbon (1–15 wt.%). Depending on the biomass type, the resulting ashes have different elemental compositions, varying from 10–50 wt.% of alkali metals, 5–20 wt.% of alkali-earth metals, 0.5–2 wt.% of Al, 2–80 wt.% of Si, and 0.5–10 wt.% of Cl, P, Mn, Fe, Zn, etc. [4]. The calcination method is carried out in a wide temperature range from 300 to 1000 °C [5]. Calcination can be performed either in the air or in an inert atmosphere, resulting in the degradation of biomass components to form carbon and inorganic compounds, such as alkali and alkali-earth metal carbonates and carbides. During calcination, temperature and biomass retention time play a critical role in the formation of the resulting carbon material morphology and structure. It is reported that a high reaction temperature and a short retention time lead to the aggregation of particles of Ca, Na, Mg, and Si oxides and, thus, reduce the specific surface area and porosity of the resulting catalysts [6,7]. Despite the simplicity of the combustion and calcination methods, the resulting materials have a low specific surface area, poor porosity, and low carbon content, leading to their limited application in catalysis.

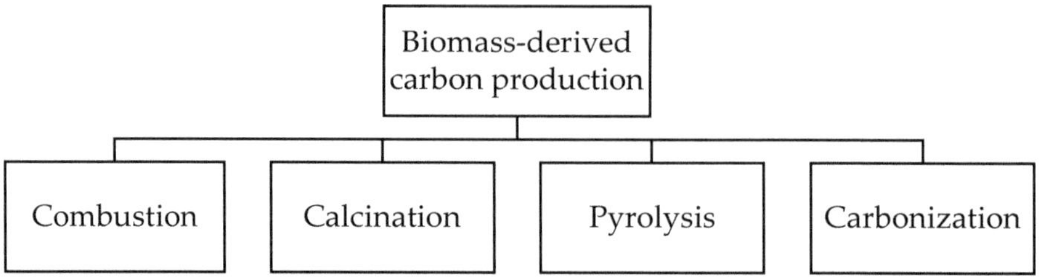

Figure 1. Main methods for biomass-derived carbon production.

Another two methods are used to prepare biochar characterized by a high porosity with a hierarchical structure. Pyrolysis is a widely used method to obtain biochar from organogenic feedstock. Depending on the temperature and the feedstock residence time, pyrolysis can be divided into slow pyrolysis occurring at 300–700 °C for several hours [8,9] and fast pyrolysis performed at 500–1000 °C at a residence time <10 s [10]. Typically, slow pyrolysis results in a biochar yield ranging from 30 to 70 wt.%, while the fast process provides a biochar yield of up to 15 wt.% [9]. Besides conventional non-catalytic pyrolysis to provide biochar production, biomass feedstock can be preliminarily treated with a metal salt solution (catalytic pyrolysis), which leads to the formation of metal-containing composites. The carbonization method seems to be close to pyrolysis and consists of heating biomass in an inert atmosphere (N_2 or Ar) at relatively low temperatures. This method can be divided into two types depending on the media used. Dry carbonization is carried out at 300–600 °C using dried feedstock [11]. Wet or hydrothermal carbonization (HTC) is performed in superheated water using either dried or wet feedstock at a temperature of 180–800 °C [12]. The latter can be classified into high-temperature (300–800 °C) and low-temperature (180–250 °C) HTC [13–15]. The high-temperature process results in the formation of carbon materials with a high surface area and a high porosity, with a hierarchical structure, while the low-temperature HTC is used for the production of colloidal carbonaceous spheres [15].

Moreover, techniques such as gasification and torrefaction can be mentioned for biochar production. Gasification is the process mainly used for synthesis gas production from biomass occurring at a temperature of 750–900 °C. Meanwhile, about 10 wt.% of biochar can be obtained by gasification [16]. Torrefaction is direct carbonization of biomass at low temperatures (up to 250–300 °C), resulting in over 80 wt.% of biochar formation [17]. The composition, structure, morphology, and textural properties of carbon materials strongly depend on the biomass type, the preparation method used, the temperature, the size of the particle, the heating rate, the residence time, etc.

Despite the different production methods, biomass-derived carbon materials find a wide range of application in catalysis. Four main directions can be highlighted: biodiesel production, organic reactions, electrocatalysis, and photocatalysis. In this review, we briefly discuss the first three directions for the use of biomass-derived carbon-based catalysts.

2. Biomass-Derived Carbon-Based Catalysts for Biodiesel Production

In recent years, biomass-carbon-based catalysts are widely used in esterification/transesterification processes to produce fatty acid methyl esters (FAMEs). Carbon produced via pyrolysis, calcination, or carbonization tends to be functionalized by sulfonic groups or alkali to form acidic or basic active sites (Figure 2). Two directions for the functionalization are used: (i) direct functionalization through the introduction of functional groups into the raw material, and (ii) post-synthesis modification through the treatment of carbonaceous materials [15].

2.1. *Acid-Functionalized Catalysts*

Acidic carbon-based catalysts are efficiently applied in esterification/transesterification processes. The most widely used ones are sulfonated carbons characterized by a high acid density. These catalysts can substitute concentrated sulfuric acid in many reactions, including biodiesel production [18]. Sulfonated catalysts are synthesized by either direct sulfonation with H_2SO_4 or by sulfonation via reductive alkylation. In contrast to concentrated sulfuric acid, carbon-based catalysts tend to cause less corrosion and can be easily separated from the reaction mixture.

A review of recent works on sulfonated carbon catalysts shows that these catalysts, regardless of the method of preparation, are consisted of amorphous or graphite-like carbon with covalently bonded $–SO_3H$ groups. The last ones provide the Brønsted acid sites with a high density. The surface area of the resulting composites mainly depends on the carbon preparation method. These catalysts have a high activity in the production of biodiesel from feedstock with high free fatty acid (HFFA) content [19,20]. Incomplete carbonization of the biomass raw material during the catalyst synthesis leads to the formation of a polycyclic carbon-ring structure with –COOH or –OH groups on the surface, which increases the catalyst acidity and facilitates the covalent bonding of sulfonic groups [21,22].

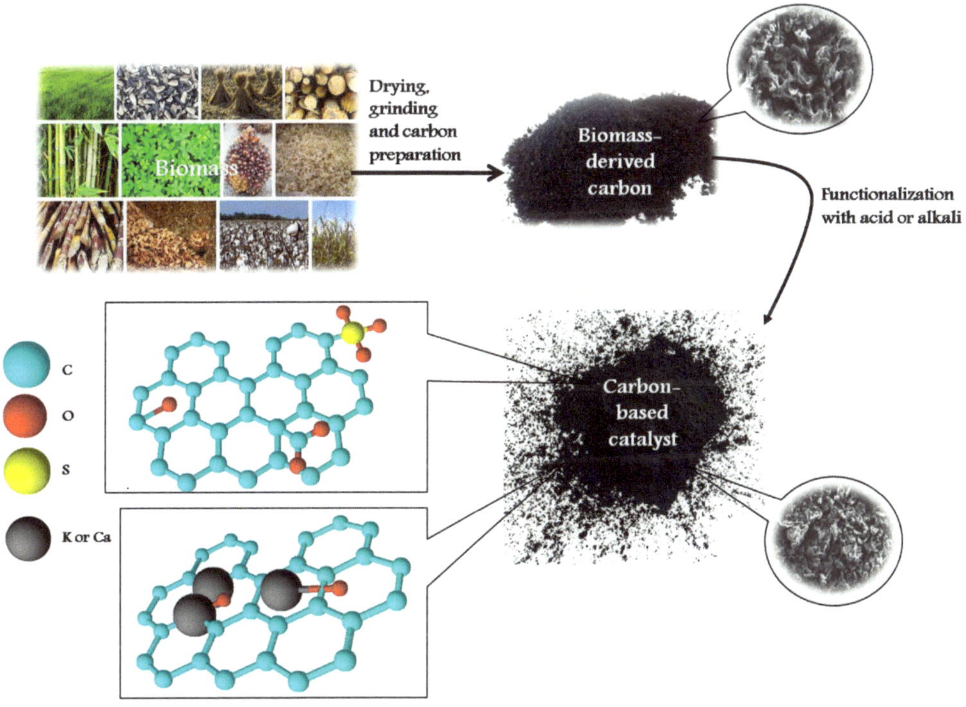

Figure 2. Biomass-derived carbons for the application in biodiesel production. Summarized from [23–46].

Table 1 presents the literature data on sulfonated carbon-based catalysts and their activity during the esterification/transesterification process.

Table 1. Sulfonated carbon-based catalysts for biodiesel production.

Biomass	Preparation Method	Oil Feedstock	Biodiesel Synthesis Conditions	Biodiesel Yield	SA *, m^2/g	Reusability	Reference
C. inophyllum seeds	Carbonization for 5 h at 400 °C, followed by sulfonation by H_2SO_4 (98 wt.%) at 150 °C.	*C. inophyllum* Oil	Oil/methanol ratio 1:30 mol, 0.3 g of catalyst, 150 °C, 2 h.	36.4	3.4	80% of initial activity at 2nd cycle	[23]
Starch	Pyrolysis for 5 h at 400 °C, followed by sulfonation by H_2SO_4 (98 wt.%) at 150 °C.	*C. inophyllum* Oil	Oil/methanol ratio 1:30 mol, 7.5 wt.% of catalyst, 180 °C, 5 h.	81.0	0.9	80% of initial activity at 4th cycle	[24]
Glucose	Carbonization for 15 h at 400 °C, followed by sulfonation by H_2SO_4 (98 wt.%) at 150 °C.	Oleic acid	Oleic acid/methanol ratio 1:50 mol, 5 wt.% of catalyst, 80 °C, 5 h.	95.0	4.1	93% of initial activity at 50th cycle	[19]
Starch	Pyrolysis for 3 h at 400 °C, followed by sulfonation by H_2SO_4 (98 wt.%) at 150 °C.	Waste oil	Oil/methanol ratio 1:30 mol, 12 wt.% of catalyst, 80 °C, 15 h.	92.0	7.2	93% of initial activity at 50th cycle	[25]
Lignin	Pyrolysis for 1 h at 400 °C, followed by sulfonation by H_2SO_4 (98 wt.%) at 150 °C.	Acidified soybean soapstock	Oil/methanol ratio 1:7 mol, 5 wt.% of catalyst, 70 °C, 5 h.	96.0	4.7	Over 90% of initial activity at 3rd cycle	[26]
Rice husk	Fast pyrolysis at 510 °C for 4 s, followed by sulfonation by H_2SO_4 (98 wt.%) at 80 °C.	Waste oil	Oil/methanol ratio 1:30 mol, 10 wt.% of catalyst, 80 °C, 15 h.	90.0	4.0	-	[27]
Mesoporous starch	Carbonization for 5 h at 400 °C, followed by sulfonation by H_2SO_4 (99 wt.%) at 80 °C.	Waste oil	Oil/methanol ratio 1:3 vol., 10 wt.% of catalyst, 80 °C, 18 h.	98.0	-	78% of initial activity at 4th cycle	[28]
Cotton stalk	Carbonization for 3 h at 450 °C, followed by sulfonation by H_2SO_4 (98 wt.%) at 80 °C.	*Madhuca indica* oil	Oil/methanol ratio 1:18 mol, 5 wt.% of catalyst, 60 °C, 5 h.	89.2	92.0	90% of initial activity at 7th cycle	[29]
Waste banana peel	Treatment by 30 wt.% H_3PO_4, carbonization for 3 h at 600 °C, and then sulfonation by H_2SO_4 (98 wt.%) at 80 °C.	Oleic acid	Oleic acid/methyl acetate ratio 1:50 mol, 12 wt.% of catalyst, 60 °C, 8 h.	52.3	-	82% of initial activity at 5th cycle	[30]
Waste cork	Pyrolysis for 2 h at 600 °C, followed by sulfonation by H_2SO_4 (98 wt.%) at 80 °C.	Waste oil	Oil/methanol ratio 1:25 mol, 1.5 wt.% of catalyst, 65 °C, 6 h.	98.0	-	80% of initial activity at 5th cycle	[31]
Sugarcane bagasse	Treatment by H_3PO_4 (1:1), calcination for 2 h at 400 °C, and then sulfonation by $ClSO_3H$ at 300 °C.	Palm fatty acid distillate (PFAD)	PFAD/methanol ratio 1:10 mol, 2 wt.% of catalyst, 60 °C, 1.5 h.	98.6	300.0	90% of initial activity at 6th cycle	[32]

* SA—surface area.

Carbonization and pyrolysis of biomass with sulfonation result in a relatively low specific surface area of the catalyst. Meanwhile, pretreatment of biomass with phosphorous acid leads to a significant increase in the surface area three times that of non-pretreated ones [33]. Sulfonated biocarbon-based catalysts mainly show a FAME yield over 80 wt.%. It is observed that the catalyst activity slightly depends on the specific surface area of the catalysts, but the acid sites' density on the carbon surface plays an important role. However, a high surface area can be crucial for the reactions with large molecules (i.e., triglycerides). Thus, the catalyst surface area, acid site density, and carbon preparation method play important roles in the acid-catalyzed esterification/transesterification process.

However, one of the main problems for sulfonated carbon-based catalysts is the loss in activity after 5–7 cycles. Moreover, sulfonated catalysts tend to be deactivated at high temperatures because of the loss of sulfonic groups from the surface. Thus, future works should focus on the improvement of catalyst stability. Another challenge associated with oil transesterification in the presence of biomass-derived catalysts is the investigation of reaction kinetics and mechanisms.

2.2. Base-Functionalized Catalysts

Basic carbon-based catalysts are less used in the transesterification process in comparison to acidic ones. This is due to the high saponification activity of bases, especially for HFFA oils [33]. Moreover, a critical disadvantage of a basic catalyst is its sensitivity to water [34]. Meanwhile, several studies on the application of base-functionalized carbon-based catalysts can be mentioned. Two directions in the preparation of base-functionalized carbon-based catalysts can be highlighted. The first one is calcination of biomass feedstock. As it is known, plant biomass contains a high amount of alkali and alkali-earth metals [4]. Calcination leads to a decrease in the concentration of H, O, and C atoms, producing alkali or alkali-metal oxides or hydroxides along with alumina or silicon oxides. The second method for base-functionalized catalyst preparation is preliminary treatment of raw biomass with solutions of KOH, potassium, or calcium salts (i.e., KF, $Ca(Ac)_2$), followed by hydrothermal or thermal carbonization. This method results in a biochar containing metal oxides on the surface. The last ones catalyze the transesterification process and show high effectiveness. Table 2 presents the results of recent studies on transesterification of oils over a base-functionalized catalyst.

Table 2. Basic carbon-based catalysts for biodiesel production.

Biomass	Preparation Method	Oil Feedstock	Biodiesel Synthesis Conditions	Biodiesel Yield	SA *, m^2/g	Reusability	Reference
Red banana peduncle	Calcination for 4 h at 700 °C.	*Ceiba pentandra* oil	Oil/methanol ratio 1:12 mol, 2.5 wt.% of catalyst, 65 °C, 2 h.	98.7	46.0	90% of initial activity at 3rd cycle	[35]
Banana peels	Calcination for 4 h at 700 °C.	*Bauhinia monandra* seed oil	Oil/methanol ratio 1:7.6 mol, 2.75 wt.% of catalyst, 65 °C, 70 min.	93.9	4.4	-	[36]
Coconut husk ash	Calcination for 1 h at 500 °C.	Jatropha oil	Oil/methanol ratio 1:12 mol, 7 wt.% of catalyst, 45 °C, 30 min.	90.0	-	-	[37]
Sugarcane bagasse	Treatment with 1M HCl, calcination for 4 h at 700 °C, treatment with 1 M NaOH, and carbonization at 300 °C for 1 h.	Waste oil	Oil/methanol ratio 1:2 vol., 10 wt.% of catalyst, 65 °C, 2 h.	99.0	-	-	[38]
Waste date seeds	Carbonization for 4 h at 450 °C, followed by impregnation by 1 M KOH and drying at room temperature for 48 h.	Waste oil	Oil/methanol ratio 1:10 mol, 3 wt.% of catalyst, 65 °C, 90 min.	93.0	260	-	[39]
Empty fruit bunch	HTC for 3 h at 250 °C, followed by impregnation by K_2CO_3 and $Cu(NO_3)_2$.	Waste oil	Oil/methanol ratio 1:12 mol, 5 wt.% of catalyst, 70 °C, 2 h.	97.1	4056.2	80% of initial activity at 5th cycle	[40]
Coffee grounds	Carbonization for 3 h at 600 °C, followed by impregnation by 1 M KOH.	Waste oil	Oil/methanol ratio 1:9 mol, 5 wt.% of catalyst, 90 °C, 2 h.	91.6	-	80% of initial activity at 5th cycle	[41]
Raw coffee husks	Carbonization for 3 h at 700 °C, followed by impregnation by 1 M KOH.	Soybean oil	Oil/methanol ratio 1:10 mol, 10 wt.% of catalyst, 65 °C, 2 h.	74.0	-	90% of initial activity at 2nd cycle	[42]
Potato peel	Pyrolysis for 5 h at 500 °C, followed by calcination.	Waste oil	Oil/methanol ratio 1:9 mol, 3 wt.% of catalyst, 60 °C, 2 h.	97.5	-	90% of initial activity at 5th cycle	[43]
Hyacinth biomass	Carbonization for 8 h at 600 °C, followed by impregnation by K_2CO_3.	Palm oil	Oil/methanol ratio 1:12 mol, 15 wt.% of catalyst, 65 °C, 3 h.	97.6	-	-	[44]
Waste passion fruit peel	Calcination for 1 h at 500 °C	Palm oil	Oil/methanol ratio 1:15 mol, 7 wt.% of catalyst, room temperature, 30 min	95.4	-	78% of initial activity at 2nd cycle	[45]

* SA—surface area.

The activity of base-functionalized catalysts is determined by both the base concentration and catalyst surface area. In this case, the surface area seems to be a significant factor, providing good triglyceride adsorption on the catalyst as well as the distribution and availability of basic active sites. For this reason, the carbonization method (in particular, HTC) can be one of the most promising methods. Additionally, the use of additional alkali during the catalyst preparation process provides higher activity and biodiesel yield in comparison with catalysts obtained without impregnation.

Generally, base-functionalized carbon catalysts prepared through the calcination method are less stable and lose their activity after 3–5 consecutive cycles because of the leaching of alkali via water or methanol or because the interaction of alkali with free fatty acids. Meanwhile, catalysts prepared by the carbonization method, followed by activation by a base, maintain their effectiveness in terms of oil conversion of FAME yield for a minimum of five cycles. In spite of the lower stability, alkali catalysts allow transesterification to be performed at lower temperatures (even at room temperature) with a high biodiesel yield. This can be significant from an economical point of view.

In summary, basic biocarbon catalysts have proven to have a higher conversion of oil into biodiesel at a lower reaction temperature in comparison with sulfonated ones. However, the use of these composites in the transesterification of waste cooking oil is limited because of the high free fatty acid content in the oil.

3. Biomass-Derived Carbon-Based Catalysts for Biomass Conversion and Organic Synthesis

Sulfonated biomass-derived carbons have been effectively used for biomass conversion into valuable chemicals, such as furfural, 5-hydroxymethylfurfural, 5-ethoxymethylfurfural, and hexitols. The presence of $-SO_3H$ groups on the carbon surface effectively catalyzes the hydrolysis and dehydration of polysaccharides of biomass (cellulose and hemicelluloses) (Figure 3) [15]. These catalysts are mainly produced by biomass pyrolysis with or without templates, followed by sulfonation by concentrated sulfuric acid. As for biodiesel synthesis, the resulting composites have the same structure and are characterized by high acidity. The summarized results are presented in Table 3.

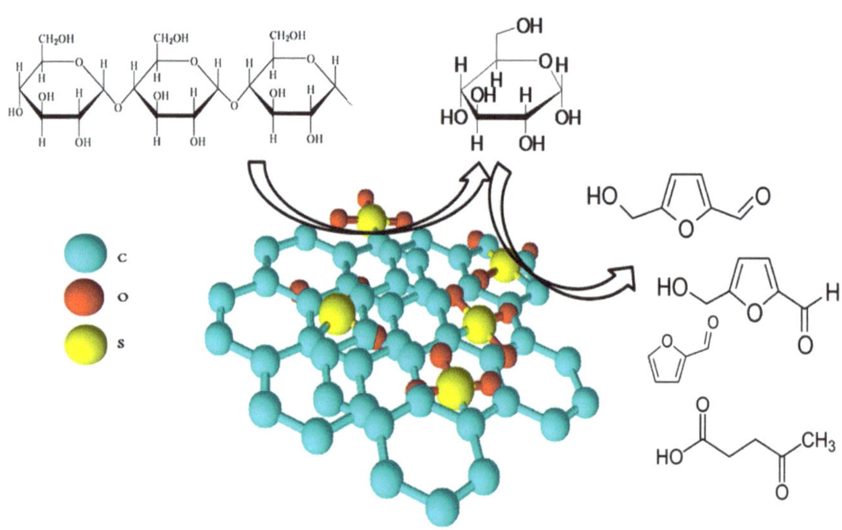

Figure 3. Application of acidic carbon-based catalysts for biomass conversion. Summarized from [46–55].

Table 3. Sulfonated carbon-based catalysts for biomass conversion.

Carbon Source	Feedstock	Reaction Conditions	Product	Product Yield, wt.%	Number of Cycles without Loss in Activity	Reference
Sucrose	Cellulose	110 °C, 4 h	Glucose	71.5	-	[46]
Microcrystalline cellulose	Eucalyptus flakes	120 °C, 3 h	Glucose	76.0	3	[47]
Glucose	Cellulose	110 °C, 4 h	Reducing sugars	72.7	5	[48]
Sucrose	Cellulose	120 °C, 4 h	Glucose	59.0	4	[49]
Cellulose	Cellulose	130 °C, 3 h	Reducing sugars	68.9	3	[50]
Glucose	Fructose	130 °C, 10 min	5-hydroxymethylfurfural	91.2	5	[51]
Cellulose	Fructose	80 °C, 10 min	5-hydroxymethylfurfural	83.0	5	[52]
Cornstalk	Cornstalk	170 °C, 30 min	Furfural	68.2	-	[53]
Waste *Camelia oleifera* shells	Fructose	100 °C, 24 h	5-ethoxymethylfurfural	25.0	4	[54]
Chestnut shell	Waste lignocellulose	180 °C, 15 min	Furfural	68.7	7	[55]

High catalytic activity of sulfonated carbons in biomass hydrolysis and dehydration reactions can be attributed to the high ability of SO_3H-groups to adsorb carbohydrates, as well as their tolerance to hydration. Moreover, the catalyst effect in hydrolysis depends significantly on the water concentration and reaches a maximum when the water content is equal to the catalyst acidity [15,56]. In comparison with traditionally used catalysts containing Lewis and Brønsted acid sites (e.g., zeolites, Amberlyst, and Nafion), sulfonated carbons show higher effectiveness in biomass hydrolysis and dehydration and are considered promising for obtaining platform molecules from biomass.

Besides the functionalization of biocarbon with acid or basic sites, the production of carbon-based composites containing different metal nanoparticles is widely used. Carbon-supported catalysts have great potential in many organic reactions, such as hydrogenation, oxidation, alkylation, and dehydrohalogenation. Biomass-derived carbonaceous supports for metal-containing catalysts are effectively prepared through hydrothermal or dry carbonization, thus providing a high surface area for the resulting materials (200–1000 m^2/g). The active metal-containing phase is traditionally formed by impregnation of the carbon supports with solutions of metal salts. Acetate and nitrates are widely used as metal precursors because of their easy decomposition by heating or calcination (Figure 4). A summary of the reactions using biomass-derived carbon-based catalysts is shown in Table 4.

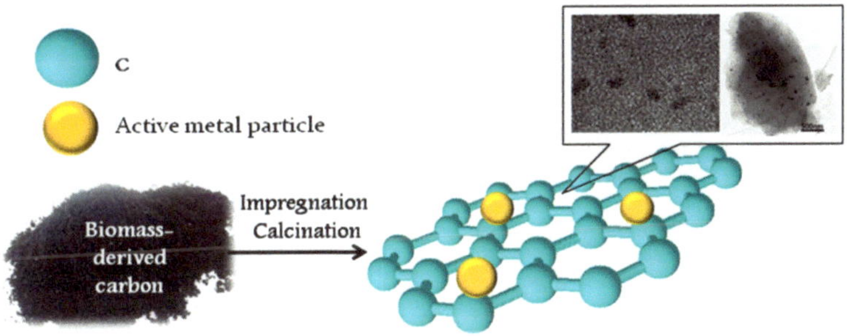

Figure 4. Preparation of biochar-based metal-containing catalysts. Summarized from [57,58].

Table 4. Summary of biomass-derived carbon-based catalysts for organic reactions.

Carbon Source and Preparation Conditions	Active Phase	Substrate	Reaction Conditions	Product	Product Selectivity, wt.%	Number of Cycles without Loss in Activity	Reference
				Hydrogenation			
Shrimp shell and pyrolysis at 600–800 °C for 2 h.	Core–shell $Co@Co_3O_4$	Nitroarenes	110 °C, 40 bar H_2, 6 h	Amines	99.0	3	[57]
Bamboo shoots and HTS at 850 °C.	Core–shell $Co@Co_3O_4$	Nitroarenes	110 °C, 5 MPa H_2, 5 h	Amines	>90.0	6	[58]
Bamboo shoots and HTS at 850 °C.	Pd	Alkynes	Room temperature, 1 atm H_2, 7 h	Alkenes	95.0	-	[58]
Bamboo shoots and HTS at 850 °C.	Co/P	Nitroarenes	170 °C, formic acid as H-donor, 7 h	Amines	>97.0	-	[58]
Cornstalks and carbonization at 300 °C for 3 h.	Pt	Cinnamaldehyde	120 °C, 40 bar H_2, 6 h	Hydrocinamyl alcohol	95.0	-	[59]
Starch and HTC at 500 °C.	Ru	Levulinic acid	150 °C, 5 MPa H_2, 3 h	γ-valerolactone	99.0	5	[60]
Sucrose and HTS with poly(ionic liquid).	Au-Pd	Phenylacetylene	Room temperature, 1 atm H_2, 7 h	Styrene	99.0	-	[61]
				Other reactions			
Starch and gelation, followed by carbonization.	Fe_3O_4	Benzyl alcohol	130 °C, MW, 1 atm air	Benzaldehyde	74.0	5	[62]
Tannic acid and gel formation, followed by carbonization at 900 °C.	Cu/Ni core-shell	Furfural		Methylfurane	48.0	4	[63]
Sugarcane bagasse and HTC at 600 °C.	Ni/NiO, MgO	$CH_4 + CO_2$	750 °C, 40 h	Synthesis gas	Methane conversion > 80%	5	[64]
Shrimp shell and pyrolysis at 600–800 °C for 2 h.	Core–shell $Co@Co_3O_4$	Arylhallides	130 °C, 30 bar H_2, 24 h	Cycloalkenes	>90.0	4	[65]
Bamboo shoots and HTC at 850 °C.	Pd	Alkynes	Room temperature, 1 atm H_2, 2.5 h	Vinylsilanes	96.0	-	[58]
Bamboo shoots and HTC at 850 °C.	Core–shell $Co@Co_3O_4$	Nitroarenes	110 °C, 5 MPa H_2, 5 h	Structurally complex imines	81.0	-	[58]
Starch and carbonization at 600 °C.	Pd	Iodobenzene + methyl acrylate	130 °C, MW, 2 min, 0.1 g catalyst	Biaryles	95.0	2	[66]
Starch and carbonization at 600 °C.	Pd	Benzeneboronic acid + bromobenzene	130 °C, MW, 2 min, 0.1 g catalyst	Biaryles	99.0	3	[67]
Chitosan and carbonization at 750 °C.	Au	Phenylboronic acid	70 °C, 7 h, 0.01 g catalyst	biphenyl	86.0	-	[68]

Along with traditional carbon-based catalysts, biomass-derived ones exhibit high activity in different reactions. Moreover, the "green" nature of these systems makes them preferable for use in sustainable technologies. Carbon materials derived from biomass are considered to provide a high dispersion of metals deposited on their surface, as well as a narrow metal particle size distribution, leading to a higher reducibility of the active phase. Moreover, the presence of dopped atoms (i.e., N and O) on the carbon surface increases the catalyst activity in comparison with traditional carbon-based catalysts. The higher surface area of the biocarbon-derived catalysts compared to those supported on inorganic oxides (such as SiO_2 or Al_2O_3) also provides a good availability of active sites and higher hydrogen adsorption, resulting in their higher activity [15].

4. Biomass-Derived Carbon-Based Catalysts for Electrocatalysis

Carbon-based materials are widely used as catalyst supports in electrocatalysis. Biomass-based carbon is a cheap and structured material that can be easily synthesized for large-scale applications. These supports can effectively replace existing electrocatalysts in the electrooxidation processes, hydrogen or oxygen revolution reactions (HRR and ORR), and oxygen reduction reactions [15]. Electrocatalysts based on biomass-derived carbons have shown high activity and stable onset potential.

The structure of carbonaceous supports plays a great role in electrocatalysis. A higher activity has been shown by catalysts with a graphite- or grapheme-like structure. Biomass-based carbon catalysts as electrocatalysts are usually synthesized through pyrolysis or carbonization of biomass feedstock. The control over the synthesis process temperature, time of carbonization, and pyrolysis atmosphere are the key factors that allow the required structure of carbon-based catalysts to be obtained [15]. Moreover, the introduction of heteroatoms in the crystal lattice of carbon is a key factor for these electrocatalysts since it forms the active sites [69]. Doping usually serves as a charge transfer and modification of the electronic structure of carbon atoms (Figure 5). Among the doping heteroatoms, nitrogen, phosphorus, sulfur, and oxygen should be mentioned.

An overview of carbon-based electrocatalysts, their synthesis methods, structure, and onset potential is presented in Table 5.

Table 5. Summary of biomass-derived carbon-based electrocatalysts.

Catalyst	Carbon Source	Synthesis Conditions	Crystal Structure	Porosity	Onset Potential, V	Reaction	Catalyst Stability	Reference
Pd/N-C	Mildew-starting orange waste	Juice extraction, mixing with K_2PdCl_4, hydrothermal carbonization at 180 °C for 4 h, and pyrolysis in nitrogen atmosphere at 800 °C for 2 h.	Graphite-like structure with pyridinic and graphite N	Microporous, SA * = 26 m^2/g	0.5	Methanol oxidation	90% in 50 cycles	[70]
N-C	N-rich yuba	Carbonization at 850 °C, followed by impregnation.	Graphite-like structure with pyridinic and pyrrolic N	Mesoporous, SA * = 740–1000 m^2/g	0.97	Fuel cells	85% after 5000 s	[71]
Fe_3C/C	Cellulose fiber	Impregnation with an iron salt, addition of dicyandiamide, carbonization, and etching.	Fe_3C crystals over a graphite-like structure	-	0.98	Oxygen reduction	85% after 10,000 s	[72]
FeO_x/N-C	Acorn shell	Impregnation with melanin, $(MgOH)_2CO_3$, and NaCl, carbonization at 900 °C, treatment with hemin, and calcination in nitrogen atmosphere.	Graphite-like structure with pyridinic and pyrrolic N, Fe_3O_4, and Fe_2O_3 coated with C	SA * = 819 m^2/g	0.88 in ORR 0.72 in HRR	ORR and HRR	73% after 36,000 s	[73]
N-C	Wood char	Activation with alkali and doping with nitrogen.	Graphite-like structure with pyridinic and pyrrolic N	Microporous, SA * = 1924 m^2/g	0.2	ORR	95% after 1200 min	[74]
Co/C	Cotton carbon fiber	Impregnation with $CoCl_2$, ultrasonification, and carbonization at 900 °C for 3 h.	Graphite-like structure with amorphous carbon	Mesoporous, SA * = 358 m^2/g	-	Lie-SeS_2 cells	70% after 45 cycles	[75]
Ru/N-C	Yeasts	Impregnation with $RuCl_3$, calcinations at 950 °C for 5 h, and doping with nitrogen.	Graphite-like structure	-	0.2	HRR	85% after 60 h	[76]
Fe_3O_4/N-C	Yeasts	Impregnation with $Fe(NO_3)_3$, calcinations at 950 °C for 5 h, and doping with nitrogen.	Graphite-like structure	-	0.3	ORR	85% after 60 h	[76]
Mo/C	Sugar beet, corn stover, pine, and miscanthus	Impregnation with Mo salt, pyrolysis at 600 °C for 6 h, and NaCl/NaF salt flux.	Graphite-like structure	-	0.3–0.5 V	HRR	-	[77]
N-C	Cocoon silk	Carbonization followed by activation by KOH.	Graphite-like structure with pyridinic and pyrrolic N	-	0.6	ORR	-	[78]
N-C	Cocoon silk	Pyrolysis followed by activation by $ZnCl_2$.	1D structure	-	0.85	ORR	95% after 12,000 s	[79]
N-C	Keratin	Precarbonization, activation with KOH, and treatment in ammonia at 1000 °C.	Grapheme-like 2D structure	-	0.8	ORR	92% after 300 s	[80]
N-C	Coconut shell	Carbonization followed by activation by H_3PO_4.	Graphite-like structure	Mesoporous, SA * = 1260 m^2/g			90% after 12 h	[81]
N-C	Ophiopogon japonicus	Hydrothermal carbonization at 180 °C.	Carbon nanodot/nanosheet aggregates	-	0.88	ORR	-	[82]
N-C	Basswood	Pyrolysis at 600 °C and activation in ammonia atmosphere.	Graphite-like structure	SA * = 1438 m^2/g	0.98	ORR	95% after 20 h	[83]
N-C	Pinecone	Precarbonization at 800 °C and activation in ammonia.	Graphite-like structure	-	0.95	ORR	-	[84]
N-C	Spent coffee grounds	Pyrolysis followed by activation by KOH.	Graphite-like structure	SA * = 1018 m^2/g	0.94	ORR	95% after 10,000 s	[85]
N-C	Banana peels	Carbonization followed by activation by KOH, and treatment with ammonia	Graphite-like structure	SA * = 1756 m^2/g	0.98	ORR	95% after 10,000 s	[86]

* SA—surface area.

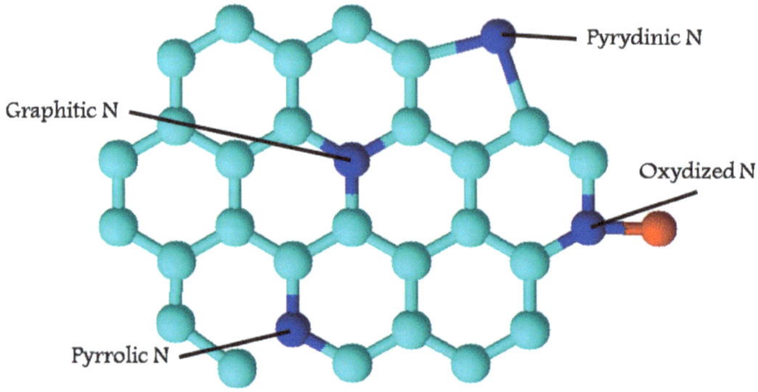

Figure 5. Overall structure of N-doped carbon-based electrocatalyst. Summarized from [69,70,73,77–85].

For electrocatalysts, the distribution of metal-containing particles and heteroatoms, the crystal structure of carbonaceous support, and its porosity play a great role. It has been shown that the presence of heteroatoms exhibits a higher catalytic activity than that of non-doped carbons, as well as a higher potential stability. The hierarchical porous structure can also improve the activity of these electrocatalysts in most reactions because of the facilitation of the diffusion of reactants and products [69,87]. A high surface area of the electrodes obtained from biocarbons is crucial for the energy storage and ion/electron transport. Thus, biomass-derived carbonaceous materials can be effectively applied as supercapacitors. Biomass-derived carbon electrodes can be considered a promising alternative to traditional electrodes [87].

5. Conclusions and Future Prospective

The key requirements that catalysts should meet are high activity and selectivity toward the target products, stability in the reaction medium, mechanical effects, and temperature. In terms of structure, heterogeneous catalysts should have a uniform distribution of the active phase, thus avoiding its aggregation; high porosity to allow reagent molecules to easily reach the active sites; and a proper structure of the active sites to catalyze the target reactions. Therefore, the catalyst design is of great interest.

Carbon-based catalysts have attracted much attention in recent years. Their low cost, simple preparation methods, and "green" nature make carbon materials a promising support for catalysts. Biomass can be successfully used for the preparation of carbon-based catalysts through different methods, mainly based on thermochemical decomposition. Different types of biomasses, e.g., starch, lignin, cellulose, chitosan, wood, agricultural or food waste, and marine waste, can be promising carbon feedstock. Expanding biomass sources, development of new methods for direct catalyst synthesis, expanding the choice of synthesis conditions, and development of "green" preparation methods are the main directions for future research for biomass-derived catalyst synthesis. Moreover, wider applications of biomass-derived catalysts, including electrocatalysis, fine and organic synthesis, and flow processes, should be developed to provide sustainable and environmentally friendly processes.

Author Contributions: Conceptualization, A.A.S. and M.G.S.; data collection, Y.V.L. and Y.Y.K.; writing—original draft preparation, A.A.S. and V.G.M.; writing—review and editing, M.E.M.; project administration, A.A.S. All authors have read and agreed to the published version of the manuscript.

Funding: The financial support for this work was provided by the Russian Science Foundation (grant 22-79-10096).

Data Availability Statement: Not applicable.

Conflicts of Interest: The authors declare no conflict of interest.

References

1. Elliott, D.C.; Hart, T.R. Catalytic hydroprocessing of chemical models for bio-oil. *Energy Fuels* **2008**, *23*, 631–637. [CrossRef]
2. Sharma, S.; Pollet, B.G. Support materials for PEMFC and. DMFC electrocatalysts—A review. *J. Power Sources* **2012**, *208*, 96–119. [CrossRef]
3. Lam, E.; Luong, J.H.T. Carbon Materials as Catalyst Supports and Catalysts in the Transformation of Biomass to Fuels and Chemicals. *ACS Catal.* **2014**, *4*, 3393–3410. [CrossRef]
4. Maroa, S.; Inambao, F. A review of sustainable biodiesel production using biomass-derived heterogeneous catalysts. *Eng. Life Sci.* **2021**, *21*, 790–824. [CrossRef] [PubMed]
5. Alrobaian, A.; Rajasekar, V.; Alagumalai, A. Critical insight into biowaste-derived biocatalyst for biodiesel production. *Environ. Prog. Sustain. Energy* **2020**, *39*, e13391. [CrossRef]
6. Smith, S.M.; Oopathum, C.; Weeramongkhonlert, V.; Smith, C.B. Transesterification of soybean oil using bovine bone waste as new catalyst. *Bioresour. Technol.* **2013**, *143*, 686–690. [CrossRef]
7. Roschat, W.; Kacha, M.; Yoosuk, B.; Sudyoadsuk, T. Biodiesel production based on heterogeneous process catalyzed by solid waste coral fragment. *Fuel* **2012**, *98*, 194–202. [CrossRef]
8. Pang, S. Advances in thermochemical conversion of woody biomass to energy, fuels and chemicals. *Biotechnol. Adv.* **2019**, *37*, 589–597. [CrossRef]
9. Yaashikaa, P.R.; Kumar, P.S.; Varjani, S.; Saravanan, A. A critical review on the biochar production techniques, characterization, stability and applications for circular bioeconomy. *Biotechnol. Rep.* **2020**, *28*, e00570. [CrossRef]
10. Cantrell, K.B.; Hunt, P.G.; Uchimiya, M.; Novak, J.M.; Ro, K.S. Impact of pyrolysis temperature and manure source on physicochemical characteristics of biochar. *Bioresour. Technol.* **2012**, *107*, 419–428. [CrossRef]
11. Nunoura, T.; Wade, S.R.; Bourke, J.P.; Antal, M.J. Studies of the flash carbonization process. 1. Propagation of the flaming pyrolysis reaction and performance of a catalytic afterburner. *Ind. Eng. Chem. Res.* **2006**, *45*, 585–599. [CrossRef]
12. Funke, A.; Ziegler, F. Hydrothermal carbonization of biomass: A summary and discussion of chemical mechanisms for process engineering. *Biofuels Bioprod. Biorefin.* **2010**, *4*, 160–177. [CrossRef]
13. Titirici, M.-M.; Thomas, A.; Antonietti, M. Back in the black: Hydrothermal carbonization of plant material as an efficient chemical process to treat the CO_2 problem? *New J. Chem.* **2007**, *31*, 787–789. [CrossRef]
14. Titirici, M.-M.; White, R.J.; Brun, N.; Budarin, V.L.; Su, D.S.; del Monte, F.; Clark, J.H.; MacLachlan, M.J. Sustainable carbon materials. *Chem. Soc. Rev.* **2015**, *44*, 250–290. [CrossRef]
15. De, S.; Balu, A.M.; van der Waal, J.C.; Lique, R. Biomass-Derived Porous Carbon Materials: Synthesis and Catalytic Applications. *ChemCatChem* **2015**, *7*, 1608–1629. [CrossRef]
16. Klinghoffer, N.B.; Castaldi, M.J.; Nzihou, A. Influence of char composition and inorganics on catalytic activity of char from biomass gasification. *Fuel* **2015**, *157*, 37–47. [CrossRef]
17. Bergman, P.C.A.; Boersma, A.R.; Zwart, R.W.R.; Kiel, J.H.A. *Torrefaction for Biomass Co-Firing in Existing Coal-Fired Power Stations*; Report No. ECNC05013 Energy Research Centre of The Netherlands (ECN); Energy Research Centre of The Netherlands: Petten, The Netherlands, 2005; p. 71.
18. Shokrolahi, A.; Zali, A.; Keshavarz, M.H. Wet carbon-based solid acid/$NaNO_3$ as a mild and efficient reagent for nitration of aromatic compound under solvent free conditions. *Chin. Chem. Lett.* **2007**, *18*, 1064–1066. [CrossRef]
19. Zong, M.H.; Duan, Z.Q.; Lou, W.Y.; Smith, T.J.; Wu, H. Preparation of a sugar catalyst and its use for highly efficient production of biodiesel. *Green Chem.* **2007**, *5*, 434–437. [CrossRef]
20. Lokman, I.M.; Rashid, U.; Yunus, R.; Taufiq-Yap, Y.H. Carbohydrate-derived Solid Acid Catalysts for Biodiesel Production from Low-Cost Feedstocks: A Review. *Catal. Rev. Sci. Eng.* **2014**, *56*, 187–219. [CrossRef]
21. Hara, M.; Yoshida, T.; Takagaki, A.; Takata, T.; Kondo, J.N.; Domen, K.; Hayashi, S. A carbon material as a strong protonic acid. *Angew. Chem. Int. Ed.* **2004**, *43*, 2955–2958. [CrossRef]
22. Shu, Q.; Zhang, Q.; Xu, G.; Nawaz, Z.; Wang, D.; Wang, J. Synthesis of biodiesel from cottonseed oil and methanol using a carbon-based solid acid catalyst. *Fuel Process. Technol.* **2009**, *90*, 1002–1008. [CrossRef]
23. Dawodu, F.A.; Ayodele, O.O.; Xina, J.; Zhanga, S. Application of solid acid catalyst derived from low value biomass for a cheaper biodiesel production. *J. Chem. Technol. Biotechnol.* **2014**, *89*, 1898–1909. [CrossRef]
24. Ayodele, O.O.; Dawodu, F.A. Conversion of Calophyllum inophyllum Oil with a High Free Fatty Acid Content to Biodiesel using a Starch-Derived Catalyst. *Energy Technol.* **2014**, *2*, 912–920. [CrossRef]
25. Lou, W.Y.; Zong, M.H.; Duan, Z.Q. Efficient production of biodiesel from high free fatty acid-containing waste oils using various carbohydrate-derived solid acid catalysts. *Bioresour. Technol.* **2008**, *99*, 8752–8758. [CrossRef]
26. Guo, F.; Xiu, Z.L.; Liang, Z.X. Synthesis of biodiesel from acidified soybean soapstock using a lignin-derived carbonaceous catalyst. *Appl. Energy* **2012**, *98*, 47–52. [CrossRef]
27. Li, M.; Zheng, Y.; Chen, Y.; Zhu, X. Biodiesel production from waste cooking oil using a heterogeneous catalyst from pyrolyzed rice husk. *Bioresour. Technol.* **2014**, *154*, 345–348. [CrossRef]
28. Luque, R.; Clark, J.H. Biodiesel-Like Biofuels from Simultaneous Transesterification/Esterification of Waste Oils with a Biomass-Derived Solid Acid Catalyst. *ChemCatChem.* **2011**, *3*, 594–597. [CrossRef]
29. Choksi, H.; Pandian, S.; Gandhi, Y.H.; Deepalakshmi, S. Studies on production of biodiesel from Madhuca indica oil using a catalyst derived from cotton stalk. *Energy Sources Part A Recovery Util. Environ. Eff.* **2019**, *43*, 3424–3433. [CrossRef]

30. Wong, W.-Y.; Lim, S.; Pang, Y.-L.; Chen, W.-H.; Lam, M.-K.; Tan, I.-S. Synthesis of glycerol-free fatty acid methyl ester using interesterification reaction based on solid acid carbon catalyst derived from low-cost biomass wastes. *Int. J. Energy Res.* **2020**, *46*, 147–162. [CrossRef]
31. Bhatia, S.K.; Gurav, R.; Choi, T.-R.; Kim, H.J.; Yang, S.-Y.; Song, H.-S.; Park, J.Y.; Park, Y.-L.; Han, Y.-H.; Choi, Y.-K. Conversion of waste cooking oil into biodiesel using heterogenous catalyst derived from cork biochar. *Bioresour. Technol.* **2020**, *302*, 122872. [CrossRef]
32. Akinfalabi, S.-I.; Rashid, U.; Ngamcharussrivichai, C.; Nehdi, I.A. Synthesis of reusable biobased nano-catalyst from waste sugarcane bagasse for biodiesel production. *Environ. Technol. Innov.* **2020**, *18*, 100788–100801. [CrossRef]
33. Meher, L.; Churamani, C.; Arif, M.; Ahmed, Z.; Naik, S. *Jatropha curcas* as a renewable source for bio-fuels—A review. *Renew. Sustain. Energy Rev.* **2013**, *26*, 397–407. [CrossRef]
34. Leung, D.; Guo, Y. Transesterification of neat and used frying oil: Optimization for biodiesel production. *Fuel Process. Technol.* **2006**, *87*, 883–890. [CrossRef]
35. Balajii, M.; Niju, S. A novel biobased heterogeneous catalyst derived from Musa acuminata peduncle for biodiesel production—Process optimization using central composite design. *Energy Convers. Manag.* **2019**, *189*, 118–131. [CrossRef]
36. Betiku, E.; Akintunde, A.M.; Ojumu, T.V. Banana peels as a biobase catalyst for fatty acid methyl esters production using Napoleon's plume (*Bauhinia monandra*) seed oil: A process parameters optimization study. *Energy* **2016**, *103*, 797–806. [CrossRef]
37. Vadery, V.; Narayanan, B.N.; Ramakrishnan, R.M.; Cherikkallinmel, S.K.; Sugunan, S.; Narayanan, D.P.; Sasidharan, S. Room temperature production of jatropha biodiesel over coconut husk ash. *Energy* **2014**, *70*, 588–594. [CrossRef]
38. Niju, S.; Kanna, S.K.A.; Ramalingam, V.; Kumar, M.S.; Balajii, M. Sugarcane bagasse derived biochar—A potential heterogeneous catalyst for transesterification process. *Energy Sources Part A Recovery Util. Environ. Eff.* **2019**, *20*, 1–12. [CrossRef]
39. Inayat, A.; Jamil, F.; Raza, F.; Khurram, S.; Ghenai, C.; Al-Muhatseb, A.H. Upgradation of waste cooking oil to biodiesel in the presence of green catalyst derived from date seeds. *Biofuels* **2021**, *12*, 1245–1250. [CrossRef]
40. Abdullah, R.F.; Rashid, U.; Hazmi, B.; Ibrahim, M.L.; Tsubota, T.; Alharthi, F.A. Potential heterogeneous nano-catalyst via integrating hydrothermal carbonization for biodiesel production using waste cooking oil. *Chemosphere* **2022**, *286*, 131913. [CrossRef]
41. Atelge, M. Production of biodiesel and hydrogen by using a double-function heterogeneous catalyst derived from spent coffee grounds and its thermodynamic analysis. *Renew. Energy* **2022**, *198*, 1–15. [CrossRef]
42. Fernández, J.V.; Faria, D.N.; Santoro, M.C.; Mantovaneli, R.; Cipriano, D.F.; Brito, G.M.; Carneiro, M.T.W.; Schettino, M.A.; Gonzalez, J.L.; Freitas, J.C. Use of Unmodified Coffee Husk Biochar and Ashes as Heterogeneous Catalysts in Biodiesel Synthesis. *BioEnergy Res.* **2022**. [CrossRef]
43. Daimary, N.; Eldiehy, K.S.; Boruah, P.; Deka, D.; Bora, U.; Kakati, B.K. Potato peels as a sustainable source for biochar, bio-oil and a green heterogeneous catalyst for biodiesel production. *J. Environ. Chem. Eng.* **2022**, *10*, 107108. [CrossRef]
44. Yani, F.; Ulhaqi, R.; Pratiwi, W.; Pontas, K.; Husin, H. Utilization of water hyacinth-based biomass as a potential heterogeneous catalyst for biodiesel production. *J. Phys. Conf. Ser.* **2019**, *1402*, 055005. [CrossRef]
45. Tarigan, J.B.; Singh, K.; Sinuraya, J.S.; Supeno, M.; Sembiring, H.; Tarigan, K.; Rambe, S.M.; Karo-Karo, J.A.; Sitepu, E.K. Waste Passion Fruit Peel as a Heterogeneous Catalyst for Room-Temperature Biodiesel Production. *ACS Omega* **2022**, *7*, 7885–7892. [CrossRef]
46. Pang, J.; Wang, A.; Zheng, M.; Zhang, T. Hydrolysis of cellulose into glucose over carbons sulfonated at elevated temperatures. *Chem. Commun.* **2010**, *46*, 6935–6937. [CrossRef]
47. Suganuma, S.; Nakajima, K.; Kitano, M.; Yamaguchi, D.; Kato, H.; Hayashi, S.; Hara, M. Hydrolysis of Cellulose by Amorphous Carbon Bearing SO_3H, COOH, and OH Groups. *J. Am. Chem. Soc.* **2008**, *130*, 12787–12793. [CrossRef]
48. Guo, H.; Qi, X.; Li, L.; Smith, R.L., Jr. Hydrolysis of cellulose over functionalized glucose-derived carbon catalyst in ionic liquid. *Bioresour. Technol.* **2012**, *116*, 355–359. [CrossRef]
49. Liu, M.; Jia, S.; Gong, Y.; Song, C.; Guo, X. Effective hydrolysis of cellulose into glucose over sulfonated sugar-derived carbon in an ionic liquid. *Ind. Eng. Chem. Res.* **2013**, *52*, 8167–8173. [CrossRef]
50. Guo, H.; Lian, Y.; Yan, L.; Qi, X.; Smith, R.L., Jr. Cellulose-derived superparamagnetic carbonaceous solid acid catalyst. *Green Chem.* **2013**, *15*, 2167–2174. [CrossRef]
51. Wang, J.J.; Xu, W.J.; Ren, J.W.; Liu, X.H.; Lu, G.Z.; Wang, Y.Q. Efficient catalytic conversion of fructose into hydroxymethylfurfural by a novel carbon-based solid acid. *Green Chem.* **2011**, *13*, 2678–2681. [CrossRef]
52. Qi, X.H.; Guo, H.X.; Li, L.Y.; Smith, R.L., Jr. Acid-Catalyzed Dehydration of Fructose into 5-Hydroxymethylfurfural by Cellulose-Derived Amorphous Carbon. *ChemSusChem* **2012**, *5*, 2215–2220. [CrossRef]
53. Ji, L.; Tang, Z.; Yang, D.; Ma, C.; He, Y.-C. Improved one-pot synthesis of furfural from corn stalk with heterogeneous catalysis using corn stalk as biobased carrier in deep eutectic solvent–water system. *Bioresour. Technol.* **2021**, *340*, 125691–125699. [CrossRef]
54. Wen, Y.; Yu, Z.; Li, K.; Guo, H.; Dai, Y.; Yan, L. Fabrication of biobased heterogeneous solid Brønsted acid catalysts and their application on the synthesis of liquid biofuel 5-ethoxymethylfurfural from fructose. *Green Energy Environ.* **2018**, *3*, 384–391. [CrossRef]
55. Liang, J.; Zha, J.; Zhao, N.; Tang, Z.; He, Y.; Ma, C. Valorization of Waste Lignocellulose to Furfural by Sulfonated Biobased Heterogeneous Catalyst Using Ultrasonic-Treated Chestnut Shell Waste as Carrier. *Processes* **2021**, *9*, 2269. [CrossRef]

56. Kang, S.; Ye, J.; Chang, J. Recent Advances in Carbon-Based Sulfonated Catalyst: Preparation and Application. *Int. Rev. Chem. Eng.* **2013**, *5*, 133–144.
57. Sahoo, B.; Formenti, D.; Topf, C.; Bachmann, S.; Scalone, M.; Junge, K.; Beller, M. Sustainable Biomass-Derived Catalysts for Selective Hydrogenation of Nitroarenes. *ChemSusChem* **2017**, *10*, 3035–3039. [CrossRef]
58. Song, T.; Yang, Y. Metal Nanoparticles Supported on Biomass-Derived Hierarchical Porous Heteroatom-Doped Carbon from Bamboo Shoots Design, Synthesis and Applications. *Chem. Rec.* **2018**, *19*, 1283–1301. [CrossRef]
59. Zhang, F.; Li, G.-D.; Chen, J.-S. Preparation and gas storage of high surface area microporous carbon derived from biomass source cornstalks. *J. Colloid Interface Sci.* **2008**, *327*, 108–114. [CrossRef]
60. Luque, R.; Clark, J.H. Water-tolerant Ru-Starbon (R) materials for the hydrogenation of organic acids in aqueous ethanol. *Catal. Commun.* **2010**, *11*, 928–931. [CrossRef]
61. Zhang, P.; Yuan, J.; Fellinger, T.-P.; Antonietti, M.; Li, H.; Wang, Y. Improving hydrothermal carbonization by using poly(ionic liquid)s. *Angew. Chem. Int. Ed.* **2013**, *52*, 6028–6032. [CrossRef]
62. Ojeda, M.; Balu, A.M.; Romero, A.A.; Esquinazi, P.; Ruokolainen, J.; Sixta, H.; Luque, R. MAGBONS: Novel Magnetically Separable Carbonaceous Nanohybrids from Porous Polysaccharides. *ChemCatChem* **2014**, *6*, 2847–2853. [CrossRef]
63. Varila, T.; Makela, E.; Kupila, R.; Romar, H.; Hu, T.; Karinen, R.; Puurunen, R.L.; Lassi, U. Conversion of furfural to 2-methylfuran over CuNi catalysts supported on biobased carbon foams. *Catal. Today* **2021**, *367*, 16–27. [CrossRef]
64. Liang, Y.; Han, J.; Wang, H.; Zhao, B.; Qin, L.; Wang, Y. Synthesis of Ni-Mg@HC catalyst derived from sugarcane bagasse and its application for producing syngas via CO_2 dry reforming. *Energy Sources Part A Recovery Util. Environ. Eff.* **2019**, *18*, 1–14. [CrossRef]
65. Sahoo, B.; Surkus, A.-E.; Pohl, M.-M.; Radnik, J.; Schneider, M.; Bachmann, S.; Scalone, M.; Junge, K.; Beller, M. A Biomass-Derived Non-Noble Cobalt Catalyst for Selective Hydrodehalogenation of Alkyl and (Hetero)Aryl Halides. *Angew. Chem.* **2017**, *129*, 11394–11399. [CrossRef]
66. Gronnow, M.J.; Luque, R.; Macquarrie, D.J.; Clark, H. A novel highly active biomaterial supported palladium catalyst. *Green Chem.* **2005**, *7*, 552–557. [CrossRef]
67. Budarin, V.L.; Clark, J.H.; Luque, R.; Macquarrie, D.J.; White, R.J. Versatile mesoporous carbonaceous materials for acid catalysis. *Green Chem.* **2008**, *10*, 382–387. [CrossRef]
68. Sk, M.P.; Jana, C.K.; Chattopadhyay, A. A gold–carbon nanoparticle composite as an efficient catalyst for homocoupling reaction. *Chem. Commun.* **2013**, *49*, 8235–8237.
69. Zhang, Z.; Yang, S.; Li, H.; Zan, Y.; Li, X.; Zhu, Y.; Dou, M.; Wang, F. Sustainable Carbonaceous Materials Derived from Biomass as Metal-Free Electrocatalysts. *Adv. Mater.* **2019**, *31*, 1805718. [CrossRef]
70. Da, Z.; Niu, X.; Li, X.; Zhang, W.; He, Y.; Pan, J.; Qui, F.; Yan, Y. From mildewy orange waste to natural reductant and catalyst support: Active palladium/biomass-derived carbonaceous hybrids for promoted methanol electro-oxidation. *ChemElectroChem* **2017**, *4*, 1372–1377. [CrossRef]
71. Zhang, J.J.; Sun, Y.; Guo, L.K.; Sun, X.N.; Huang, N.B. Ball-Milling Effect on Biomass-Derived Nanocarbon Catalysts for the Oxygen Reduction Reaction. *ChemistrySelect* **2021**, *6*, 6019–6028. [CrossRef]
72. Wei, J.; Liang, Y.; Hu, Y.; Kong, B.; Simon, G.P.; Zhang, J.; Jiang, S.P.; Wang, H. A Versatile Iron–Tannin-Framework Ink Coating Strategy to Fabricate Biomass-Derived Iron Carbide/Fe-N-Carbon Catalysts for Efficient Oxygen Reduction. *Agnew. Chem. Int. Ed.* **2016**, *55*, 1355–1359. [CrossRef]
73. Lu, Z.; Chen, J.; Wang, W.L.; Li, W.; Sun, M.; Wang, Y.; Wang, X.; Ye, J.; Rao, H. Electrocatalytic, Kinetic, and Mechanism Insights into the Oxygen-Reduction Catalyzed Based on the Biomass-Derived FeO_x@N-Doped Porous Carbon Composites. *Small* **2021**, *17*, 2007326. [CrossRef]
74. Kaare, K.; Yu, E.; Käämbre, T.; Volperts, A.; Dobele, G.; Zhurinsh, A.; Niaura, G.; TamasauskaiteTamasiunaite, L.; Norkus, E.; Kruusenberg, I. Biomass-derived Graphene-like Catalyst Material for Oxygen Reduction Reaction. *ChemNanoMat* **2021**, *7*, 307–313. [CrossRef]
75. Zhang, Y.; Wang, B.; Jing, P.; Guo, Y.; Zhang, Y.; Wei, Y.; Wang, Q.; Zhang, Y.; Wu, H. Bioderived carbon fiber conductive networks with inlaid electrocatalysts as an ultralight freestanding interlayer for working Lie-SeS_2 pouch cells. *Carbon* **2022**, *189*, 10–20. [CrossRef]
76. Tiwari, J.N.; Dang, N.K.; Sultan, S.; Thangavel, P.; Jeong, H.Y.; Kim, K.S. Multi-heteroatom-doped carbon from waste-yeast biomass for sustained water splitting. *Nature Sust.* **2020**, *3*, 556–563. [CrossRef]
77. Harris, D.; Budhi, S.; She, Y.; Henry, J.; Leonard, B.M. Biomass derived metal carbide catalysts formed using a salt flux synthesis. *Mater. Res. Express* **2019**, *6*, 115519. [CrossRef]
78. Wang, Y.; Lei, Y.; Wang, H. Astridia velutina-like S, N-codoped hierarchical porous carbon from silk cocoon for superior oxygen reduction reaction. *RSC Adv.* **2016**, *6*, 73560–73565. [CrossRef]
79. Fu, P.; Zhou, L.; Sun, L.; Huang, B.; Yuan, Y. Nitrogen-doped porous activated carbon derived from cocoon silk as a highly efficient metal-free electrocatalyst for the oxygen reduction reaction. *RSC Adv.* **2017**, *7*, 13383–13389. [CrossRef]
80. Zhang, J.; Zhou, H.; Liu, X.; Zhang, J.; Peng, T.; Yang, J.; Huang, Y.; Mu, S. Keratin-derived S/N co-doped graphene-like nanobubble and nanosheet hybrids for highly efficient oxygen reduction. *J. Mater. Chem. A* **2016**, *4*, 15870–15879. [CrossRef]

81. Borghei, M.; Laocharoen, N.; Kibena-Põldsepp, E.; Johansson, L.S.; Campbell, J.; Kauppinen, E.; Tammeveski, K.; Rojas, O.J. Porous N,P-doped carbon from coconut shells with high electrocatalytic activity for oxygen reduction. *Appl. Catal. B* **2017**, *204*, 394–402. [CrossRef]
82. Oh, J.; Park, S.; Jang, D.; Shin, Y.; Lim, D.; Park, S. Metal-free N-doped carbon blacks as excellent electrocatalysts for oxygen reduction reactions. *Carbon* **2019**, *145*, 481–487. [CrossRef]
83. Tang, Z.; Pei, Z.; Wang, Z.; Li, H.; Zeng, J.; Ruan, Z.; Huang, Y.; Zhu, M.; Xue, Q.; Yu, J.; et al. Highly anisotropic, multichannel wood carbon with optimized heteroatom doping for supercapacitor and oxygen reduction reaction. *Carbon* **2018**, *130*, 532–538. [CrossRef]
84. Lei, X.; Wang, M.; Lai, Y.; Hu, L.; Wang, H.; Fang, Z.; Li, J.; Fang, J. Nitrogen-doped micropore-dominant carbon derived from waste pine cone as a promising metal-free electrocatalyst for aqueous zinc/air batteries. *J. Power Sources* **2017**, *365*, 76–82. [CrossRef]
85. Srinu, A.; Peera, S.G.; Parthiban, V.; Bhuvaneshwari, B.; Sahu, A.K. Heteroatom Engineering and Co-Doping of N and P to Porous Carbon Derived from Spent Coffee Grounds as an Efficient Electrocatalyst for Oxygen Reduction Reactions in Alkaline Medium. *ChemistrySelect* **2018**, *3*, 690–702. [CrossRef]
86. Zhang, J.; Zhang, C.; Zhao, Y.; Amiinu, I.S.; Zhou, H.; Liu, X.; Tang, Y.; Mu, S. Three dimensional few-layer porous carbon nanosheets towards oxygen reduction. *Appl. Catal. B* **2017**, *211*, 148–156. [CrossRef]
87. Manasa, P.; Sambasivam, S.; Ean, F. Recent progress on biomass waste derived activated carbon electrode materials for supercapacitors applications—A review. *J. Energy Storage* **2022**, *54*, 105290. [CrossRef]

catalysts

MDPI

Article

Doped Ruthenium/Hypercrosslinked Polystyrene (HPS) Catalysts in the Modification of Fatty Acid Methyl Esters

Romulo C. Dutra [1], Thatiane V. S. Martins [2], Delma da G. Rocha [2], Mario R. Meneghetti [2], Simoni M. P. Meneghetti [2], Mikhail G. Sulman [3,4], Valentina G. Matveeva [3,4] and Paulo A. Z. Suarez [1,*]

1 Laboratory of Materials and Fuels, Chemistry Institute, University of Brasilia, Brasilia 70910-000, Brazil
2 Group of Catalysis and Chemical Reactivity, Chemistry and Biotechnology Institute, Federal University of Alagoas, Maceio 57072-970, Brazil
3 Department of Biotechnology, Chemistry and Standardization, Tver Technical University, Tver 170026, Russia
4 Regional Technological Centre, Tver State University, Tver 170100, Russia
* Correspondence: psuarez@unb.br

Abstract: The modification of fatty substrates using heterogeneous catalysts have extensive industrial application and has an emphatic position in a sustainable context. Herein ruthenium, nickel, cobalt and copper-doped ruthenium, supported on hypercrosslinked polystyrene (HPS) catalysts were prepared by wet impregnation, characterized and applied on the modification of canola fatty acid methyl esters (FAME). The characterization showed a dispersive effect of doping metals over the ruthenium particles, the presence of acid sites and overall surface morphology, which allows targeting potential applications. The chosen modifications consisted of hydrogenation, hydroformylation, oxidative cleavage and deoxygenation, resulting in superb catalytic activities of over 99% conversions for hydrogenation and deoxygenation. The simplicity of the canola FAME composition allows the understanding of the catalytic processes and allows the upscale of more complex FAME matrixes. Finally, the obtained data stimulates further optimization studies for each application with a variation on the catalysts and the usage of residual fatty substrates, greatly enhancing the sustainability profile of the systems.

Keywords: ruthenium; doped; catalyst; hypercrossed polystyrene; heterogenized; FAME

1. Introduction

Heterogeneous catalysis plays a prominent role due to its great advantages compared to homogeneous ones, because they are usually easy to regenerate and reuse, exhibiting thermal, chemical and structural stability in addition to being low cost. However, some disadvantages must be considered, such as the complexity of active sites, leaching and limited surface-substrate contact [1–3].

One area in which the use of heterogeneous catalysts has emerged is that of functionalization of fatty substrates, mainly by their simple operation, requiring less unitary operations for product isolation, and results in products exhibiting high purity with increased value, which may be characterized as a circular economy and biorefinery application [4–6].

The unsaturations in fatty acids chains are active sites for a large range of reactions, which could be employed to modify biodiesel characteristics such as the cold-flow properties, besides making it possible to prepare value-added derivatives with a wide range of applications such as additives, plasticizers, lubricants, detergents, surfactants, polymer precursors, etc. In addition, since allylic hydrogens are susceptible to radical attack and consequent oxidative degradation, the removal of unsaturations permits to improve the product's oxidative stability.

Some modifications plausible for fatty substrates (Scheme 1) include, among others, hydrogenations (A), hydroformylations (B), oxidative cleavages (C), deoxygenations (D), metathesis (E) and ethenolysis (F) [7–10]. The hydrogenation process is still used for the production of hydrogenated fats using nickel Raney, but can also be accomplished using, bimetallic ruthenium-nickel supported over molecular sieves and others [11–13]. The hydroformylation processes normally occur by homogeneous catalysis but may proceed using rhodium or ruthenium heterogenized over MOFs, functionalized silica and others [14,15]. The oxidative cleavage process uses ozonolysis, but may also be accomplished using ruthenium catalysts and composites of ruthenium-vanadium heterogenized and others [16–18]. The deoxygenation of fatty compounds is an efficient route for green diesel production using heterogenized catalysts supported on zeolites and others [19–21].

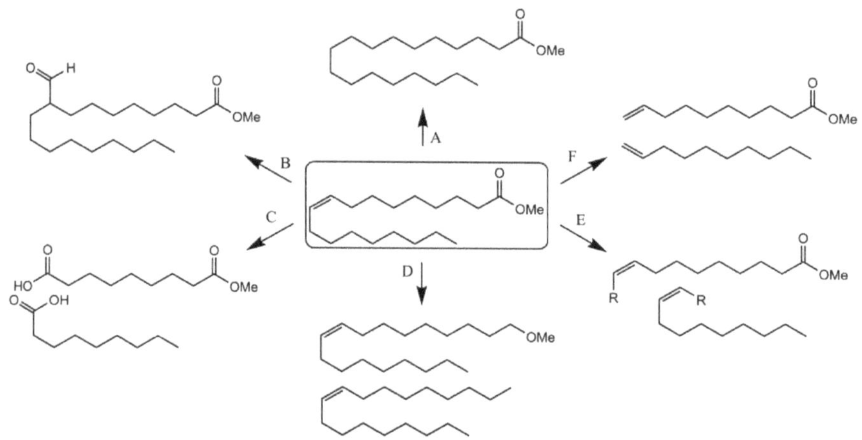

Scheme 1. Products from different modification routes of the methyl oleate: (A) hydrogenation; (B) hydroformylation; (C) oxidative cleavage; (D) deoxygenation; (E) metathesis and (F) ethenolysis.

Hypercrosslinked polystyrenes (HPS), or functionalized polystyrenes (such as amine-functionalized), are useful in the stabilization of metallic nanoparticles and clusters [22–25]. The stabilization of nanoparticles is vital for the production of ruthenium mono and bimetallic catalysts applicable in biorefinery processes, such as the oxidation and hydrogenation of glucose, the hydrogenolysis of cellulose, on Suzuki cross-coupling and on the lactic acid hydrogenation [26–28].

In this work, the synthesis and characterization data of the ruthenium, and doped ruthenium/HPS (5%Ru/HPS and 3%Ru-0.1%Me/HPS; Me = Co, Ni and Cu) catalysts allow focus on the modification of canola fatty acid methyl ester (FAME), a mixture containing 60–70% of oleic, 17–20% of linoleic and 6–10% of linolenic methyl esters [29]. In this way, 5%Ru/HPS catalyst was tested on hydroformylation, the 3%Ru-0.1%Ni/HPS on hydrogenation, the 3%Ru-0.1%Co/HPS on hydroformylation and deoxygenation, and the 3%Ru-0.1%Cu/HPS on oxidative cleavage.

2. Results and Discussion

2.1. Catalyst Characterization

The compiled thermogravimetry for the catalysts 3%Ru/HPS and 3%Ru-0.1%Me/HPS; Me = Co, Ni and Cu is presented in Figure 1. The first weight loss between 35–105 °C is attributed to water chemically and physically adsorbed on the catalyst's surfaces. The second weight loss between 400–500 °C is attributed to the decomposition of the polymeric matrix. However, on every sample, it was possible to verify a subtle onset of weight loss from 250 °C and, therefore, this temperature was chosen as the limit temperature for the catalytic tests.

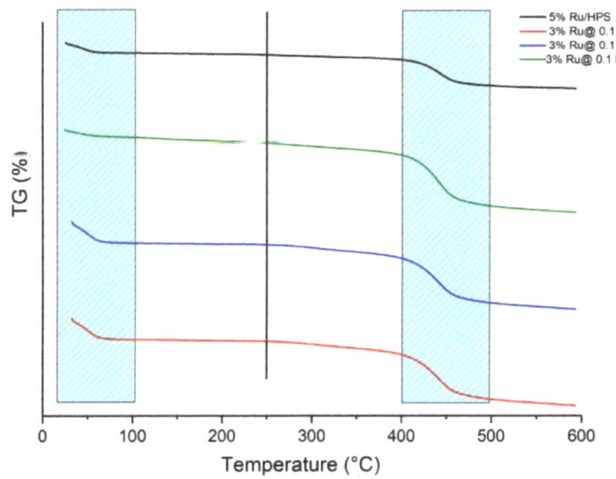

Figure 1. Ruthenium and doped ruthenium/HPS catalysts thermogravimetric profiles.

The catalysts' FT-IR spectrums are compiled in Figure 2. It is possible to verify a similar profile for every sample, with bands between 3100–2900 cm^{-1} and 1600–1400 cm^{-1} typical of bond stretching for sp^3 C-H and C-C, as well as vibration and folding for C-H bonds, all of which are associated with the polymeric matrix of HPS.

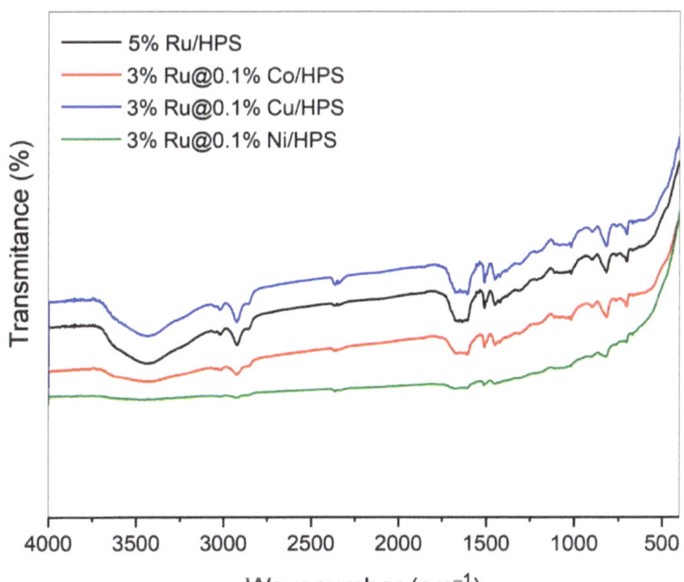

Figure 2. Ruthenium and doped ruthenium/HPS catalysts FT-IR compilation.

To investigate the nature and the amount (mmol g^{-1}) of the acid sites present in these materials, they were analyzed by infrared spectroscopy using pyridine as a probe molecule. Furthermore, the acid strength of the sites was investigated following the thermal stability of the interaction pyridine-acid sites by the acquisition of infrared spectra at 100, 200 and 300 °C [30,31].

The results (Table 1) point out just the presence of a low amount of Lewis acid sites (LAS) in the case of 3%Ru-0.1%Ni/HPS, 3%Ru-0.1%Cu and 3%Ru-0.1%Zn. Bronsted acid sites (BAS) were not detected for any catalysts and no LAS were observed for 3%Ru/HPS and 3%Ru-0.1%Co/HPS; however, it is not possible to ignore the presence of this type of site in very low amounts as they may not be detectable for the analytical strategy employed. It is important to note that even submitted to 300 °C, the detectable acid sites still remain in the material's structure, showing the temperature resistance of the catalysts.

Table 1. Lewis acid sites (LAS) at different temperatures obtained from the infrared spectra using pyridine adsorbed on catalysts.

T (°C)	5%Ru/HPS	3%Ru-0.1%Co/HPS	3%Ru-0.1%Ni/HPS	3%Ru-0.1%Cu/HPS
25	nd	nd	164.2	258.0
100	nd	nd	78.8	140.4
200	nd	nd	61.4	132.5
300	nd	nd	52.9	127.5

nd = non-detected.

The X-ray diffraction patterns of the materials are presented in Figure 3 and just two diffraction bands at 38.35° and 44.50° are observed which can be associated with planes (010) and (011) of metallic ruthenium [32]. The traces of copper, nickel and cobalt present on doped ruthenium catalysts were not present due to their low abundance over the catalysts' surface and equipment detection limit.

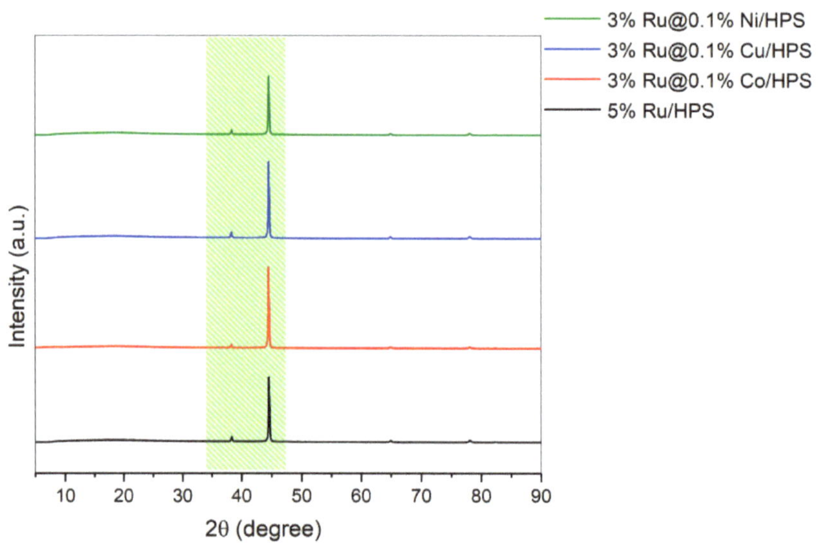

Figure 3. Ruthenium and doped ruthenium/HPS catalysts XRD compilation.

Nitrogen physisorption results for the catalysts (Figure 4 and Table 2) show a slight tendency to decrease on specific surface area and pore volume, in comparison with monometallic 5%Ru/HPS, maybe due to deposition of metallic species in the material walls. All isotherms are of type I, as shown in Figure 4, which are associated with the presence of the hysteresis loop of type H4, indicating mainly the presence of micropores. This observation is confirmed by pore diameter distribution and DBJH values, shown in Figure 5.

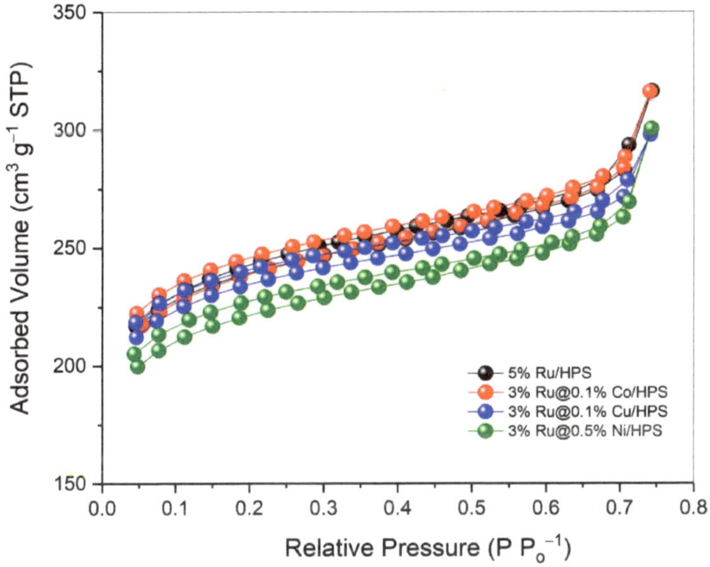

Figure 4. Ruthenium and doped ruthenium/HPS catalysts adsorption–desorption isotherms compilation.

Table 2. BET specific surface area (SBET), pore volume (V) and average pore diameter (DBJH) for doped and non-doped ruthenium/HPS catalysts.

Sample	S_{BET} ($m^2\ g^{-1}$)	V ($cm^3\ g^{-1}$)	D_{BJH} (Å)
5% Ru/HPS	726	0.125	32.5
3% Ru@ 0.1% Co/HPS	719	0.121	32.6
3% Ru@ 0.1% Cu/HPS	712	0.098	36.0
3% Ru@ 0.1% Ni/HPS	672	0.124	32.4

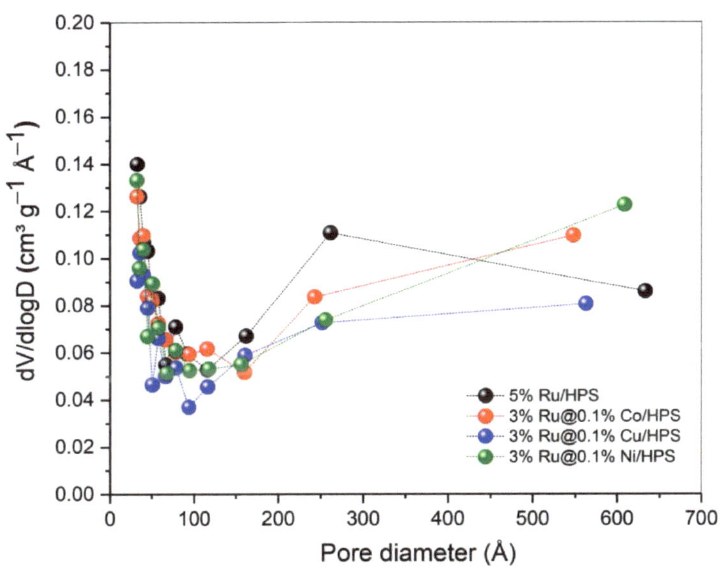

Figure 5. Ruthenium and doped ruthenium/HPS catalysts pore volume distribution compilation.

Previous studies involving 5%Ru/HPS and 3%Ru-0.1%Co/HPS have revealed that for 5%Ru/HPS material hydrated ruthenium (IV) oxide was the predominant form of Ru on the catalyst surface (about 69% of all the ruthenium-containing compounds) and Ru-containing particles are located non-uniformly, mainly closer to the outer surface of the polymer granule. In the case of 3%Ru-0.1%Co/HPS, no products of chemical interaction of cobalt and ruthenium were detected and cobalt is present as $Co(OH)_2$ [28,33].

The dispersion parameters observed in previous studies were obtained, as shown on the SEM compilation in Figure 6. The introduction of cobalt, nickel and copper changes the distribution of ruthenium-containing particles, increasing their uniformity. Copper promoted the greatest dispersion followed by nickel and cobalt. Other SEM images in different zooms are available in the supporting information, Supplementary Material Figures S1–S16.

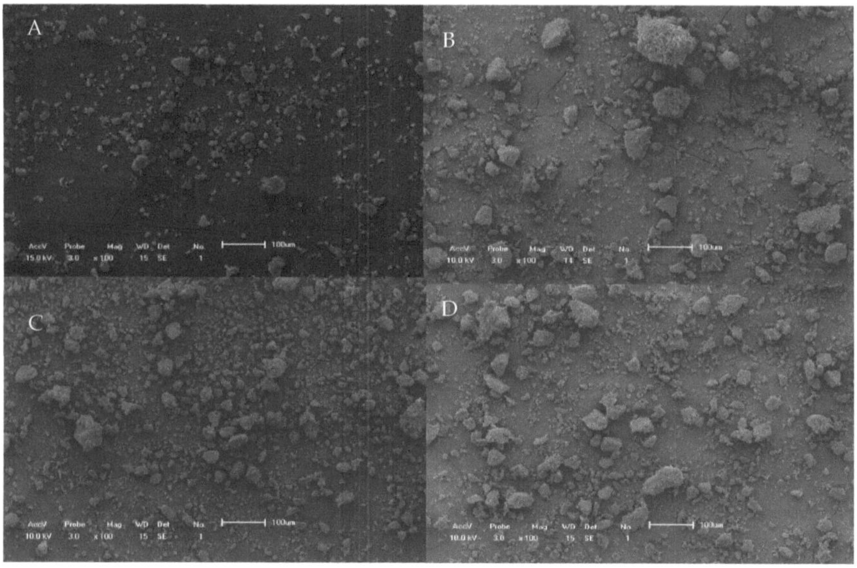

Figure 6. SEM 100× magnitude compilation. (**A**) 5%Ru/HPS; (**B**) 3%Ru@0.1%Co/HPS; (**C**) 3%Ru@0.1%Cu/HPS e; (**D**) 3%Ru@0.1%Ni/HPS.

2.2. Catalyst Activity Tests

The canola FAME mixture was chosen as substrate due to its major composition of 70% methyl oleate [29], which is less expensive if compared to pure oleic acid which is obtained by fraction, facilitating the application of this catalytic system on greater scales. The usage of FAME mixtures is therefore friendly to circular economic applications while contributing to green chemistry, enhancing the sustainability of a process. On the other hand, the methyl oleate has also only one unsaturation per fatty chain which facilitates the interpretation of activity data [4,34,35].

The FT-IR of both canola oil and canola FAME shows only the shifting of carbonyl stretch from 1743 cm^{-1} on oil to 1743 cm^{-1} on FAME, as demonstrated in Supplementary Material Figure S17. The ^{1}H NMR for the canola FAME (Supplementary Material Figure S18) demonstrates the disappearance of internal hydrogens of glycerol over 4.40 and 4.15 ppm, while also showing the appearance of a methyl peak at 3.67 ppm, indicating almost total conversion of oil into FAME. The GC-MS chromatogram of canola FAME and main component list are also available in Supplementary Material Figure S19 and Table S1.

The hydrogenation methodology chosen, Scheme 2, was typical of heterogenized nickel catalytic systems. On the heating ramp, over 175 °C, it was possible to verify a

10 bar fall, indicative of activity at temperatures lower than the selected and also effectivity at the oxidative addition step, considered the limiting catalytic step for hydrogenations. The obtained white and solid product was extracted with petroleum ether to avoid HPS solubilization [36].

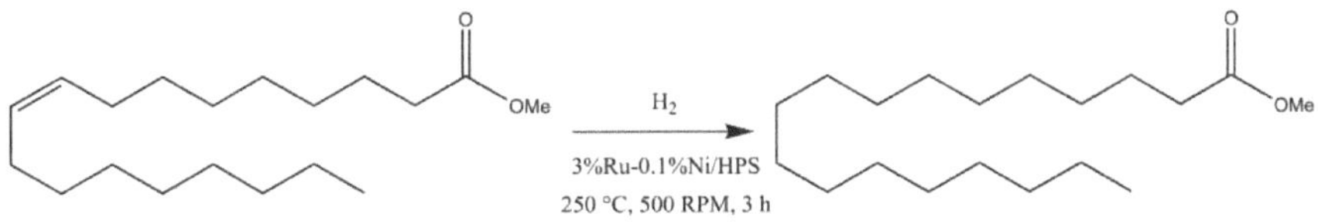

Scheme 2. Hydrogenation of canola FAME.

The hydrogenation of canola FAME with 3%Ru@0.1%Ni/HPS showed an outstanding result with the apparent total conversion of starting unsaturations, using 50 bar of H_2, at 250 °C and 500 rpm, during 180 min. The FT-IR spectrum of the hydrogenated canola FAME product, Figure 7, shows the disappearance of the C-H sp^2 stretch band, while the ^{1}H NMR spectrum, Figure 8, allows the conversion calculation at 99%. The calculation of selectivity and yield was not possible due to the superposition of fatty chain hydrogens. The GC-MS chromatogram and main component list, Supplementary Material Figure S20 and Table S2, corroborate the full conversion of canola FAME into saturated fatty chains. The high activity justification is due to synergic effects between nickel and ruthenium allowing a greater dispersion of particles over the catalyst surface, as shown on the SEM and greater acid stability at high temperatures if compared to 5%Ru/HPS.

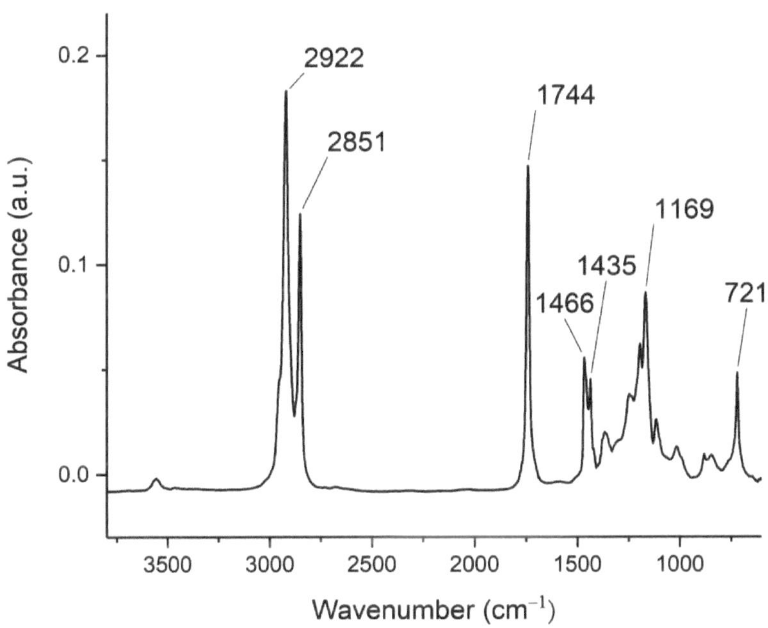

Figure 7. FT-IR spectrum for hydrogenated canola FAME.

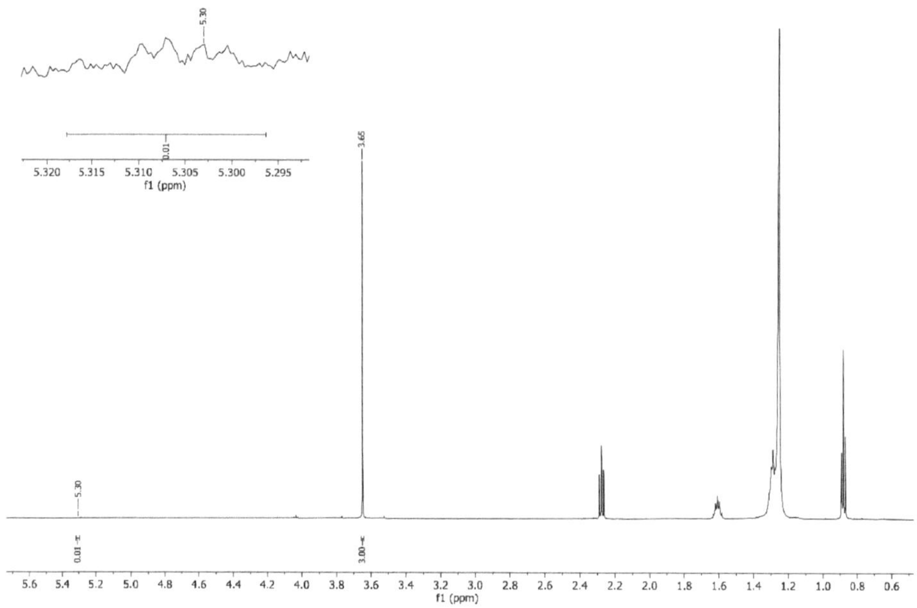

Figure 8. ^{1}H NMR spectrum for hydrogenated canola FAME.

An example from the literature points out that ruthenium supported over carbon was studied on hydrogenation of soybean oil (2 bar of H_2 at 165 °C) resulting in moderate non-selective yields (18.8% of linoleic (C18:2), 53.3% of oleic (C18:1) and 14.4% of stearic (C18:0) fatty chains) [37]. In another study, the hydrogenation of glicerol trioleate (1.60 mmol of the double bond) was described in the presence of ruthenium nanoparticles supported over ion exchange resin DOWEX-300 using water:heptane at 30 °C, 10 bar of H_2 and 1000 rpm, during 120 min, yielding 45% of unsaturations conversion [38]. Nanocatalysts containing nickel, paladium and ruthenium were also applied on the selective hydrogenation of canola oil attaining 75.3% of cis-C18:1 and 2.46% of C18:0, at 100 °C, 10 bar of H_2 and 500 rpm, during 60 min [39].

The first hydroformylation followed a classical heterogenized ruthenium methodology with syn-gas (1:1), toluene and 5%Ru/HPS as catalyst, under 150 °C for 48 h. In contrast with the hydrogenation, no pressure fall was detected, indicating no intense activity. The obtained product ^{1}H NMR (Figure 9) allows the calculations of conversion and selectivity for aldehyde and yield, 44.23%, 1.8% and 0.8% respectively. As there were no signs of alcohol, most of the conversions were into hydrogenated products, while the mild temperature of 150 °C might not be enough for the oxidative addition step, stimulating the selection of another methodology.

The second hydroformylation test, Scheme 3, followed a slightly more drastic methodology with syn-gas (1:1), no solvent, 3%Ru@0.1%Co/HPS as catalyst, under 250 °C for 4 h. The cobalt-doped ruthenium catalyst was chosen due to the famous cobalt synergic effect over ruthenium aiming for higher activities. Indeed, as expected, a pressure fall was verified and the obtained product was partially solidified.

The hydroformylated canola FAME FT-IR spectrum, Figure 10, shows intensity reduction on the C-H sp^2 stretch band, while the ^{1}H NMR spectrum, Figure 11, allows the conversion calculation at 89%, but with no signals of aldehyde or alcohol, indicating the materials hydrogenation. The calculation of selectivity and yield was not possible due to the superposition of fatty chain hydrogens. The GC-MS chromatogram and main component list, Supplementary Material Figure S21 and Table S3, corroborate the conversion of canola FAME into saturated fatty chains. In comparison with the nickel-doped ruthenium/HPS, the lower activity was either due to the addition of carbon monoxide or to the catalyst acid properties, lower than the nickel-doped.

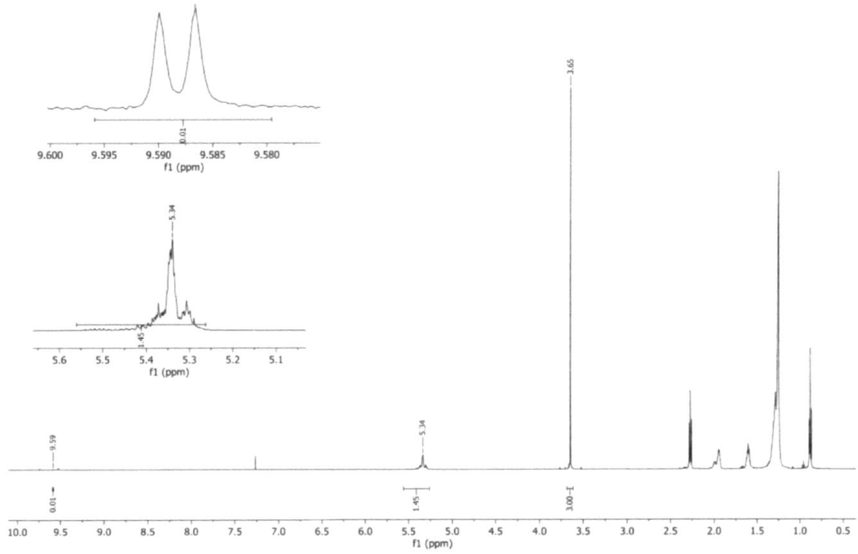

Figure 9. H NMR spectrum for hydroformylated product following the first methodology.

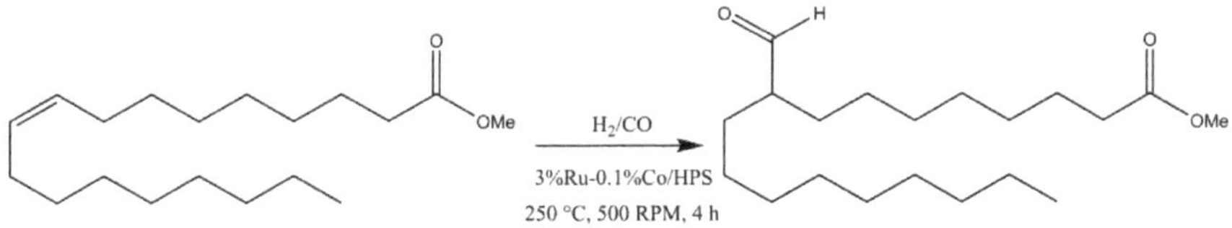

Scheme 3. Hydroformylation methodology was chosen for the canola FAME modification.

The canola FAME oxidative cleavage methodology, Scheme 4, was adapted from classical heterogenized ruthenium on the oxidation of unsaturations, with sodium periodate as co-oxidant, 3%Ru@0.1Cu/HPS as catalyst and a solvent mixture in a glass flask at room temperature during 48 h. The oxidative cleavage was simpler in comparison with the hydrogenations, requiring no gas pressure and no product extraction. The copper-doped ruthenium catalyst was chosen due to its promoter properties on oxidative cleavage reactions. The oxidized canola FAME FT-IR spectrum, Figure 12, show the total disappearance of the C-H sp^2 band stretch and asymmetry on the carbonyl C=O stretch band indicating the formation of other carbonylated species.

The ^{1}H NMR spectrum for the oxidized canola FAME, Figure 13, allows the conversion of 85%. Values for selectivity and yield were not calculated due to the hydrogen labile character on the oxidized major products azelaic acid and pelargonic acid, while the ester variations of such overlap with the starter fatty chain hydrogens. The ^{1}H NMR allowed the identification of an aldehyde triplet at 9.7 ppm, coupled with hydrogens β to the carbonyl position, with calculated selectivity for aldehyde calculated at 17%. The aldehyde presence is explained by a possible water addition during the oxidative cleavage process. The ^{13}C NMR, Figure 14, shows the characteristic peaks of aldehyde at 200 ppm, carboxylic acid at 180 ppm, ester at 172 ppm and unsaturation at 130 ppm, indicating the partial conversion. The corroborative GC-MS chromatogram and main component list are available in Supplementary Material Figure S22 and Table S4. Among the tested catalysts, the copper-doped ruthenium/HPS was the most acidic and presented the greater pore volume and greater dispersion on the support surface and these characteristics explain the elevated activity at room temperature.

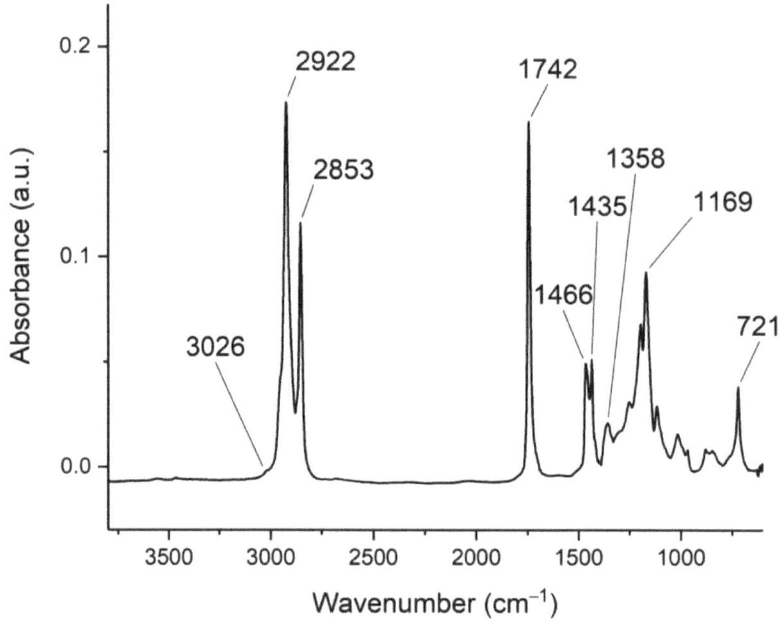

Figure 10. FT-IR spectrum for hydroformylated canola FAME product on the second methodology.

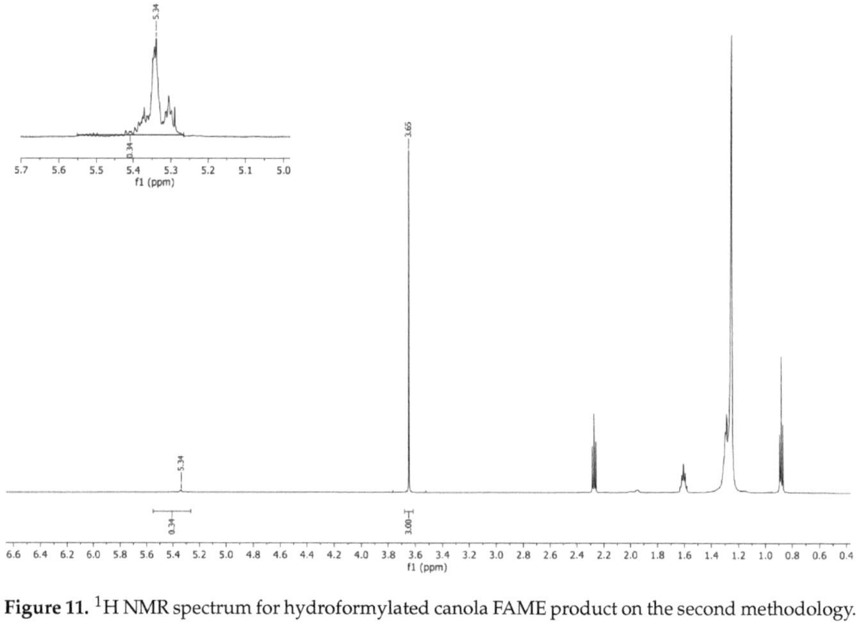

Figure 11. ^{1}H NMR spectrum for hydroformylated canola FAME product on the second methodology.

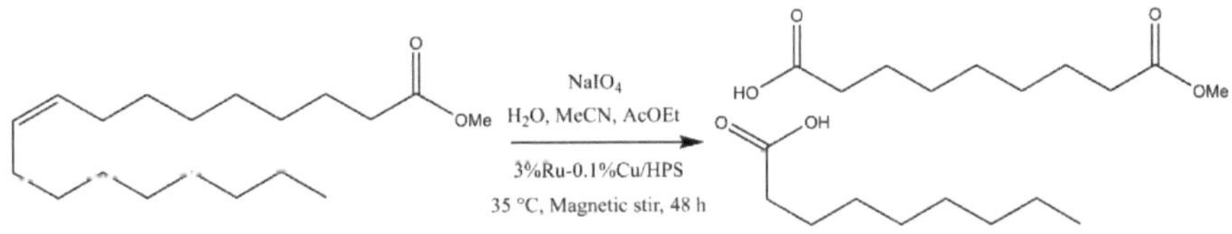

Scheme 4. Oxidative cleavage of canola FAME.

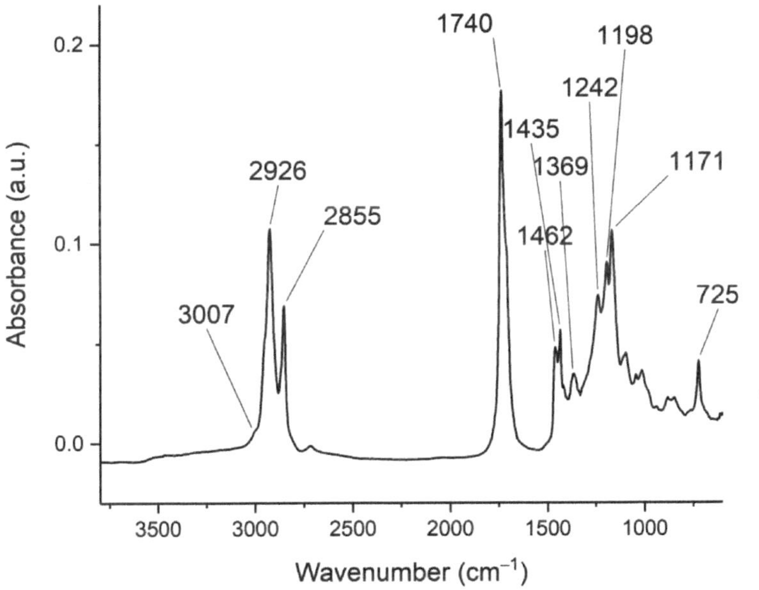

Figure 12. FT-IR spectrum for oxidized canola FAME product.

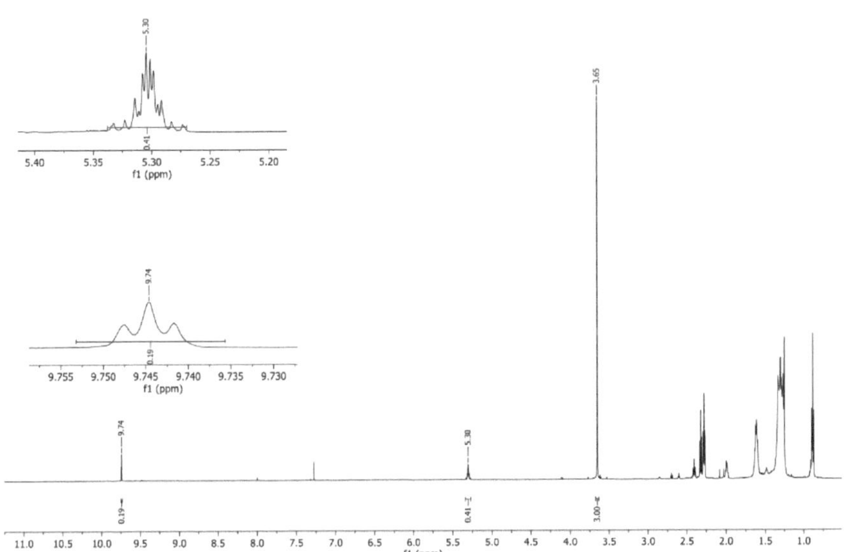

Figure 13. ^{1}H NMR spectrum for oxidized canola FAME product.

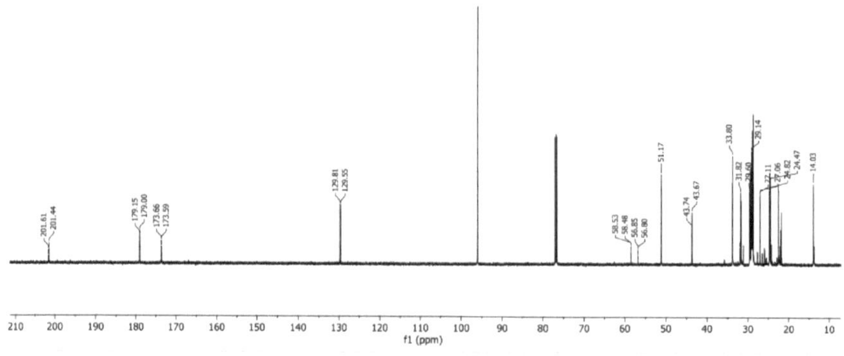

Figure 14. ^{13}C NMR spectrum for oxidized canola FAME product.

Few examples of oxidative cleavage of fatty compounds using supported ruthenium catalysts are available in the literature. A catalytic system based on ruthenium nanoparticles supported on hydroxyapatite was used for the oxidative cleavage of oleic acid and 16% of unsaturation conversion and 84% of pelargonic acid yield were detected, after 12 h at room temperature [40]. In another study, a functionalized carbon black support for the complexation of ruthenium was tested on oxidative cleavage of oleic acid, yielding almost complete conversions to pelargonic and azelaic acids, after 24 h and 1500 rpm at room temperature [41].

The deoxygenation methodology of canola FAME with 3%Ru@0.1%Co/HPS, Scheme 5, was the same as for hydrogenation, but using milder conditions (30 bar of H_2, 200 °C and 500 rpm), in the presence of water and longer reaction time. The cobalt-doped ruthenium/HPS catalyst was selected due to the cobalt synergic properties over the ruthenium clusters. The obtained product was a white solid associated with the occurrence of hydrogenation and decarbonylation. The FT-IR of the product, Figure 15, shows the total disappearance of the C-H sp^2 band stretch and decreased carbonyl C=O stretch band, indicating the formation of hydrogenated and decarbonylated species.

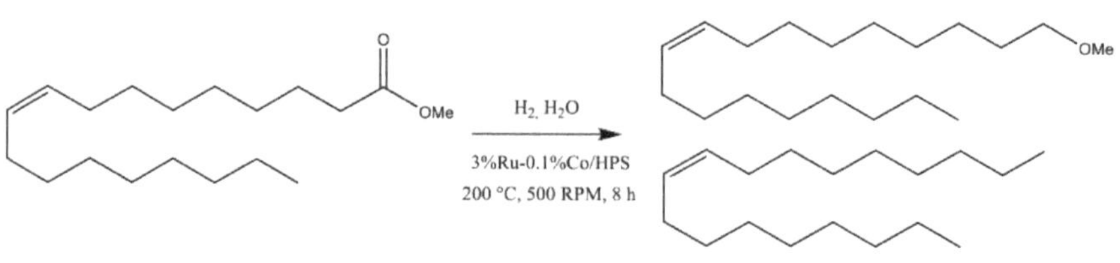

Scheme 5. Deoxygenation of canola FAME.

The ^1H NMR for the deoxygenated product, Figure 16, shows almost the total disappearance of unsaturation and methoxyl peaks. The conversion calculations were possible by normalization based on the chain methyl group, at 0.88 ppm, which was similarly done to the starter FAME ^1H NMR. The calculated conversions for unsaturation and methoxyl were 99%. On the ^{13}C NMR, Figure 17, is possible to verify the total disappearance of unsaturations peaks and the presence, although at a very low intensity, of ester at 173 ppm, carboxylic acid at 179 ppm and ether terminal carbon at 64 ppm. The corroborative GC-MS chromatogram and main compoment list are available in Supplementary Material Figure S23 and Table S5.

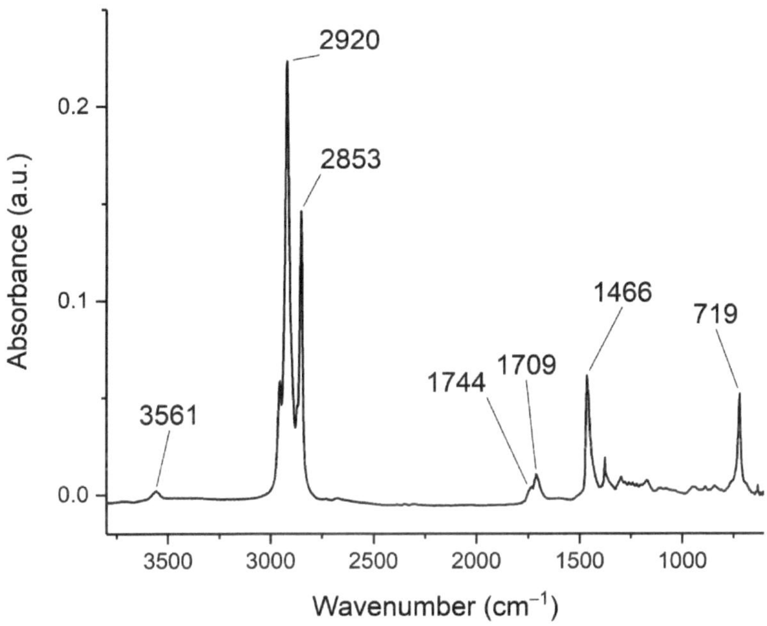

Figure 15. FT-IR spectrum for deoxygenated canola FAME product.

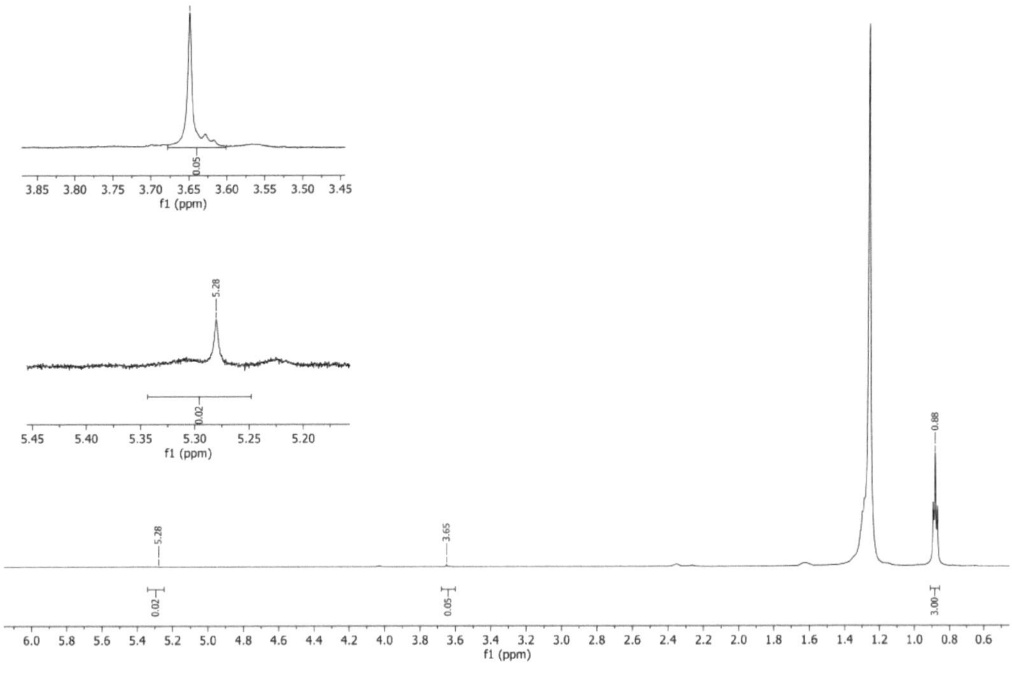

Figure 16. ^{1}H NMR spectrum for deoxygenated canola FAME product.

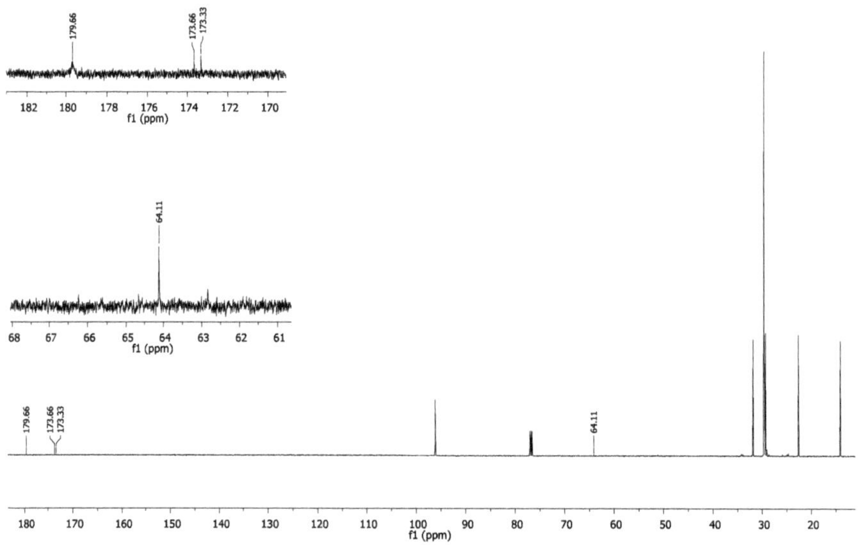

Figure 17. ^{13}C NMR spectrum for deoxygenated canola FAME product.

It is important to mention some examples from the literature. Stearic acid was deoxygenated using hexane, 68 bar of H_2 and 50 mg of 10%Co/HPS, at 250 °C for 1 h with total conversion [42]. The deoxygenation of waste cooking oil using water and ruthenium supported over hydroxyapatite catalyst was studied. In this case, conversions over 100% and an yield of 89% of alkane long chain were attained at 180 °C, 20 bar of H_2 and 1000 rpm, for 4.5 h. Additionally, the catalyst could be recycled [43]. Deoxygenation of tall oil fatty acids was tested using carbon-supported ruthenium nanoparticles in the presence of aqueous formic acid at 326 °C and 2 h, leading to high conversion (~100%) and selectivity of the 82% to C5–C20 alkane chains [44]. Ruthenium nanoparticles on a porous organic network were employed on the deoxygenation of stearic acid (30 bar of H_2, 180 °C, 8 h) with high conversion and selectivity [45]. However, herein, using canola FAME, a not hydrogenated substrate, was possible to hydrogenate and deoxygenate under milder conditions of pressure and temperature, and with fewer metallic sites, all indications of great catalyst activity.

In summary, the applications tested and discussed here, many of them with extremely promising results, demonstrated the potential of catalysts based on ruthenium heterogenized species. From these encouraging results, subsequent optimization works can be carried out, in which distinct catalysts and residual fatty substrates can be studied at several reaction conditions.

3. Materials and Methods

3.1. Materials

The reagents used for the catalyst preparation and characterization were analytical grade and used as received. The canola oil, used for canola FAME synthesis, was food-grade and used as received.

3.2. Catalyst Synthesis

The doped and non-doped ruthenium/HPS catalysts were prepared by wet impregnation methodologies described elsewhere [26,28,33,42]. The simpler non-doped ruthenium catalyst (5%Ru/HPS) was prepared by a wash of the hypercrossed polystyrene (3 g) with water and acetone, followed by impregnation of $Ru(OH)Cl_3$ (on adequate molarity considering wt% of metal) dissolved in tetrahydrofuran, methanol and water (5:1:1). After the impregnation, the mixture was dried at 70 °C for 1 h, refluxed with 21 mL of a 0.1 mol L^{-1} NaOH solution and strongly agitated with 2 mL of hydrogen peroxide. The impregnated catalyst was washed, dried and reduced with a H_2 flux of 100 mL min^{-1} at 300 °C for 2 h.

Similarly, on the doped ruthenium/HPS (3%Ru-0.1%Me/HPS, Me = Co, Ni and Cu), the appropriate molarities of acetates of cobalt(II), copper(II) and nickel(II) were dissolved with the $Ru(OH)Cl_3$ on the mixture of tetrahydrofuran, methanol and water. The materials were analyzed without previous treatment by the technique described below.

3.3. Catalyst Characterization

Thermogravimetric characterizations were performed using a Shimadzu TGA-50 apparatus under a nitrogen atmosphere in the temperature range from 25 to 1000 °C, at a heating rate of 10 °C/min.

The FTIR spectra were recorded on a Shimadzu IR Prestige 21 spectrometer using the KBr pellet method. The following operational parameters were used: spectral range of 400–4000 cm^{-1}, 65 scans in transmittance mode and a resolution of 4 cm^{-1}. For pyridine (Py)-adsorption infrared spectroscopy tests, a KBr pellet (sample and KBr were weighed) was placed in a closed receptacle containing liquid pyridine (none in direct contact with the pellet) and a vacuum was formed to obtain pyridine vapor. The system remained under this condition for 48 h so that pyridine vapor could interact with the acid sites of the samples. After this procedure, the infrared spectra were acquired at 25 °C and after an increase in temperature of 100, 200 and 300 °C.

In all cases, the number of acid sites was evaluated by integrating the area of the absorption bands related to different acid sites (Bronsted or Lewis) using Equation (1), in which D = pellet diameter (cm), w = mass of the sample (g), and $A_{B,L}$ = integration of the Lewis and Bronsted characteristic bands: approximately 1537 cm^{-1} for Py-Bronsted site and 1445 cm^{-1} (Py-Lewis site). The coefficient of extinctions of the Py-Bronsted interaction site and of the Py-Lewis site are 1.67 ± 0.12 cm μmol^{-1} and 2.22 ± 0.21 cm mol^{-1}, respectively.

$$q_{B,L} = \left(A_{B,L} \cdot \pi \cdot D^2\right)(4w \cdot E_{B,L}) \quad (1)$$

X-ray diffraction (DRX) measurements on powder samples, were performed on a Rigaku Multiflex instrument equipped with nickel filtered Cu K (0.15418 nm) radiation source and a scintillation counter detector. The data were selected in the range of 2θ from 2 to 80 ° at a goniometer speed of 2°/min and a step of 0.02 ° with a counting time of 1 s at each step. The crystallite sizes were determined by the Scherrer equation (Equation (2)).

$$D_{hkl} = \frac{K\lambda}{\beta \cos(\theta)} \quad (2)$$

The nitrogen physisorption isotherms at 77 K were obtained using a Quantachrome Nova 2200e Instrument. The specific surface area was determined by the Brunauer–Emmett–Teller method.

Microscopy images of the samples were obtained using a Hitachi scanning electron microscope (SEM), model S-3400N (all the samples were previously metalized with gold), and an FEI Tecnai G2 Spirit TWIN transmission electron microscope (TEM) (120 kV).

3.4. FAME Synthesis and Activity Tests

This canola FAME mixture was obtained by alkaline transesterification of canola oil following established methodologies [46]. The activity tests for 3%Ru-0.1%Me/HPS, Me = Co, Ni and Cu consisted of modifications of FAME by hydrogenation, hydroformylation, oxidative cleavage and deoxygenation. The obtained products were analyzed by FT-IR and $^1H/^{13}C$ NMR.

The hydrogenation methodology was adapted from the literature [47], using a 100 mL Parr reactor model 4590 HP. Before usage, the Parr reactor was meticulously cleaned with steel wool and tuned for each mixture's temperature without the catalysts (canola

FAME and hydrogen, canola FAME water and hydrogen, etc.), and no pressure fall was observed. To the reactor were added 10 mL of dry canola FAME and 200 mg of catalyst (3%Ru-0.1%Ni/HPS). The reactor was purged and pressurized to 20 bar of N_2. After the pressurization, the system was almost totally depressurized and pressurized with H_2 to 50 bar. The pressurized system was heated to 250 °C and stirred under 500 RPM for 3 h. Past the reaction time, the system was cooled, quenched by depressurization, the obtained mixture dissolved in petroleum ether, filtered in alumina and the product dried under reduced pressure.

For the hydroformylation, two adapted methodologies were applied. First, [48–50], to a 150 mL autoclave reactor, with glass cup and magnetic stirring, was added 1.76 mL of dry canola FAME, 168 mg of catalyst (5%Ru/HPS) and 5 mL of dry toluene. The reactor was purged and pressurized to 20 bar of N_2. After the pressurization, the system was almost totally depressurized and pressurized with CO to 25 bar and H_2 to 50 bar. The pressurized system was heated to 150 °C and magnetically stirred for 48 h. Past the reaction time, the system was cooled, quenched by depressurization, the obtained mixture dissolved in toluene, filtered in alumina and the product dried under reduced pressure.

Second, hydroformylation methodology was applied, adapted from literature [47], using a 100 mL Parr reactor model 4590 HP. To the reactor was added 10 mL of dry canola FAME and 140 mg of catalyst (3%Ru-0.1%Co/HPS). The reactor was purged and pressurized to 20 bar of N_2. After the pressurization, the system was almost totally depressurized and pressurized with CO to 25 bar and H_2 to 50 bar. The pressurized system was heated to 250 °C and stirred under 500 RPM for 4 h. After the reaction time, the system was cooled, quenched by depressurization, the obtained mixture dissolved in petroleum ether, filtered in alumina and the product dried under reduced pressure.

The oxidative cleavage methodology was adapted from applications for ruthenium complexes methodologies found elsewhere [51]. To a glass flask was added 4 mL of canola FAME, 12 g of sodium periodate, 30 mL of distilled water, 20 mL of acetonitrile, 20 mL of ethyl acetate and 78 mg of catalyst (3%Ru-0.1%Cu/HPS). The mixture was magnetically stirred at room temperature for 48 h. After the reaction time, the products were extracted with petroleum ether, filtered in alumina and dried under reduced pressure.

The deoxygenation methodology was adapted from the literature [52,53] for a 100 mL Parr reactor model 4590 HP. To the reactor was added 0.7 mL of canola FAME, 10 mL of mili-Q water and 150 mg of catalyst (3%Ru-0.1%Co/HPS). The reactor was purged and pressurized to 20 bar of N_2. After the pressurization, the system was almost totally depressurized and pressurized with H_2 to 30 bar. The pressurized system was heated to 200 °C and stirred under 500 RPM for 8 h. After the reaction time, the system was cooled, quenched by depressurization, the obtained mixture dissolved in petroleum ether, filtered in alumina and the product dried under reduced pressure.

The FT-IR were recorded using a Shimadzu IR Prestige 21 spectrometer with an ATR Miracle cell. The following operational parameters were used: spectral range of 650–4000 cm^{-1}, 32 scans in absorbance mode and a resolution of 4 cm^{-1}. The ^{1}H/ ^{13}C NMR spectra were obtained from a Bruker 600 MHz NMR spectrometer, equipped with a broad-band probe Bruker 5 mm (BBFO). Spectrums were obtained for ^{1}H and ^{13}C on deuterated chloroform, referenced with TMS, using 16 scans for ^{1}H and 2000 scans for ^{13}C and 5 s delay.

The GC-MS chromatograms were obtained using a Shimadzu GC-MS model QP-2020. The samples were eluded through a Zebron Phase 0.5 µm thickness × 0.25 mm diameter × 50 m length. Samples were diluted in hexane (1:10). The column initial temperature was 60 °C, which was maintained for 5 min, and then ramped from 60 °C to 270 °C at 2 °C min^{-1}. At 270 °C, the temperature was maintained for 5 min. The injection temperature was 280 °C, the ion source temperature was 200 °C and the interface temperature was 280 °C. The injected volume was 1 µL with a split of 1:100. The helium analytical grade was used on a flow rate of 0.95 mL min^{-1}.

The activity quantification method was found in the literature, using the integrated areas from ^{1}H NMR [54,55]. The Double Bond Number (DBN) was determined using Equation (3), where B is the unsaturation area over 5.3 ppm and A is the methyl group over 3.6 ppm. The variation of DBN before and after is possible to determine the conversion, as stated in Equation (4).

$$\mathrm{DBN} = \frac{\frac{B}{2}}{\frac{A}{3}} \tag{3}$$

$$\mathrm{Conversion}(\%) = \frac{\mathrm{DBN_i} - \mathrm{DBN_f}}{\mathrm{DBN_i}} \times 100 \tag{4}$$

The selectivity values for the hydroformylated samples were calculated using Equation (5), where C is the aldehyde hydrogen duplet over 7.1 ppm and A is the methyl group over 3.6 ppm. The yield values for hydroformylated samples were determined using Equation (6).

$$\mathrm{Aldehyde\ selectivity}(\%) = \frac{\frac{C}{\frac{A}{3}}}{\mathrm{DBN_i} - \mathrm{DBN_f}} \times 100 \tag{5}$$

$$\mathrm{Yield}(\%) = \frac{\mathrm{Conversion} \times \mathrm{Selectivity}}{100} \tag{6}$$

4. Conclusions

Different catalysts based on ruthenium and doped ruthenium supported on the surface of hypercrosslinked polystyrene were prepared by wet impregnation. The catalysts were characterized for acidity, morphology, surface area and thermic stability to select possible applications. Canola FAME was chosen as the substrate due to its major composition of oleic acid. The catalysts were highly active for hydrogenation, with a conversion of 99%, and effective on oxidative cleavage reactions, with a conversion of 85%. The catalyst was also extremely active for deoxygenation reactions, on which double behavior was observed, simultaneously deoxidizing and hydrogenating the canola FAME with conversions of 99%. The obtained data stimulates further collaborative studies for optimization using distinct catalysts and residual fatty substrates at several reaction conditions.

Supplementary Materials: The following supporting information can be downloaded at: https://www.mdpi.com/article/10.3390/catal13030630/s1, Figure S1. 5%Ru/HPS 100× magnitude SEM; Figure S2. 5%Ru/HPS 3000× magnitude SEM; Figure S3. 5%Ru/HPS 5000× magnitude SEM; Figure S4. 5%Ru/HPS 10,000× magnitude SEM; Figure S5. 3%Ru@0.1%Co/HPS 100× magnitude SEM; Figure S6. 3%Ru@0.1%Co/HPS 3000× magnitude SEM; Figure S7. 3%Ru@0.1%Co/HPS 5000× magnitude SEM; Figure S8. 3%Ru@0.1%Co/HPS 10,000× magnitude SEM; Figure S9. 3%Ru@0.1%Cu/HPS 100× magnitude SEM; Figure S10. 3%Ru@0.1%Cu/HPS 3000× magnitude SEM; Figure S11. 3%Ru@0.1%Cu/HPS 5000× magnitude SEM; Figure S12. 3%Ru@0.1%Cu/HPS 10,000× magnitude SEM; Figure S13. 3%Ru@0.1%Ni/HPS 100× magnitude SEM; Figure S14. 3%Ru@0.1%Ni/HPS 3000× magnitude SEM; Figure S15. 3%Ru@0.1%Ni/HPS 5000× magnitude SEM; Figure S16. 3%Ru@0.1%Ni/
HPS 10,000× magnitude SEM; Figure S17. FT-IR spectrum of canola oil (red) and canola FAME (black); Figure S18. ^{1}H NMR spectrum of canola FAME in $CDCl_3$; Figure S19. Canola FAME GC-MS chromatogram; Table S1. List of identified compounds for the canola FAME CG-MS; Figure S20. Hydroformylated canola FAME GC-MS chromatogram; Table S2. List of identified compounds for the hydroformylated canola FAME CG-MS; Figure S21. Hydrogenated canola FAME GC-MS chromatogram; Table S3. List of identified compounds for the hydroformylated canola FAME CG-MS., Figure S22. Oxidized canola FAME GC-MS chromatogram; Table S4. List of identified compounds for the oxidized canola FAME CG-MS; Figure S23. Deoxygenated canola FAME GC-MS chromatogram. And Table S5. List of identified compounds for the deoxygenated canola FAME CG-MS.

Author Contributions: Conceptualization, M.G.S., V.G.M., S.M.P.M., M.R.M. and P.A.Z.S.; methodology, R.C.D., T.V.S.M., D.d.G.R., V.G.M., S.M.P.M., M.R.M. and P.A.Z.S.; investigation, R.C.D., T.V.S.M., D.d.G.R., V.G.M., S.M.P.M., M.R.M. and P.A.Z.S.; resources, M.G.S., V.G.M., S.M.P.M., M.R.M. and P.A.Z.S.; writing—review and editing, V.G.M., S.M.P.M., M.R.M. and P.A.Z.S.; supervision, S.M.P.M. and P.A.Z.S.; project administration, V.G.M., S.M.P.M. and P.A.Z.S.; funding acquisition, M.G.S., V.G.M., S.M.P.M., M.R.M. and P.A.Z.S. All authors have read and agreed to the published version of the manuscript.

Funding: This research was supported by the National Council for Scientific and Technological Development (CNPq)—BRICS project 442095/2017-1-Call N. 2 BRICS-STI/CNPQ, the Brazilian Federal Agency for the Improvement of Higher Education (CAPES), the Brazilian Innovation Agency (FINEP), Alagoas Research Foundation (FAPEAL) and the DF Research Foundation (FAP-DF). R.C.D., T.V.S.M. and D.G.R. express their appreciation for fellowships granted by CAPES and CNPq. S.M.P.M., M.R.M. and P.A.Z.S. thank CNPq for research fellowships.

Acknowledgments: The authors thank the LSCAT/CTEC/UFAL and GON/IF/UFAL teams for their contributions.

Conflicts of Interest: The authors declare no conflict of interest.

References

1. Poovan, F.; Chandrashekhar, V.G.; Natte, K.; Jagadeesh, R.V. Synergy between Homogeneous and Heterogeneous Catalysis. *Catal. Sci. Technol.* **2022**, *12*, 6623–6649. [CrossRef]
2. Khan, H.M.; Iqbal, T.; Yasin, S.; Irfan, M.; Abbas, M.M.; Veza, I.; Soudagar, M.E.M.; Abdelrahman, A.; Kalam, M.A. Heterogeneous Catalyzed Biodiesel Production Using Cosolvent: A Mini Review. *Sustainability* **2022**, *14*, 5062. [CrossRef]
3. Farnetti, E.; Di Monte, R.; Kašpar, J. Homogeneous and Heterogeneous Catalysis. In *Inorganic and Bio-Inorganic Chemistry*; Bertini, I., Ed.; EOLSS Publications: Abu Dhabi, United Arab Emirates, 2009; Volume 2.
4. Velenturf, A.P.M.; Purnell, P. Principles for a Sustainable Circular Economy. *Sustain. Prod. Consum.* **2021**, *27*, 1437–1457. [CrossRef]
5. Sekoai, P.T.; Ghimire, A.; Ezeokoli, O.T.; Rao, S.; Ngan, W.Y.; Habimana, O.; Yao, Y.; Yang, P.; Fung, A.H.Y.; Yoro, K.O. Valorization of Volatile Fatty Acids from the Dark Fermentation Waste Streams-A Promising Pathway for a Biorefinery Concept. *Renew. Sustain. Energy Rev.* **2021**, *143*, 110971. [CrossRef]
6. Kahar, P.; Rachmadona, N.; Pangestu, R.; Palar, R.; Adi, D.T.N.; Juanssilfero, A.B.; Manurung, I.; Hama, S.; Ogino, C. An Integrated Biorefinery Strategy for the Utilization of Palm-Oil Wastes. *Bioresour. Technol.* **2022**, *344*, 126266. [CrossRef]
7. Awogbemi, O.; Von Kallon, D.V.; Aigbodion, V.S. Trends in the Development and Utilization of Agricultural Wastes as Heterogeneous Catalyst for Biodiesel Production. *J. Energy Inst.* **2021**, *98*, 244–258. [CrossRef]
8. Behr, A.; Seidensticker, T. *Chemistry of Renewables: An Introduction*; Springer Nature: London, UK, 2020; ISBN 3662614308.
9. Jayakumar, M.; Karmegam, N.; Gundupalli, M.P.; Gebeyehu, K.B.; Asfaw, B.T.; Chang, S.W.; Ravindran, B.; Awasthi, M.K. Heterogeneous Base Catalysts: Synthesis and Application for Biodiesel Production–A Review. *Bioresour. Technol.* **2021**, *331*, 125054. [CrossRef]
10. Lin, L.; Han, X.; Han, B.; Yang, S. Emerging Heterogeneous Catalysts for Biomass Conversion: Studies of the Reaction Mechanism. *Chem. Soc. Rev.* **2021**, *50*, 11270–11292. [CrossRef]
11. Behr, A.; Gomes, J.P. The Refinement of Renewable Resources: New Important Derivatives of Fatty Acids and Glycerol. *Eur. J. Lipid Sci. Technol.* **2010**, *112*, 31–50. [CrossRef]
12. Luo, S.; Yao, J.; Wang, R.; Wang, L.; Chen, X.; Yu, D.; Elfalleh, W. Effect of Nickel Modification on Ru–Ni/NaY Catalyst Structure and Linoleic Acid Isomerization Selectivity. *J. Food Meas. Charact.* **2021**, *15*, 5584–5598. [CrossRef]
13. Strekalova, A.; Shesterkina, A.; Kustov, L. Recent Progress in Hydrogenation of Esters on Heterogeneous Bimetallic Catalysts. *Catal. Sci. Technol.* **2021**, *11*, 7229–7238. [CrossRef]
14. Liu, B.; Wang, Y.; Huang, N.; Lan, X.; Xie, Z.; Chen, J.G.; Wang, T. Heterogeneous Hydroformylation of Alkenes by Rh-Based Catalysts. *Chem* **2022**, *8*, 2630–2658. [CrossRef]
15. Rodrigues, F.M.S.; Carrilho, R.M.B.; Pereira, M.M. Reusable Catalysts for Hydroformylation-Based Reactions. *Eur. J. Inorg. Chem.* **2021**, *2021*, 2294–2324. [CrossRef]
16. Kerenkan, A.E.; Béland, F.; Do, T.-O. Chemically Catalyzed Oxidative Cleavage of Unsaturated Fatty Acids and Their Derivatives into Valuable Products for Industrial Applications: A Review and Perspective. *Catal. Sci. Technol.* **2016**, *6*, 971–987. [CrossRef]
17. Gámez, S.; de la Torre, E.; Gaigneaux, E.M. Carbon Black-Polydopamine-Ruthenium Composite as a Recyclable Boomerang Catalyst for the Oxidative Cleavage of Oleic Acid. *Chem. Eng. J.* **2022**, *427*, 131820. [CrossRef]
18. Upadhyay, R.; Rana, R.; Sood, A.; Singh, V.; Kumar, R.; Srivastava, V.C.; Maurya, S.K. Heterogeneous Vanadium-Catalyzed Oxidative Cleavage of Olefins for Sustainable Synthesis of Carboxylic Acids. *Chem. Commun.* **2021**, *57*, 5430–5433. [CrossRef] [PubMed]
19. Aiamsiri, P.; Tumnantong, D.; Yoosuk, B.; Ngamcharussrivichai, C.; Prasassarakich, P. Biohydrogenated Diesel from Palm Oil Deoxygenation over Unsupported and γ-Al_2O_3 Supported Ni–Mo Catalysts. *Energy Fuels* **2021**, *35*, 14793–14804. [CrossRef]

20. Mahdi, H.I.; Bazargan, A.; McKay, G.; Azelee, N.I.W.; Meili, L. Catalytic Deoxygenation of Palm Oil and Its Residue in Green Diesel Production: A Current Technological Review. *Chem. Eng. Res. Des.* **2021**, *174*, 158–187. [CrossRef]
21. da Costa, A.A.F.; de Oliveira Pires, L.H.; Padrón, D.R.; Balu, A.M.; da Rocha Filho, G.N.; Luque, R.; do Nascimento, L.A.S. Recent Advances on Catalytic Deoxygenation of Residues for Bio-Oil Production: An Overview. *Mol. Catal.* **2022**, *518*, 112052. [CrossRef]
22. Alsalahi, W.; Trzeciak, A.M. Rhodium-Catalyzed Hydroformylation under Green Conditions: Aqueous/Organic Biphasic,"on Water", Solventless and Rh Nanoparticle Based Systems. *Coord. Chem. Rev.* **2021**, *430*, 213732. [CrossRef]
23. Cheng, S.; Martínez-Monteagudo, S.I. Hydrogenation of Lactose for the Production of Lactitol. *Asia-Pac. J. Chem. Eng.* **2019**, *14*, e2275. [CrossRef]
24. Sulman, M.; Doluda, V.; Grigoryev, M.; Manaenkov, O.; Filatova, A.; Molchanov, V.; Sidorov, A.; Bykov, A.; Shkileva, I.; Sulman, A. Influence of the Mesoporous Polymer Matrix Nature on the Formation of Catalytically Active Ruthenium Nanoparticles. *Bull. Chem. React. Eng. Catal.* **2015**, *10*, 313–323. [CrossRef]
25. Xu, Y.; Hua, M.; Dong, X.; Chen, C.; Duan, Y.; Tang, H. From Scrap Polystyrene Foam to Efficient Demercurizer: In-Situ Synthesis of Fe-Embedded Hyper-Cross-Linked Polymers. *Appl. Catal. B Environ.* **2021**, *285*, 119791. [CrossRef]
26. Grigorev, M.E.; Mikhailov, S.P.; Bykov, A.V.; Sidorov, A.I.; Tiamina, I.Y.; Vasiliev, A.L.; Nikoshvili, L.Z.; Matveeva, V.G.; Meneghetti, S.M.P.; Sulman, M.G. Mono-and Bimetallic (Ru-Co) Polymeric Catalysts for Levulinic Acid Hydrogenation. *Catal. Today* **2021**, *378*, 167–175. [CrossRef]
27. Nikoshvili, L.; Bakhvalova, E.S.; Bykov, A.V.; Sidorov, A.I.; Vasiliev, A.L.; Matveeva, V.G.; Sulman, M.G.; Sapunov, V.N.; Kiwi-Minsker, L. Study of Deactivation in Suzuki Reaction of Polymer-Stabilized Pd Nanocatalysts. *Processes* **2020**, *8*, 1653. [CrossRef]
28. Protsenko, I.I.; Abusuek, D.A.; Nikoshvili, L.Z.; Bykov, A.V.; Matveeva, V.G.; Sulman, E.M. The Use of the Ru-Containing Catalyst Based on Hypercrosslinked Polystyrene in the Hydrogenation of Levulinic Acid to γ-Valerolactone. *Catal. Ind.* **2018**, *10*, 301–312. [CrossRef]
29. Chew, S.C. Cold-Pressed Rapeseed (Brassica Napus) Oil: Chemistry and Functionality. *Food Res. Int.* **2020**, *131*, 108997. [CrossRef] [PubMed]
30. Noda, L.K.; de Almeida, R.M.; Gonçalves, N.S.; Probst, L.F.D.; Sala, O. TiO_2 with a High Sulfate Content—Thermogravimetric Analysis, Determination of Acid Sites by Infrared Spectroscopy and Catalytic Activity. *Catal. Today* **2003**, *85*, 69–74. [CrossRef]
31. Noda, L.K.; de Almeida, R.M.; Probst, L.F.D.; Gonçalves, N.S. Characterization of Sulfated TiO_2 Prepared by the Sol–Gel Method and Its Catalytic Activity in the n-Hexane Isomerization Reaction. *J. Mol. Catal. A Chem.* **2005**, *225*, 39–46. [CrossRef]
32. Rojas, J.V.; Toro-Gonzalez, M.; Molina-Higgins, M.C.; Castano, C.E. Facile Radiolytic Synthesis of Ruthenium Nanoparticles on Graphene Oxide and Carbon Nanotubes. *Mater. Sci. Eng. B* **2016**, *205*, 28–35. [CrossRef]
33. Nikoshvili, L.Z.; Protsenko, I.I.; Abusuek, D.A.; Zaykovskaya, A.O.; Bykov, A.V.; Matveeva, V.; Sulman, E. Hydrogenation of Biomass-Derived Levulinic Acid to Gamma-Valerolactone Using Polymer-Based Metal-Containing Catalysts. *Chem. Eng. Trans.* **2017**, *61*, 895–900. [CrossRef]
34. Dunn, R.O. Correlating the Cloud Point of Biodiesel with Its Fatty Acid Methyl Ester Composition: Multiple Regression Analyses and the Weighted Saturation Factor (WSF). *Fuel* **2021**, *300*, 120820. [CrossRef]
35. Zuin, V.G.; Eilks, I.; Elschami, M.; Kümmerer, K. Education in Green Chemistry and in Sustainable Chemistry: Perspectives towards Sustainability. *Green Chem.* **2021**, *23*, 1594–1608. [CrossRef]
36. Barton, A.F.M. *Handbook of Polymer-Liquid Interaction Parameters and Solubility Parameters*; Routledge: Oxfordshire, UK, 2018; Volume 1, p. 297, ISBN 0203752619.
37. Cizmeci, M.; Musavi, A.; Tekin, A.; Kayahan, M. Catalytic Behavior of Ruthenium in Soybean Oil Hydrogenation. *Eur. J. Lipid Sci. Technol.* **2009**, *111*, 607–611. [CrossRef]
38. Madureira, A.; Noël, S.; Léger, B.; Ponchel, A.; Monflier, E. Catalytic Hydrogenation of Derived Vegetable Oils Using Ion-Exchange Resin-Supported Ruthenium Nanoparticles: Scope and Limitations. *ACS Sustain. Chem. Eng.* **2022**, *10*, 16588–16597. [CrossRef]
39. Sarno, M.; Iuliano, M.; Viscusi, G.; Zarli, A.; Ciambelli, P. A Nickel/Palladium/Ruthenium-Graphene Based Nanocatalyst for Selective Catalytic Hydrogenation of Vegetable Oils. *Ind. Crop. Prod.* **2021**, *170*, 113815. [CrossRef]
40. Ho, C.-M.; Yu, W.-Y.; Che, C.-M. Ruthenium Nanoparticles Supported on Hydroxyapatite as an Efficient and Recyclable Catalyst Forcis-Dihydroxylation and Oxidative Cleavage of Alkenes. *Angew. Chem. Int. Ed.* **2004**, *43*, 3303–3307. [CrossRef]
41. Gámez, S.; Magerat, A.; de la Torre, E.; Gaigneaux, E.M. Functionalization of Carbon Black for Ru Complexation Towards the Oxidative Cleavage of Oleic Acid. *SSRN Electron. J.* **2022**. [CrossRef]
42. Stepacheva, A.A.; Sidorov, A.I.; Matveeva, V.G.; Sulman, M.G.; Sulman, E.M. Fatty Acid Deoxygenation in Supercritical Hexane over Catalysts Synthesized Hydrothermally for Biodiesel Production. *Chem. Eng. Technol.* **2019**, *42*, 780–787. [CrossRef]
43. Xu, G.; Zhang, Y.; Fu, Y.; Guo, Q. Efficient Hydrogenation of Various Renewable Oils over Ru-HAP Catalyst in Water. *ACS Catal.* **2017**, *7*, 1158–1169. [CrossRef]
44. Konwar, L.J.; Oliani, B.; Samikannu, A.; Canu, P.; Mikkola, J.-P. Efficient Hydrothermal Deoxygenation of Tall Oil Fatty Acids into N-Paraffinic Hydrocarbons and Alcohols in the Presence of Aqueous Formic Acid. *Biomass Convers. Biorefinery* **2022**, *12*, 51–62. [CrossRef]
45. Mondal, S.; Singuru, R.; Chandra Shit, S.; Hayashi, T.; Irle, S.; Hijikata, Y.; Mondal, J.; Bhaumik, A. Ruthenium Nanoparticle-Decorated Porous Organic Network for Direct Hydrodeoxygenation of Long-Chain Fatty Acids to Alkanes. *ACS Sustain. Chem. Eng.* **2018**, *6*, 1610–1619. [CrossRef]

46. Ramalho, H.F.; di Ferreira, K.M.C.; Machado, P.M.A.; Oliveira, R.S.; Silva, L.P.; Prauchner, M.J.; Suarez, P.A.Z. Biphasic Hydroformylation of Soybean Biodiesel Using a Rhodium Complex Dissolved in Ionic Liquid. *Ind. Crop. Prod.* **2014**, *52*, 211–218. [CrossRef]
47. Corradini, S.A.d.S.; Lenzi, G.G.; Lenzi, M.K.; Soares, C.M.F.; Santos, O.A.A. Characterization and Hydrogenation of Methyl Oleate over Ru/TiO_2, Ru–Sn/TiO_2 Catalysts. *J. Non. Cryst. Solids* **2008**, *354*, 4865–4870. [CrossRef]
48. Alvila, L.; Pakkanen, T.A.; Krause, O. Hydroformylation of Olefins Catalysed by Supported $Ru_3(CO)_{12}$ with 2, 2′-Bipyridine or with Other Heterocyclic Nitrogen Base. *J. Mol. Catal.* **1993**, *84*, 145–156. [CrossRef]
49. Oresmaa, L.; Moreno, M.A.; Jakonen, M.; Suvanto, S.; Haukka, M. Catalytic Activity of Linear Chain Ruthenium Carbonyl Polymer [Ru (CO) 4] n in 1-Hexene Hydroformylation. *Appl. Catal. A Gen.* **2009**, *353*, 113–116. [CrossRef]
50. Ungvary, F. Application of Transition Metals in Hydroformylation. Annual Survey Covering the Year 1995. *Coord. Chem. Rev.* **1997**, *160*, 129–159. [CrossRef]
51. Zimmermann, F.; Meux, E.; Mieloszynski, J.-L.; Lecuire, J.-M.; Oget, N. Ruthenium Catalysed Oxidation without CCl_4 of Oleic Acid, Other Monoenic Fatty Acids and Alkenes. *Tetrahedron Lett.* **2005**, *46*, 3201–3203. [CrossRef]
52. Chen, J.; Xu, Q. Hydrodeoxygenation of Biodiesel-Related Fatty Acid Methyl Esters to Diesel-Range Alkanes over Zeolite-Supported Ruthenium Catalysts. *Catal. Sci. Technol.* **2016**, *6*, 7239–7251. [CrossRef]
53. Krishnapriya, R.; Gupta, U.; Soni, V.K.; Sharma, R.K. Catalytic Conversion of Methyl Oleate to Hydrocarbons: Impact of Cobalt Oxide Species Integration in SiO_2–Al_2O_3. *Sustain. Energy Fuels* **2020**, *4*, 3308–3317. [CrossRef]
54. Mendes, A.N.F.; Gregório, J.R.; da Rosa, R.G. Studies on the Experimental Variables Effects on Rhodium Catalyzed Hydroformylation of Unsaturated Fatty Esters and Comparison of [RhH(CO)(PPh3)3] and [RhCl3.3H_2O] as Starting Catalytic Precursors. *J. Braz. Chem. Soc.* **2005**, *16*, 1124–1129. [CrossRef]
55. Ramalho, H.F.; Ferreira, K.M.C.; Machado, P.M.A.; Silva, T.B.; Rangel, E.T.; Prauchner, M.J.; Suarez, P.A.Z. Production of Additives with Antimicrobial Activity via Tandem Hydroformylation-Amine Condensation of Soybean FAME Using an Ionic Liquid-Based Biphasic Catalytic System. *J. Braz. Chem. Soc.* **2015**, *27*, 321–333. [CrossRef]

catalysts

MDPI

Article

Hydrogen Production by N-Heterocycle Dehydrogenation over Pd Supported on Aerogel-Prepared Mg-Al Oxides

Danil M. Shivtsov, Anton P. Koskin *, Sergey A. Stepanenko, Ekaterina V. Ilyina, Artem B. Ayupov, Alexander F. Bedilo and Vadim A. Yakovlev

Federal Research Center, Boreskov Institute of Catalysis SB RAS, Lavrentyev Av. 5, Novosibirsk 630090, Russia
* Correspondence: koskin@catalysis.ru

Abstract: Tetradecahydrophenazine (14HP) is a nitrogen-containing heterocycle compound with a high content of hydrogen that can be released during its dehydrogenation to phenazine (P). The high stability of the 14HP/P pair and relatively low dehydrogenation temperature make 14HP a promising organic hydrogen carrier. This manuscript is devoted to the investigation of hydrogen production by 14HP dehydrogenation over Pd supported on a series of magnesium-aluminum oxides prepared by the aerogel method. This technique made it possible to synthesize catalyst supports characterized by a high surface area and high concentration of surface active sites where active transition metals could be stabilized in a finely dispersed state. The synthesized aerogels had high specific surface areas and pore volumes. A surface area as high as 600 m^2/g after calcination at 500 °C was observed for the mixed aerogel with an Mg:Al ratio of 1:4. An increase in the concentration of acidic electron-acceptor sites determined by EPR on the surface of the mixed magnesium-aluminum oxide supports with a high surface area prepared by the aerogel method was found to result in higher hydrogen production due to the faster dehydrogenation of sterically hindered nitrogen-containing tetradecahydrophenazine heterocycles.

Keywords: hydrogen storage; N-heterocycles; Pd catalysts; acceptorless dehydrogenation; aerogel supports; porous structure; nitrogen adsorption; surface active sites

1. Introduction

The widespread use of fossil and other carbon-containing fuels has led to a significant increase in the content of greenhouse gases in the atmosphere [1]. Today, hydrogen, which is an environmentally friendly and renewable energy source, is considered to be a promising alternative fuel. Hydrogen has the highest known gravimetric energy density (143 MJ/kg, which is three times higher than that of gasoline). However, its volumetric energy density is very low (0.0108 MJ/L, which is 3000 times lower than that of gasoline) [2,3]. This is the main aspect hindering the widespread introduction of hydrogen energy. The development of methods for its energy-efficient storage and transportation is an important step in the progress of hydrogen energy implementation [4].

In the last decade, various methods of H_2 accumulation and storage have been studied. It has been suggested that hydrogen should be stored in compressed [5], cryo-compressed [6], liquefied [7], and adsorbed states [8], as well as in the form of metal hydrides [9], complex hydrides [10], and organic hydrogen carriers [2,3,11]. According to a number of feasibility studies [12,13], the storage and transportation of hydrogen using organic carriers is currently considered to be one of the most promising technologies.

Since the 2000s, an intensive search for optimal pairs of substrate molecules acting as organic H_2 carriers has been carried out, which would allow one to carry out reversible catalytic hydrogenation/dehydrogenation reactions with high selectivity at moderate temperatures [14–16]. In this regard, the optimization of the hydrogen evolution process (the selection of reaction conditions for dehydrogenation, catalytic systems, and organic substrates) is very important, since this high-temperature endothermic process could be associated with the decomposition of an organic H_2 carrier. It has been repeatedly shown that H_2-rich nitrogen-containing heterocyclic molecules are characterized by low dehydrogenation temperatures (140–250 °C) in comparison with those of cycloalkanes (300–400 °C). This fact accounts for the decrease in the decomposition rate of the H_2-rich substrate. In addition, heterocycles are less volatile, ensuring the production of high-purity hydrogen [17,18]. One of the most promising N-heterocyclic compounds for hydrogen production is tetradecahydrophenazine (perhydrophenazine, 14HP) [19]. Tetradecahydrophenazine is characterized by a high gravimetric hydrogen capacity, good thermal stability, and a high purity of the hydrogenation process, and 14HP precursors can be obtained from the degradation products of lignocellulose [19]. In our recent study, it was shown that the acceptorless dehydrogenation of 14HP made it possible to achieve higher rates of hydrogen evolution compared to decahydroquinoline [20]. In that study, and in the literature on the acceptorless dehydrogenation of N-heterocyclic compounds in general, Pd-containing systems were shown to be the most efficient catalysts [15,21,22]. Using the dehydrogenation of 12H-NEC as an example, it was shown that the catalytic activity of Pd was much higher than that of Pt, Ru, and Rh [23]. This result was confirmed by the study of catalysts containing these metals under identical reaction conditions.

An equally important aspect during the selection of an effective catalyst for the dehydrogenation of N-heterocycles is the choice of a support for the dispersed metal particles. One of the most promising approaches for improving the properties of the supports is the use of the aerogel method for their synthesis [24–27]. The aerogel method was successfully used for the preparation of V-Mg-containing catalysts for the oxidative dehydrogenation of propane. This method made it possible to obtain nanocrystalline samples characterized by a layered structure, high specific surface area, and uniform distribution of vanadium in the magnesium oxide matrix [28,29]. The aerogel-prepared two-component catalyst exhibited a high catalytic activity that exceeded the activity of catalysts previously described in the literature [28]. In addition to their bifunctionality, an important advantage of aerogel supports is their ability to reduce the active component loading. It was shown that the use of an aerogel support for Pd/Al_2O_3 and Rh/Al_2O_3 made it possible to significantly reduce the loading of noble metals in three-way catalysts due to the uniform distribution of the metals on the surface [30].

Recent studies have shown that the use of $MgAl_2O_4$ as a support improves the properties of dehydrogenation catalysts. For example, $Pd/MgAl_2O_4$ was shown to have high catalytic activity in the dehydrogenation of dodecahydro-N-ethylcarbazole [31]. This activity was explained by the strong interaction between Pd and $MgAl_2O_4$, leading to the stabilization of small particles. It was also noted that the $MgAl_2O_4$ spinel suppressed the sintering of PdO particles during the low-temperature combustion of methane due to the strong interaction of PdO with the support [32,33]. It was shown that the presence of a positive charge on the Pd atom promoted the rapid desorption of dehydrogenation products and made the active sites more accessible [34].

In summary, the development of Pd catalysts supported on mixed Mg-Al-O_x aerogels for the dehydrogenation of nitrogen-containing heterocycles seems to be promising. In this case, the aerogel structure of the support provides advanced characteristics of a porous structure, whereas the MgO addition contributes to the stabilization of the Pd nanoparticles in the ultrafine state and changes their charge. In the present study, several Mg-Al-O_x aerogel supports with different compositions were synthesized. The structure of the prepared materials was investigated by various physical methods. The influence of the surface properties of the mixed aerogel supports on the catalytic properties of these Pd-containing systems in the dehydrogenation of a potential hydrogen carrier, tetradecahydrophenazine, was observed.

2. Results and Discussion

2.1. Characterization of $MgAlO_x$ Aerogel Supports and Pd Catalysts Based on Them

Aerogels are porous materials based on gels, wherein the solvent is substituted by a gas. Typical methods for the synthesis of aerogels are based on a sol–gel process followed by the drying of the gel under supercritical conditions. Conventional drying results in the substantial deformation of the gel structure due to the capillary hydrostatic pressure of the solvent evaporating from the gel pores caused by the surface tension. Under supercritical conditions, the distinction between the liquid and the gas phases vanishes, resulting in the almost complete disappearance of the surface tension. As a result, the use of supercritical conditions during the synthesis of aerogels makes it possible to preserve the original developed structure of the gel almost intact and synthesize materials with high surface areas and pore volumes [35–38].

In the first step, one-component Al_2O_3 and MgO and mixed Mg-Al-O_x supports with different Mg/Al molar ratios (4:1, 2:1, 1:1, 1:2, and 1:4) were synthesized by the aerogel method. Pd was deposited on the surface of the synthesized supports using a previously reported technique [20], which was adapted taking into account the specific features of the aerogel supports. Absolute ethanol was used as the solvent during the noble metal deposition to avoid the substantial deterioration of the textural characteristics of the aerogels in water. Additionally, reduction in hydrogen at 500 °C was used instead of reduction with sodium borohydride, as it secured the complete removal of chloride ions [39].

The porous structure of the aerogels before and after palladium supporting was studied by nitrogen adsorption–desorption at 77 K (Figure 1 and Table 1). One can see that the porous structure depended on both the chemical composition of the initial gel and the subsequent post-synthetic modification. Indeed, the isotherms of nitrogen adsorption on the MgO and Pd/MgO samples showed virtually no hysteresis loop formed due to capillary condensation. On the samples M1A1, M1A2, Pd/M1A1, and Pd/M1A2, this loop was very narrow (for example, see Figure 1c). Interestingly, on the samples M1A4 and Al_2O_3, this loop also was very narrow, while after palladium supporting the form of the loop changed and became typical of mesoporous samples with cylindrical pores (see Figure 1e). On the samples M4A1, M2A1, Pd/M4A1, and Pd/M2A1, the hysteresis loop formed due to capillary condensation was large and visible, as expected for mesoporous materials. The observed difference in the appearance of hysteresis loops suggested that the pores in the studied samples had different forms. Slit-like or wedge-like pores should show no hysteresis loop, whereas cylindrical pores usually demonstrate hysteresis loops of different forms [40]. Thus, we used the slit-like pore model for the samples that demonstrated a narrow loop or no hysteresis loop at all. The cylindrical model was used for the samples that showed a hysteresis loop.

The highest surface area and total pore volume among the synthesized aerogels were observed on the sample M1A4 (Table 1). This sample also showed the highest total pore volume among the supports. However, the porous structure of this sample changed after supporting Pd. The initial wedge-like or slit-like geometry became closer to a cylindrical geometry. Palladium deposition changed the porous structure of all the samples. Both the surface area and the total pore volume were reduced, and the maximum pore size distribution shifted to a larger value. The extent of these changes depended on the chemical composition of the aerogel sample. The MgO and Pd/MgO samples showed the most drastic transformation: the surface area decreased from 220 m^2/g to 70 m^2/g, and the maximum pore size distribution increased from 9.4 nm to 34 nm. The pore structure comprised different wedge-like and slit-like pores (Figure 1b). Most probably, these pores appeared between parallel or almost parallel platelets and crystals in this oxide. As one can see from the cumulative pore volume curves, after Pd deposition involving the dispersion of the precursor materials in liquid ethanol followed by calcination, the smallest pores disappeared due to the cohesion and sintering of the platelets after capillary contraction in solution. Larger pores could be retained after such treatment.

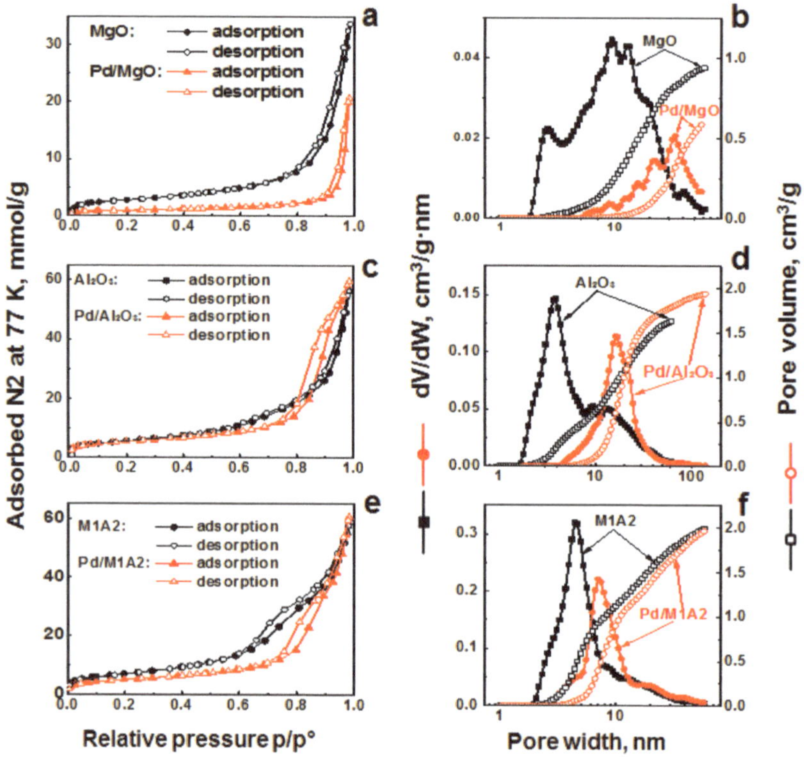

Figure 1. Nitrogen adsorption–desorption isotherms (**a**,**c**,**e**) and pore size distribution (**b**,**d**,**f**) on aerogel-prepared MgO (**a**,**b**), M1A2 (**c**,**d**), and Al_2O_3 (**e**,**f**) before and after supporting Pd. The differential pore size dV/dW distributions are given by closed symbols, the cumulative volume pore size distributions are given with open symbols. The Broekhoff–de Boer model was used for slit-like pores in MgO, Pd/MgO, M1A2, Pd/M1A2, and Al_2O_3 and for cylindrical pores in Pd/Al_2O_3; the regularization factor was 0.1.

Samples M4A1 and M2A1 showed a large hysteresis loop that could be attributed to the fact that the pores in these materials were formed by contact between particles of mixed oxides with large cavities and channels. After Pd deposition, the pore structure was mostly preserved, with the disappearance of the smallest pores to some extent. In the mixed oxides with a higher aluminum content close to that of spinel $MgAl_2O_4$, the pores were almost wedge-like or slit-like. The smallest pore size estimated as the maximum of the pore size distribution among the supports and catalysts was obtained for the samples M1A1 and Pd/M1A1, respectively. One could suggest that the cohesion and sintering capability after calcination was lower for this sample than for the others. The increase in the aluminum content made it possible to produce materials with a higher surface area (see samples M1A2 and M1A4, Table 1) and different pore structures, which became mostly cylindrical after supporting Pd (sample M1A4). In agreement with our expectations, the higher aluminum content brought the properties of the sample closer to those of pure alumina, which showed slit-like pores in its pristine form and the cylindrical pores after supporting Pd (sample Al_2O_3, Table 1).

To determine the phase composition of the synthesized systems, we performed an XRD analysis of the Pd/MgO, Pd/Al_2O_3, and Pd/M1A2 catalysts. According to the XRD data, only the pure MgO system was well-crystallized (Figure 2, 111, 200, 220). Aerogel-prepared MgO samples calcined at 500 °C are known to consist of MgO nanoparticles with average dimensions of 4–5 nm [41]. In the mixed oxide, the peaks were shifted to larger angles in comparison with MgO. This meant that the lattice parameter was smaller than in pure MgO, due to the smaller radius of Al^{3+} cations compared to Mg^{2+}. Furthermore, the mixed oxide contained cationic vacancies to compensate for the extra charge of the Al^{3+} cations. No peaks attributable to the $MgAl_2O_4$ spinel phase were observed in the XRD pattern of M1A2.

Table 1. Characteristics of the porous structure in aerogel-prepared supports and catalysts determined by nitrogen adsorption.

Sample	SSA, m^2/g	V_{pore}, cm^3/g	Pore Size [1], nm	D_{av}, nm
		Supports		
MgO	220	1.2	9.4 [2]	21
M4A1	415	1.6	12 [3]	15
M2A1	495	1.5	8.0 [3]	12
M1A1	535	1.4	3.3 [2]	11
M1A2	570	2.0	4.5 [2]	14
M1A4	600	2.3	5.2 [2]	15
Al_2O_3	545	2.3	3.8 [2]	17
		Catalysts 1 wt% Pd		
Pd/MgO	70	0.7	34 [2]	40
Pd/M4A1	310	1.6	16 [3]	20
Pd/M2A1	345	1.2	9.6 [3]	14
Pd/M1A1	340	1.4	5.1 [2]	17
Pd/M1A2	410	2.1	7.1 [2]	21
Pd/M1A4	445	1.9	14 [3]	17
Pd/Al_2O_3	435	2.1	17 [3]	19

[1] The maximum of the differential pore size distribution dV/dW, where V is the pore volume and W is the pore width for slit-like pores or the pore diameter for cylindrical pores. [2] Slit-like pores were assumed, and the pore size indicates the pore width. [3] Cylindrical pores were assumed, and the pore size indicates the pore diameter.

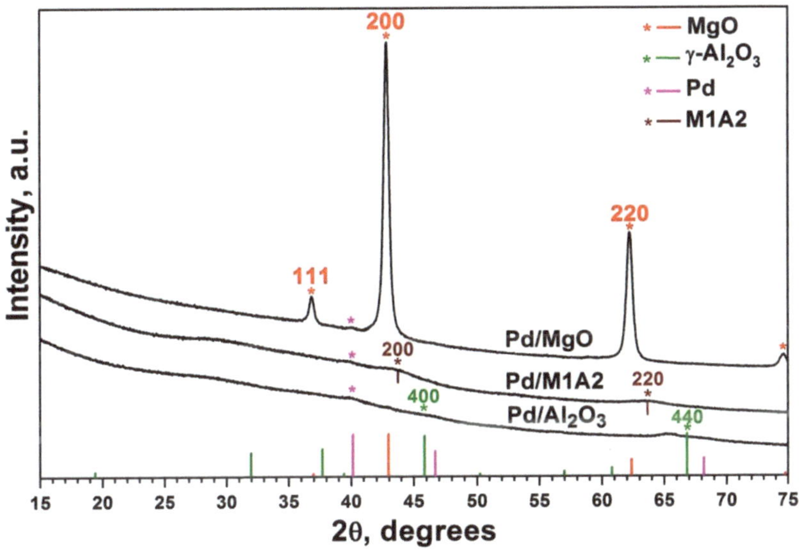

Figure 2. XRD diffractograms of MgO, M1A2, and Al_2O_3 aerogels.

The Pd/MgO, Pd/Al_2O_3, and Pd/M1A2 catalysts were studied by transmission electron microscopy. The TEM micrographs and EDX mapping of the studied catalysts are presented in Figure 3. According to the obtained data, the Pd/Al_2O_3 structure consisted of particles with a flake morphology forming a network. The catalyst-supported magnesium oxide aerogel (Pd/MgO) had a lamellar structure, which was presumably formed as a result of heat treatment during the support preparation step.

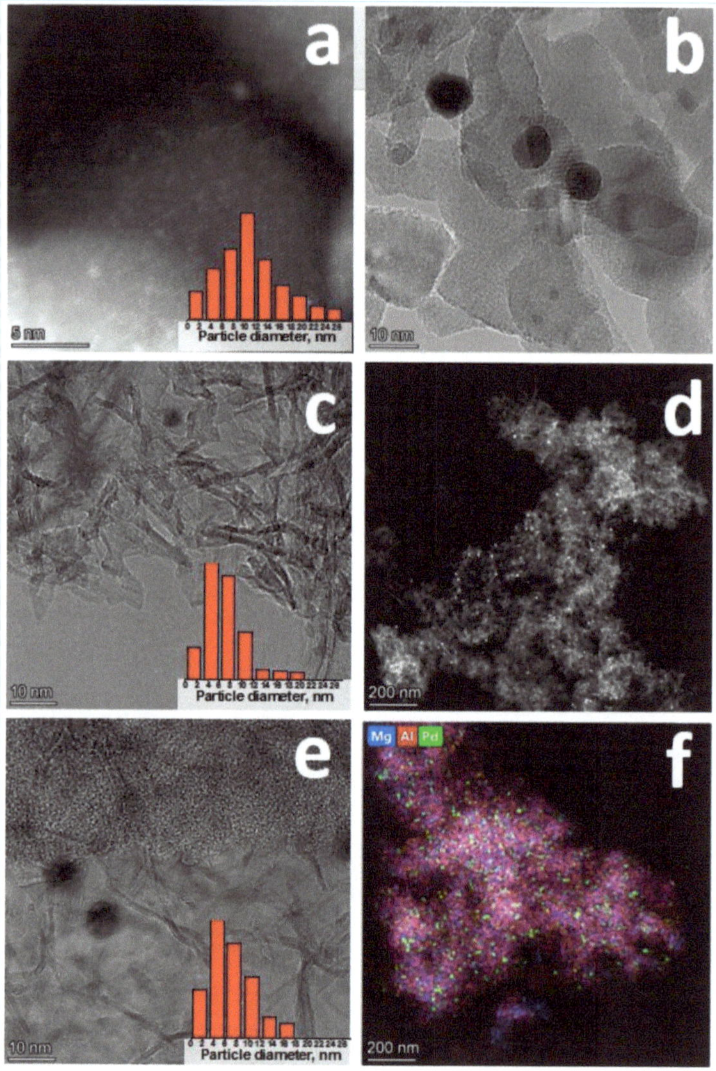

Figure 3. HR-TEM images of aerogel catalysts Pd/MgO (**a**,**b**), Pd/Al_2O_3 (**c**,**d**), Pd/M1A2 (**e**), and EDX of Pd/M1A2 (**f**).

Clusters of palladium nanoparticles about 1 nm in size and particles in the range of 2 to 26 nm with a maximum size of about 10 nm were observed on the MgO surface. The particle size of the palladium on the surface of Al_2O_3 was in the range of 2–20, with a maximum of about 6 nm. The diameter of the Pd nanoparticles on the mixed oxide M1A2 varied from 2 to 16 nm, with a maximum distribution of about 4 nm. According to the mapping data, Pd was deposited evenly on the surface of the support, whereas Al_2O_3 and MgO were uniformly mixed with each other.

To determine the possible effect of the support composition on the electronic properties of the supported palladium, the Pd/Al_2O_3 and Pd/M1A2 catalysts were studied by XPS. Core-level photoelectron peaks corresponding to Pd$3d_{5/2}$ (335.1 eV), O1s (531.3 eV), C1s (284.6 eV), and Al2p (74.5 eV) were observed in the survey spectra of both catalysts. In addition, a peak at 50.9 eV corresponding to Mg2p was observed for Pd/M1A2. The Pd*3d* core-level spectra of the studied samples are shown in Figure 4. The spectra are described well by the Pd$3d_{5/2}$-Pd$3d_{3/2}$ doublet with a 3:2 intensity ratio. The binding energy of the Pd$3d_{5/2}$ peak was 335.1 eV for both samples. This value is in an excellent agreement with the literature data for palladium in the metal state (335.0–335.2 eV) [42,43]. Thus, it was demonstrated that palladium was in the Pd^0 state on the surface of both catalysts, and the presence of basic sites associated with Mg did not have a significant effect on the electronic state of Pd. The ratios of the atomic concentrations of Pd/Al and Pd/(Al+Mg) in the near-surface catalyst layer were similar for both samples and equal to 0.0038 and 0.0035 for Pd/Al_2O_3 and Pd/M1A2, respectively. These values were close to the theoretical value of 0.005 corresponding to 1 wt.% Pd.

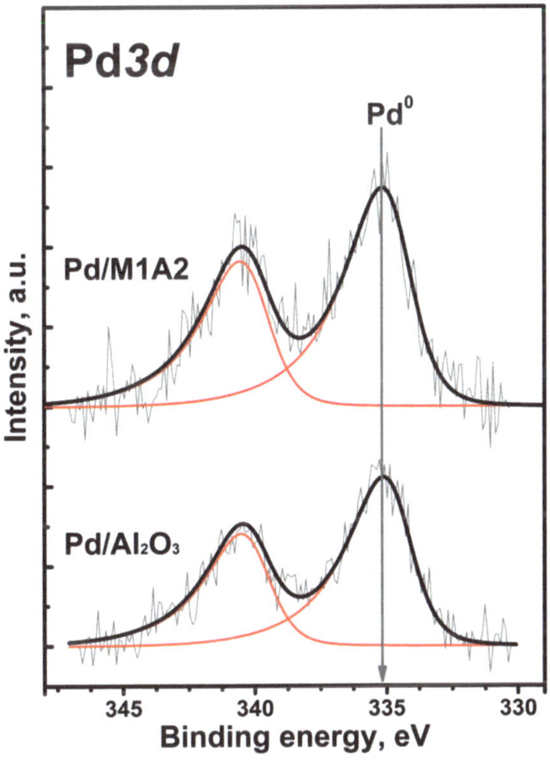

Figure 4. Core-level Pd3d XPS spectra of Pd/Al_2O_3 and Pd/M1A2 catalysts. The spectra were normalized to the integral intensity of the core-level Al2p peaks corresponding to the spectrum of Al_2O_3.

A series of palladium-containing catalysts were studied by CO pulse chemisorption. This method provided information on the average size of the supported metal species in terms of dispersity determined from CO adsorption data (D_{CO}, %; Table 2). According to the obtained results, the presence of Mg in the support favored the preservation of good Pd dispersity after reduction at 500 °C. Presumably, this resulted from the stabilization of the palladium nanoparticles on basic surface sites of the Mg-Al-O_x supports, in good agreement with the literature data [31–33].

Table 2. Dispersity of supported palladium determined by CO chemisorption and concentrations of electron-donor and electron-acceptor sites on the surface of the aerogel-prepared supports.

Aerogel Supports	Pd Dispersion by CO Chemisorption (D_{CO}), %	Active Site Concentration, 10^{18} g^{-1}	
		1,3,5-Trinitrobenzene (Electron-Donor Sites)	Phenothiazine (Electron-Acceptor Sites)
MgO	46	2.4	3.2
M4A1	50	2.4	2.4
M2A1	54	1.8	3.7
M1A1	54	1.1	4.4
M1A2	55	1.2	7.5
M1A4	49	0.7	9.6
Al_2O_3	33	0.6	16.8

Electron paramagnetic resonance (EPR) spectroscopy with the use of spin probes makes it possible to study various active sites on the surface of catalysts and sorbents and their role in chemical and catalytic reactions. This method was successfully used for the characterization of electron-acceptor sites, which are capable of abstracting electrons from aromatic probe molecules, and electron-donor sites, which are capable of donating electrons to aromatic nitro compounds [44–49]. As discussed in detail in the cited publications, electron-acceptor sites on metal oxides accounting for the formation of radical cations are associated with strong acid sites, most likely Bronsted. The presence of various metal cations has little or no positive effect. Meanwhile, the electron-donor sites responsible for the opposite process seem to be related to basic sites.

Both probe molecules had one nitrogen atom. Thus, the EPR spectra of the ion radicals formed after their adsorption on the active sites were characterized as three-component spectra with hyperfine splitting on the nitrogen atom (Figure 5). The spectra observed after the adsorption of 1,3,5-trinitrobenzene (TNB) on the aerogel supports (Figure 5a) were similar to the previously reported spectra of TNB radical anions on the Al_2O_3 surface [42]. They were characterized by a triplet EPR signal with frozen rotation with g = 2.005 and a hyperfine splitting constant A_{ZZ} of about 28 G, which did not depend much on the Mg:Al ratio. This result indicates that electron-donor sites donating a single electron to the TNB molecule, similar to those present on the Al_2O_3 surface, existed on the surface of all studied aerogel samples. Previously, such sites were shown to stabilize palladium deposited on the Al_2O_3 surface in a finely dispersed state with high CO oxidation activity [50]. The concentration of strong electron-donor sites revealed immediately after the probe adsorption generally grew with the increase in the Mg concentration in the sample (Table 2). This was quite natural because the increase in the Mg:Al ratio was expected to lead to an increase in the sample basicity. The highest concentration (2.4×10^{18} g^{-1}) was observed on M4A1 and on pure MgO, which had a substantially lower surface area. The concentration of such sites on the Al_2O_3 aerogel was approximately four times lower.

The ionization potential of phenothiazine is 6.8 eV, making it a good probe for the investigation of weak electron-acceptor sites capable of abstracting an electron from phenothiazine and stabilizing its radical cation on the surface [48,49,51]. A triplet EPR signal with hyperfine splitting on the nitrogen atom A_{ZZ} of about 18 G was observed after phenothiazine adsorption on all the studied samples. This corresponded to adsorbed phenothiazine radical cations. The concentration of such acidic sites decreased with an increase in the Mg:Al ratio, which made the sample more basic and less acidic. The highest concentration of electron-acceptor sites was observed on the Al_2O_3 aerogel: 16.8×10^{18} g^{-1} (Table 2). The M4A1 sample was characterized by the lowest concentration of such sites 2.4×10^{18} g^{-1}, which was even lower than on the MgO surface.

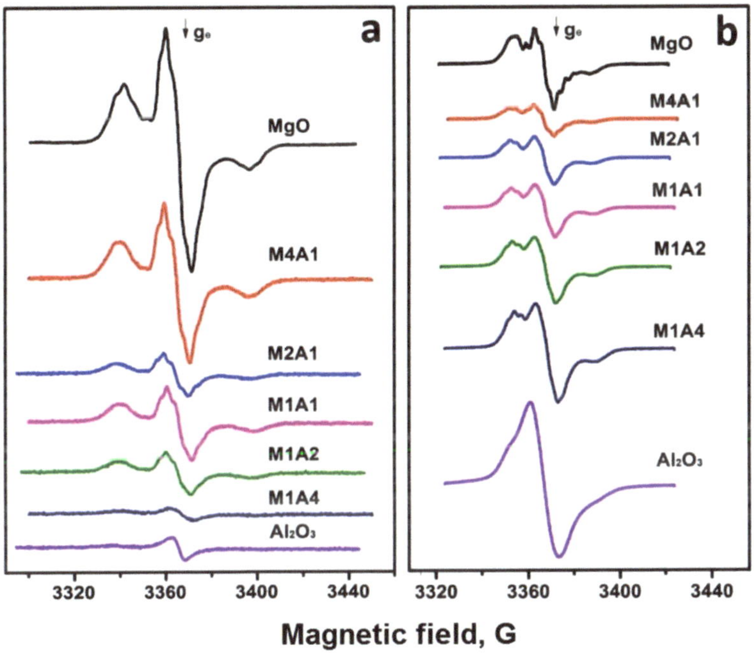

Figure 5. EPR spectra observed after adsorption on 1,3,5-trinitrobenzene (**a**) and phenothiazine (**b**) on the studied aerogel supports.

Thus, electron-donor and electron-acceptor sites were present on the surface of all the studied aerogel supports. The observed trends were in a good agreement with the expected increase in the material basicity with the growth of the Mg:Al ratio. Its growth led to an increase in the concentration of basic electron-donor sites and a decrease in the concentration of acidic electron-acceptor sites.

Overall, it was shown that the mixed oxide samples had a high surface area, a high concentration of electron-donor sites where Pd could be stabilized in a finely dispersed active state, and a low concentration of electron-acceptor sites that might lead to unwanted transformations of organic molecules initiated by their acid properties. In the next stage, the catalytic activity of the synthesized materials was studied to determine the optimal Mg:Al ratio.

2.2. Investigation of the Catalytic Activity of Aerogel-Supported Pd Catalysts

The catalytic activity of Pd/Mg-Al-O_x catalysts was studied in the dehydrogenation of tetradecahydrophenazine (14HP) using tetraglyme as the solvent. A solvent had to be used because both the reagent 14HP and the reaction product phenazine (P) are solid at the dehydrogenation temperature of 240 °C. In our previous publication [20], the effect of the solvent on 14HP dehydrogenation was studied for 1 wt% Pd/γ-Al_2O_3 catalyst. Polar tetraglyme with a high boiling temperature was shown to be the most preferable solvent. The main parameter characterizing the catalytic activity was the hydrogen yield relative to the maximum theoretically possible hydrogen release (YH_2, %). This characterizes the depth of 14HP conversion to dehydrogenation products. In addition, the concentrations of side products 5-methyl-5,10-dihydrophenazine and 5,10-dimethyl-5,10- dihydrophenazine were determined using gas chromatography (Y_{MP}, %). The formation of these compounds is related to partial tetraglyme decomposition over the surface acid sites of the support followed by the methylation of the nitrogen atom in phenazine.

The analysis of the reaction mixture probes taken at intermediate dehydrogenation times (10, 20, and 30 min) revealed significant differences in the reaction routes of dehydrogenation between the catalysts containing MgO and the catalysts based on Al_2O_3. Typical chromatograms of the reaction mixture probes taken after 30 min of dehydrogenation over Pd/M1A2 and Pd/Al_2O_3 are shown in Figure 6. 1,2,3,4,4a,5,10,10a-octahydrophenazine (8HPc) was registered as an intermediate over Pd/M1A2, whereas the reaction mixture composition over Pd/Al_2O_3 was similar to that observed previously over Pd/Pural-550 prepared using a commercial Al_2O_3 support [20].

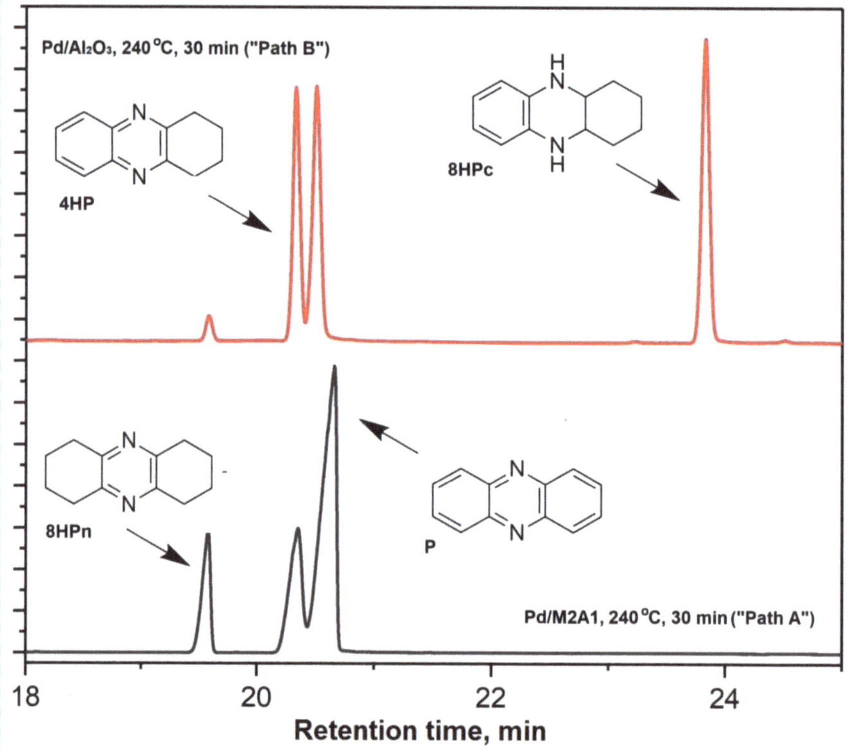

Figure 6. Typical chromatograms of the reaction mixture probes taken after 30 min of 14HP dehydrogenation over Pd/M1A2 and Pd/Al_2O_3.

As was shown previously [20], hydrogen release over Pd/Al_2O_3 begins from the side ring, despite the fact that the dehydrogenation of the central ring in 14HP is more thermodynamically favorable. This reaction route is apparently related to sterical problems for hydrogen release from the central ring in 14HP. For catalysts containing MgO, the hydrogen release began from the dehydrogenation of the central ring (Figure 7). As was shown above, the electronic state of Pd was the same in Pd/Al_2O_3 and Pd/M1A2 and corresponded to palladium metal (Figure 4). The average Pd particle size was also very similar (Figure 3). Thus, the presence of basic sites in the supports containing Mg affected the reaction pathway. Apparently, basic sites related to Mg can relieve some of the sterical hindrance, opening up a more thermodynamically favorable pathway [20].

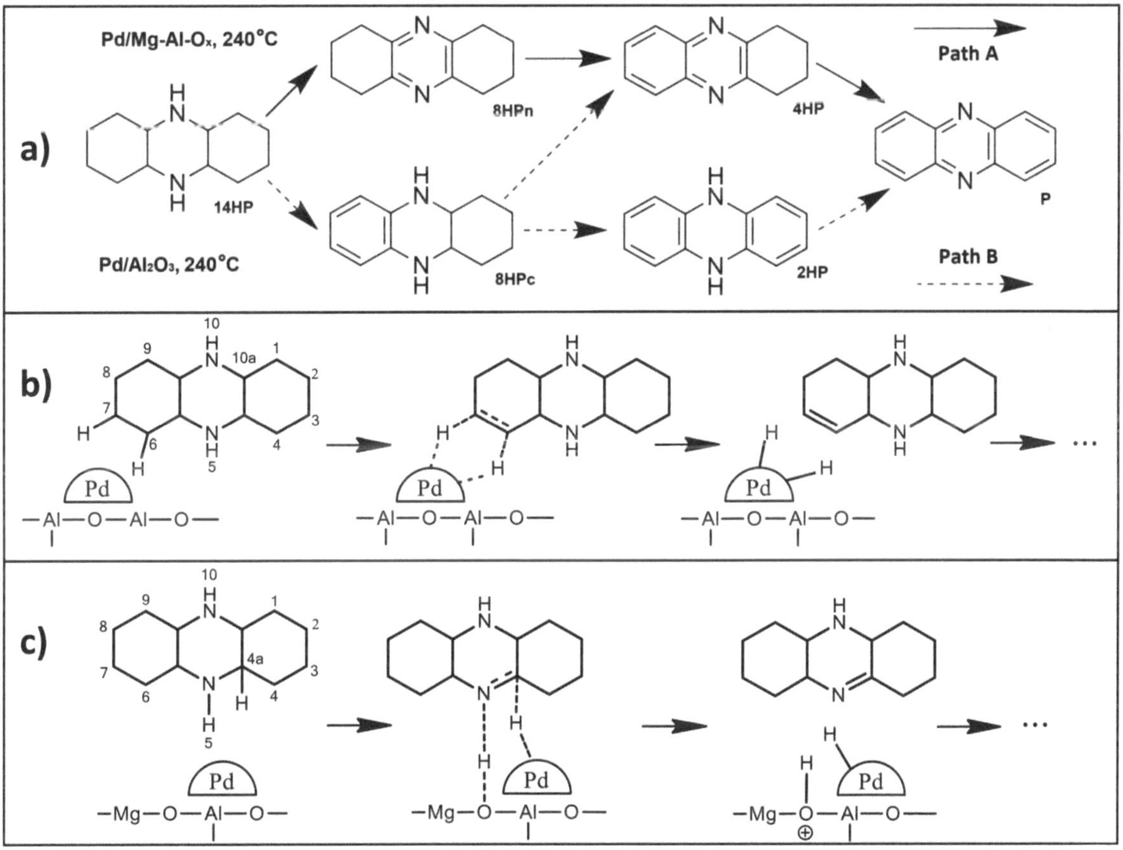

Figure 7. Differences in reaction routes of 14HP dehydrogenation over acidic and basic catalysts (**a**), and suggested reaction mechanisms over Pd/Al_2O_3 (**b**) and Pd/Mg-Al-O_x (**c**).

These observations were in a good agreement with previously obtained results showing that the basic sites of the support in addition to the noble metal play an important role in the hydrogenation and dehydrogenation of nitrogen-containing heterocycles [52–56]. When MgO was used as the support [54], the presence of strong basic sites (surface oxygen atoms) altered the mechanism of quinoline hydrogenation.

Based on the GC data and the previously reported results [50–54], the following reaction mechanisms were suggested for the first stage of 14HP dehydrogenation (Figure 7b,c). According to the proposed mechanisms, dehydrogenation over Pd/Al_2O_3 occurs in one stage via the abstraction of two hydrogen atoms by a Pd nanoparticle (Figure 7b). In this case, both hydrogen atoms must be oriented in roughly the same direction towards the Pd nanoparticle, which is apparently possible only for the side ring due to the sterical problems. Over supports containing Mg, strong basic sites can participate in the hydrogen abstraction (Figure 7c). In this case, one hydrogen atom is initially abstracted by a basic site, whereas the second atom is taken by a Pd nanoparticle in the following step. Thus, the sterical problems are removed, and the dehydrogenation process can follow the most energetically favorable pathway. In both cases, the eventual coupling of the hydrogen atoms yields molecular hydrogen.

Note that the reaction pathway "Path A" observed over mixed oxide supports is preferable, because 5,10-dihydrophenazine (2HP) is one of intermediates in "Path B"; 2HP is more stable in the desired dehydrogenation reaction and can be accumulated in the reaction mixture. Furthermore, it is easily methylated to form methylphenazines (MP) as side products.

Figure 8 illustrates the effect of the support on the catalytic properties. Catalysts containing 1 wt% Pd deposited on aerogel-prepared MgO; Al_2O_3 oxide supports; mixed oxides (M4A1, M2A1, M1A1, M1A2, and M1A4); and a γ-Al_2O_3 support from commercial Pural SB were studied under identical conditions. Hydrogen was not released when the dehydrogenation experiment was carried out in the absence of the catalyst or when the reaction was performed using pristine supports without Pd. Despite the fact that slightly poorer Pd dispersion was observed according to the CO chemisorption data over the aerogel supports (D_{CO} = 33–55%) in comparison with 1 wt% Pd/γ-Al_2O_3 (D_{CO} = 59%), their activity in 14HP dehydrogenation was significantly higher (Figure 8). The highest hydrogen production yield (YH_2 83%) was observed for the catalysts using the mixed aerogel M1A2 as the support.

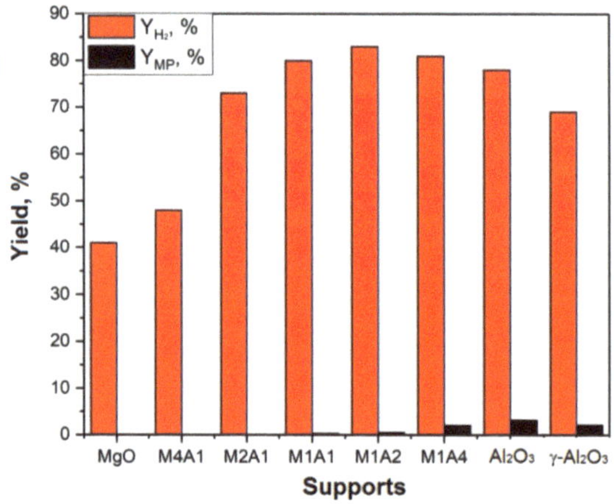

Figure 8. Effect of the support on the catalytic properties of catalysts containing 1 wt% Pd in 14HP dehydrogenation. Y_{H2} is the hydrogen yield, % of the theoretical maximum; Y_{MP} is the yield of methylphenazines, %.

The observed dependence of YH_2 on the composition of the support was presumably determined by a number of factors. In addition to a high surface area, the most efficient supports (M1A1, M1A, and M1A4), which contained magnesium, were characterized by significant concentrations of basic electron-donor sites (0.7–$1.2 \cdot 10^{18}$ g^{-1}, Table 2). These sites accounted for the efficient stabilization of the palladium nanoparticles. Additionally, their presence altered the mechanism of 14HP dehydrogenation, favoring the formation of 8HPn. Overall, this led to the higher efficiency of the dehydrogenation process and accelerated hydrogen production. Based on the catalytic activity data, Pd/M1A2 was characterized by an optimal combination of the surface area and concentration of basic sites. According to the XRF data, the Pd concentration in the Pd/M1A2 sample after the catalytic reaction was similar to that observed for the fresh catalyst. The recycled catalyst was studied again in the catalytic reaction. Its catalytic activity was preserved after recycling, remaining within the experimental error limits.

Meanwhile, the concentration of side products resulting from methylation (Y_{MP}) also substantially depended on the support used. An increase in the concentration of acidic electron-acceptor sites (Table 2) as well a surface area increase intensified the solvent decomposition reaction, resulting in an increased concentration of the methylation side products in the reaction mixture. Thus, the 1 wt% Pd/M2A1, Pd/M1A1, and Pd/M1A2 catalysts produced using aerogel-prepared mixed oxide supports were the most promising for 14HP dehydrogenation in the presence of tetraglyme, as they demonstrated both good tetraglyme stability and a high hydrogen production rate. The hydrogen evolution rate observed in this study was compared with the literature data (Table 3).

Table 3. Comparison of the hydrogen evolution rate observed in this study with the literature data.

Catalyst	Reactant	Reaction Conditions	H_2 Evolution Rate ($mmol\ g_{Met}^{-1}\ min^{-1}$)	Ref.
Pd_2Ru@SiCN	14HP	2 mmol 14HP, 70 mg Pd_2Ru@SiCN (0.36 mol. % active metal), 190 °C, 0.75 mL diglyme, 24 h.	39.31	[19]
Ir complexes (Homog.)	2,6-dimethyldecahydro-1,5-naphthyridine	0.25 mmol of reactant and the catalyst under reflux in p-xylene for 20 h, 138 °C, 5 mol% of Ir.	0.43	[57]
1 wt% Pd/γ-Al_2O_3	decahydroquinoline	10.83 mmol of reactant, 230 °C, 4 h, M/R = 0.1 mol%.	62	[58]
1 wt% Pd/C	2-[(n-methylcyclohexyl)-methyl]piperidine (MBP)	7.732 mmol of H12-MBP, M/R = 0.1 mol%, 270 °C, 4 h.	237	[59]
Pd/3.3CCA		7.732 mmol of H12-MBP, M/R = 0.1 mol%, 250 °C, 4 h.	168	[60]
3PdA		7.732 mmol of H12-MBP, 270 °C, 4 h, M/R = 0.59 mol%.	27.5	[61]
MPdA600_5h		7.3 mmol of reactant, M/R ratio of 0.1 mol%, and 250 °C for 4 h.	16.69	[62]
1 wt% Pd/Mg-Al-Ox	14HP	1 mmol 14HP in tetraglyme (3 mL), 240 °C, 45 min, 50 mg of catalyst.	252	This study

In the literature, there are very few examples of the catalytic dehydrogenation of perhydro-N-heterocycles consisting of six-membered rings. Meanwhile, this type of N-containing heterocycle is characterized by a high content of hydrogen that could be released. In addition, nitrogen-containing six-membered rings have higher stability with respect to ring opening due to hydrodenitrogenation compared to five-membered rings [63,64].

3. Materials and Methods

3.1. Catalyst Preparation

Aerogel-prepared magnesium oxide was synthesized according to a procedure adapted from the method pioneered by Klabunde et al. [65]. First, magnesium metal ribbon (1.2 g) was subjected to a reaction with methanol (75 mL). Then, 150 mL toluene was added to the reaction medium. The formed magnesium methoxide was hydrolyzed with a stoichiometric amount of distilled water (1.8 mL). The solvents (methanol and toluene) were purchased from J.T. Baker (Phillipsburg, NJ, USA) and Baza1R JSC (Staraya Kupavna, Russia), respectively. All the materials were used as received.

For the synthesis of aluminum oxide according to the earlier developed method [66], 6 g of aluminum isopropoxide (Alfa Aesar, Ward Hill, MA, USA) was dissolved in a mixture of 75 mL methanol and 150 mL toluene. Next, hydrolysis was carried out with 1.6 mL of distilled water.

For the synthesis of mixed oxide aerogels, first, magnesium metal ribbon was subjected to a reaction with methanol excess (75 mL). Then, aluminum isopropoxide and 150 mL toluene were added to the reaction mixture. The masses of magnesium metal and aluminum isopropoxide were chosen so that their molar ratios were 1:4; 1:2; 1:1; 2:1; and 4:1. After the mixture was stirred for 1 h, it was subjected to hydrolysis with a stoichiometric ratio of distilled water.

After the addition of water for hydrolysis, each resulting gel was stirred for 16 h. Then, it was placed in a 500 mL autoclave (Amar, Mumbai, India), in which it was heated for 3 h to 270 °C. The final pressure in the autoclave was 80–95 atm. Then, the gas from the autoclave was slowly released at 270 °C, and the autoclave was purged with argon for 5 min. Each synthesized aerogel sample was additionally calcined in air at 500 °C for 3 h at a heating rate of 100 °C/h. The synthesized samples of the mixed oxides were designated as MXAY, where *X* and *Y* are the amounts of Mg and Al corresponding to their molar ratio.

For the synthesis of 1 wt% Pd/Mg-Al-O_x catalysts, the adsorption precipitation technique was used with $H_2[PdCl_4]$ as the precursor of the active component. Applying this method, 1 g of the support was dispersed in 20 mL of an EDTA solution (56 mg) in ethanol using a magnetic stirrer (600 rpm). Then, liquid $H_2[PdCl_4]$ precursor at the required concentration was added dropwise and stirred for one hour (600 rpm). The resulting samples were filtered, washed with ethanol, dried at 120 °C, and calcined at 500 °C. Then, the obtained catalysts were reduced in the hydrogen flow at 500 °C for 1 h with a H_2 flow rate of 100 mL/min.

In addition, 1 wt% Pd/γ-Al_2O_3 catalyst prepared using the commercial precursor Pural SB-1 (Sasol, Hamburg, Germany) according to a previously reported procedure [20] was used for comparison.

3.2. Catalyst Characterization

X-ray diffraction patterns of the samples were recorded in the range 2θ 15–75° with a step of 0.05° and an accumulation time of 3 s using a Bruker D8 diffractometer with CuKα radiation (λ = 1.5418 A) (Bruker AXS GmbH, Karlsruhe, Germany). The dispersion of metal particles and the microstructure of the catalysts were studied by transmission electron microscopy. The images were acquired with a JEM-2010 (JEOL Ltd., Tokyo, Japan) operating at an accelerating voltage of 200 kV and a resolution of 0.14 nm. The device was equipped with an energy dispersive X-ray (EDX) spectrometer XFlash (Bruker AXS GmbH, Karlsruhe, Germany) with an energy resolution of 130 eV.

The Pd concentration in the fresh and exhaust catalysts was determined using an ARL Advant'X 2247 X-ray fluorescence spectrometer (Thermo Fisher Scientific Inc., Waltham, MA, USA).

The porous structure was studied by nitrogen adsorption–desorption at 77 K. The isotherms were measured by means of an automatic adsorption analyzer ASAP-2400 (Micromeritics Corp., Norcross, GA, USA). Before the measurements, all samples were degassed in a vacuum of less than 1 Pa at 150 °C for 16 h. All calculations were performed using ASAP 2020 Plus ver. 2.00 software. The desorption branch of the isotherms was used for the calculation of pore size distributions.

The XPS measurements were carried out using a photoelectron spectrometer (SPECS Surface Nano Analysis GmbH, Germany) equipped with a PHOIBOS-150 hemispherical electron energy analyzer and an XR-50 X-ray source with a double Al/Mg anode. The core-level spectra were obtained using AlKα radiation (hν = 1486.6 eV) under ultrahigh vacuum conditions. The binding energy (Eb) of the photoemission peaks was corrected to the Al2p peak (Eb = 74.5 eV) of alumina [67,68]. The curve fitting was carried out using CasaXPS software [69].

Active sites on the surface of the obtained materials were studied by electron paramagnetic resonance using 1,3,5-trinitrobenzene (TNB) and phenothiazine as spin probes. Before the adsorption of the spin probes, the samples were activated by heating in air at 500 °C for 3 h. The probes were adsorbed from their 2×10^{-2} M solutions in toluene. EPR spectra were recorded at room temperature with an ERS-221 spectrometer (Center of Scientific Instruments Engineering, Leipzig, Germany).

CO chemisorption was measured in a pulsed mode using a Chemosorb analyzer (MBE, Novosibirsk, Russia) equipped with a thermal conductivity detector. The catalyst (50 mg) was loaded into a U-shaped quartz reactor and treated with an H_2 flow (100 mL/min) at 20 and 100 °C with a heating rate of 10 °C/min. The reactor was kept at the final temperature for 20 min, then purged with argon and cooled to room temperature. After cooling, CO pulses (0.1 cm^3) were injected into the reactor until the sample was saturated, and the amount of chemisorbed CO was estimated.

3.3. Tetradecahydrophenzine Dehydrogenation

Tetradecahydrophenazine was synthesized from a commercially available phenazine substrate (Sigma Aldrich, St. Louis, MO, USA) according to the previously described method [20]. The 1H-NMR and GC-MS methods were used to confirm the purity of the obtained heterocyclic substrate (higher than 99.7%).

To study tetradecahydrophenazine dehydrogenation, a reaction setup similar to that described previously [20] was used. In a standard test of the catalytic activity, the catalyst (50 mg) and a suspension of 1 mmol tetradecahydrophenazine in tetraglyme (3 mL) (Acros Organics, Geel, Belgium) were loaded into the reactor. The system was purged with argon (20 mL/min, 5 min, room temperature), and the reactor was sealed and connected to a weight analysis line. Next, the reaction mixture was heated to 240 °C with stirring (500 rpm) for 15 min (0–900 s), and the dehydrogenation reaction was carried out for 45 min (900–3600 s) before the reactor was removed from the heating zone. The error of the weighing system (i.e., comparison with data on the volume of released gas) was estimated as ±2 mL H_2. When tetraglyme (3 mL) was used as a solvent, the increase in the gas volume in the reactor due to thermal expansion (up to 240 °C) was estimated as 19 ± 1 mL. Reaction mixture probes were taken at the end of the reaction and at intermediate dehydrogenation times (10, 20, and 30 min). Then, the selected aliquot of the reaction mixture was separated from the catalyst particles by centrifugation and analyzed by the GC method.

The GC analysis was performed using an Agilent-7890A gas chromatograph (Santa Clara, CA, USA) equipped with a ZB-5HT column. A catalytic experiment using a recycled catalyst was performed using the same procedure after washing the catalyst from the first experiment with acetone and drying in the argon flow at 50 °C for 1 h.

The hydrogen yield characterizing the depth of dehydrogenation was calculated as follows:

$$Y_{H_2} = Y_P + \frac{6Y_{2HP}}{7} + \frac{5Y_{4HP}}{7} + \frac{3Y_{8HPc}}{7} + \frac{3Y_{8HPn}}{7}$$

where Y is the molar fractions of the product (Y_P) and intermediates (Y_{2HP}, Y_{4HP}, Y_{8HPn}, Y_{8HPc}) obtained by gas chromatography. The dehydrogenation selectivity was estimated from the total yield of methylphenazines (Y_{MP}, %), i.e., the sum of the yields of 5-methyl-5,10-dihydrophenazine and 5,10-dimethyl-5,10-dihydrophenazine.

4. Conclusions

Mixed magnesium-aluminum oxide supports with different Mg:Al ratios (Mg:Al = 4:1, 2:1, 1:1, 1:2, 1:4) as well as pristine MgO and Al_2O_3 were synthesized by the aerogel method with high-temperature autoclave drying. The structure and properties of the synthesized aerogel samples were studied by a number of physical methods (nitrogen adsorption, XRD, HRTEM, EPR, and CO chemisorption). All the synthesized aerogels were characterized by high specific surface areas and pore volumes. Surface areas as high as 600 m^2/g after calcination at 500 °C were observed for M1A4. Electron-donor and electron-acceptor sites were observed on the surface of all the studied aerogel supports. An increase in the Mg concentration in the aerogels led to an increase in the concentration of basic electron-donor sites and a decrease in the concentration of acidic electron-acceptor sites. Catalysts containing 1 wt% Pd were prepared by palladium deposition from ethanol solution on the synthesized aerogel supports. The presence of surface active sites provided for Pd stabilization in a finely dispersed state with a nanoparticle size of about 5 nm and a narrow particle size distribution. It was demonstrated that the high dispersity of the supported metal and the high surface area of the supports had a favorable effect on the catalytic activity in the acceptorless dehydrogenation of tetradecahydrophenazine, leading to a high reaction rate. The promotion of the basic properties by the addition of magnesium to alumina was shown to change the tetradecahydrophenazine dehydrogenation mechanism, making it possible to start the process from the central ring of the heterocycle. The application of tetradecahydrophenazine as the hydrogen source required the use of a solvent such as tetraglyme. The use of catalysts with a low concentration of acidic electron-acceptor sites made it possible to suppress tetraglyme decomposition to form side products (methylphenazines). Overall, 1wt% Pd/M2A1, Pd/M1A1, Pd/M1A2 catalysts prepared with mixed oxide aerogel supports were found to be the most efficient catalysts for the acceptorless dehydrogenation of tetradecahydrophenazine.

Author Contributions: Conceptualization, A.P.K.; data curation, A.P.K. and A.F.B.; formal analysis, A.P.K., S.A.S., D.M.S. and E.V.I.; funding acquisition, A.F.B.; investigation, D.M.S., A.P.K., S.A.S., E.V.I. and A.B.A.; methodology, D.M.S., A.P.K., E.V.I. and A.B.A.; project administration, A.P.K. and A.F.B.; resources, V.A.Y. and A.F.B.; supervision, A.P.K. and A.F.B.; validation, A.P.K. and A.F.B.; visualization, D.M.S., S.A.S., E.V.I. and A.B.A.; writing—original draft, A.P.K. and D.M.S.; writing—review and editing, A.P.K. and A.F.B. All authors have read and agreed to the published version of the manuscript.

Funding: This research was funded by the Ministry of Science and Higher Education of the Russian Federation within the governmental order for the Boreskov Institute of Catalysis (projects AAAA-A21-121011390007-7 and AAAA-A21-121011390054-1).

Data Availability Statement: All data included in this study are available upon request by contacting the corresponding author.

Acknowledgments: The characterization of the samples was performed using the equipment of the Center of Collective Use "National Center of Catalysts Research". The authors are grateful to Dmitriy Y. Yermakov for CO chemisorption studies, Svetlana A. Cherepanova for XRD studies, Ekaterina I. Shuvarakova for assistance with EPR studies, Andrey A. Saraev for XPS studies and Evgeniy Y. Gerasimov for the HRTEM studies.

Conflicts of Interest: The authors declare no conflict of interest.

References

1. Yan, J. The impact of climate policy on fossil fuel consumption: Evidence from the Regional Greenhouse Gas Initiative (RGGI). *Energy Econ.* **2021**, *100*, 105333. [CrossRef]
2. Durbin, D.J.; Malardier-Jugroot, C. Review of hydrogen storage techniques for on board vehicle applications. *Int. J. Hydrogen Energy* **2013**, *38*, 14595–14617. [CrossRef]
3. Usman, M.R. Hydrogen storage methods: Review and current status. *Renew. Sustain. Energy Rev.* **2022**, *167*, 112743. [CrossRef]
4. Sharma, S.; Ghoshal, S.K. Hydrogen the future transportation fuel: From production to applications. *Renew. Sustain. Energy Rev.* **2015**, *43*, 1151–1158. [CrossRef]
5. Teichmann, D.; Arlt, W.; Schlücker, E.; Wasserscheid, P. Transport and Storage of Hydrogen via Liquid Organic Hydrogen Carrier (LOHC) Systems. In *Hydrogen Science and Engineering: Materials, Processes, Systems and Technology*; Wiley-VCH Verlag GmbH & Co. KGaA: Weinheim, Germany, 2016; pp. 811–830. [CrossRef]
6. Sordakis, K.; Tang, C.; Vogt, L.K.; Junge, H.; Dyson, P.J.; Beller, M.; Laurenczy, G. Homogeneous Catalysis for Sustainable Hydrogen Storage in Formic Acid and Alcohols. *Chem. Rev.* **2017**, *118*, 372–433. [CrossRef]
7. Dalebrook, A.F.; Gan, W.; Grasemann, M.; Moret, S.; Laurenczy, G. Hydrogen storage: Beyond conventional methods. *Chem. Commun.* **2013**, *49*, 8735–8751. [CrossRef]
8. Egeland-Eriksen, T.; Hajizadeh, A.; Sartori, S. Hydrogen-based systems for integration of renewable energy in power systems: Achievements and perspectives. *Int. J. Hydrogen Energy* **2021**, *46*, 31963–31983. [CrossRef]
9. Abdin, Z.; Khalilpour, K.R. Single and Polystorage Technologies for Renewable-Based Hybrid Energy Systems. In *Polygeneration with Polystorage for Chemical and Energy Hubs*; Academic Press: London, UK, 2019; pp. 77–131. [CrossRef]
10. Andersson, J.; Grönkvist, S. Large-scale storage of hydrogen. *Int. J. Hydrogen Energy* **2019**, *44*, 11901–11919. [CrossRef]
11. Kumar, A.; Muthukumar, P.; Sharma, P.; Kumar, E.A. Absorption based solid state hydrogen storage system: A review. *Sustain. Energy Technol. Assess.* **2022**, *52*, 102204. [CrossRef]

12. Abdin, Z.; Tang, C.; Liu, Y.; Catchpole, K. Large-scale stationary hydrogen storage via liquid organic hydrogen carriers. *iScience* **2021**, *24*, 102966. [CrossRef]
13. Niermann, M.; Drünert, S.; Kaltschmitt, M.; Bonhoff, K. Liquid organic hydrogen carriers (LOHCs)—Techno-economic analysis of LOHCs in a defined process chain. *Energy Environ. Sci.* **2019**, *12*, 290–307. [CrossRef]
14. Byun, M.; Lee, A.; Cheon, S.; Kim, H.; Lim, H. Preliminary feasibility study for hydrogen storage using several promising liquid organic hydrogen carriers: Technical, economic, and environmental perspectives. *Energy Convers. Manag.* **2022**, *268*, 116001. [CrossRef]
15. Aakko-Saksa, P.T.; Cook, C.; Kiviaho, J.; Repo, T. Liquid organic hydrogen carriers for transportation and storing of renewable energy—Review and discussion. *J. Power Sources* **2018**, *396*, 803–823. [CrossRef]
16. Singh, R.; Singh, M.; Gautam, S. Hydrogen economy, energy, and liquid organic carriers for its mobility. *Mater. Today Proc.* **2021**, *46*, 5420–5427. [CrossRef]
17. Zhu, Q.-L.; Xu, Q. Liquid organic and inorganic chemical hydrides for high-capacity hydrogen storage. *Energy Environ. Sci.* **2015**, *8*, 478–512. [CrossRef]
18. Zubar, V.; Borghs, J.C.; Rueping, M. Hydrogenation or Dehydrogenation of N-Containing Heterocycles Catalyzed by a Single Manganese Complex. *Org. Lett.* **2020**, *22*, 3974–3978. [CrossRef]
19. Forberg, D.; Schwob, T.; Zaheer, M.; Friedrich, M.; Miyajima, N.; Kempe, R. Single-catalyst high-weight% hydrogen storage in an N-heterocycle synthesized from lignin hydrogenolysis products and ammonia. *Nat. Commun.* **2016**, *7*, 13201. [CrossRef]
20. Stepanenko, S.A.; Shivtsov, D.M.; Koskin, A.P.; Koskin, I.P.; Kukushkin, R.G.; Yeletsky, P.M.; Yakovlev, V.A. N-Heterocyclic Molecules as Potential Liquid Organic Hydrogen Carriers: Reaction Routes and Dehydrogenation Efficacy. *Catalysts* **2022**, *12*, 1260. [CrossRef]
21. He, T.; Pei, Q.; Chen, P. Liquid organic hydrogen carriers. *J. Energy Chem.* **2015**, *24*, 587–594. [CrossRef]
22. Wu, Y.; Liu, X.; Bai, X.; Wu, W. Ultrasonic-assisted preparation of ultrafine Pd nanocatalysts loaded on Cl−-intercalated MgAl layered double hydroxides for the catalytic dehydrogenation of dodecahydro-N-ethylcarbazole. *Ultrason. Sonochemistry* **2022**, *88*, 106097. [CrossRef]
23. Yang, M.; Dong, Y.; Fei, S.; Ke, H.; Cheng, H. A comparative study of catalytic dehydrogenation of perhydro-N-ethylcarbazole over noble metal catalysts. *Int. J. Hydrogen Energy* **2014**, *39*, 18976–18983. [CrossRef]
24. Koper, O.B.; Lagadic, I.; Volodin, A.; Klabunde, K.J. Alkaline-Earth Oxide Nanoparticles Obtained by Aerogel Methods. Characterization and Rational for Unexpectedly High Surface Chemical Reactivities. *Chem. Mater.* **1997**, *9*, 2468–2480. [CrossRef]
25. Mishakov, I.; Heroux, D.; Chesnokov, V.; Koscheev, S.; Melgunov, M.; Bedilo, A.; Buyanov, R.; Klabunde, K. Reaction of nanocrystalline MgO with 1-iodobutane. *J. Catal.* **2005**, *229*, 344–351. [CrossRef]
26. Ilyina, E.V.; Gerus, Y.Y.; Cherepanova, S.V.; Bedilo, A.F. Synthesis of C12A7 calcium aluminate aerogels. *Mater. Lett.* **2021**, *293*, 129699. [CrossRef]
27. Ilyina, E.V.; Bedilo, A.F.; Veselov, G.B.; Gerus, Y.Y.; Shuvarakova, E.I.; Stoyanovskii, V.O.; Vedyagin, A.A. Comparative Study of Pd-Mayenite Catalysts Prepared via Aerogel Approaches. *Gels* **2022**, *8*, 809. [CrossRef]
28. Vedyagin, A.A.; Mishakov, I.V.; Ilyina, E.V. A step forward in the preparation of V–Mg–O catalysts for oxidative dehydrogenation of propane. *J. Sol-Gel Sci. Technol.* **2020**, *97*, 117–125. [CrossRef]
29. Ilyina, E.V.; Mishakov, I.V.; Vedyagin, A.A.; Cherepanova, S.V.; Nadeev, A.N.; Bedilo, A.F.; Klabunde, K.J. Synthesis and characterization of mesoporous VOx/MgO aerogels with high surface area. *Microporous Mesoporous Mater.* **2012**, *160*, 32–40. [CrossRef]
30. Anderson, A.M.; Bruno, B.A.; Santos, J.; Barry, P.J.; Carroll, M.K. PGM nanoparticle-based alumina aerogels for three-way catalyst applications. *Catal. Commun.* **2022**, *172*, 106547. [CrossRef]
31. Li, J.; Tong, F.; Li, Y.; Liu, X.; Guo, Y.; Wang, Y. Dehydrogenation of dodecahydro-N-ethylcarbazole over spinel supporting catalyst in a continuous flow fixed bed reactor. *Fuel* **2022**, *321*, 124034. [CrossRef]
32. Song, J.; Wang, S.; Xu, Y.; Liu, Q.; Zhao, Y. LDH derived $MgAl_2O_4$ spinel supported Pd catalyst for the low-temperature methane combustion: Roles of interaction between spinel and PdO. *Appl. Catal. A Gen.* **2021**, *621*, 118211. [CrossRef]
33. Liu, Y.; Wang, S.; Gao, D.; Sun, T.; Zhang, C.; Wang, S. Influence of metal oxides on the performance of Pd/Al_2O_3 catalysts for methane combustion under lean-fuel conditions. *Fuel Process. Technol.* **2013**, *111*, 55–61. [CrossRef]
34. Tuo, Y.; Meng, Y.; Chen, C.; Lin, D.; Feng, X.; Pan, Y.; Li, P.; Chen, D.; Liu, Z.; Zhou, Y.; et al. Partial positively charged Pt in $Pt/MgAl_2O_4$ for enhanced dehydrogenation activity. *Appl. Catal. B Environ.* **2021**, *288*, 119996. [CrossRef]
35. Schwarz, J.A.; Contescu, C.; Contescu, A. Methods for Preparation of Catalytic Materials. *Chem. Rev.* **1995**, *95*, 477–510. [CrossRef]
36. Bedilo, A.F.; Klabunde, K.J. Synthesis of Catalytically Active Sulfated Zirconia Aerogels. *J. Catal.* **1998**, *176*, 448–458. [CrossRef]
37. Rechberger, F.; Niederberger, M. Synthesis of aerogels: From molecular routes to 3-dimensional nanoparticle assembly. *Nanoscale Horiz.* **2017**, *2*, 6–30. [CrossRef]
38. Yorov, K.E.; Baranchikov, A.E.; Kiskin, M.A.; Sidorov, A.A.; Ivanov, V.K. Functionalization of Aerogels with Coordination Compounds. *Russ. J. Coord. Chem.* **2022**, *48*, 89–117. [CrossRef]
39. Lin, Q.; Ji, Y.; Jiang, Z.-D.; Xiao, W.-D. Effects of Precursors on Preparation of Pd/α-alumina Catalyst for Synthesis of Dimethyl Oxalate. *Ind. Eng. Chem. Res.* **2007**, *46*, 7950–7954. [CrossRef]
40. Gregg, S.J.; Sing, K.S.W. *Adsorption, Surface Area and Porosity*, 2nd ed.; Academic Press: London, UK, 1982; pp. 126–132. [CrossRef]

41. Mishakov, I.V.; Bedilo, A.F.; Richards, R.M.; Chesnokov, V.V.; Volodin, A.M.; Zaikovskii, V.I.; Buyanov, R.A.; Klabunde, K.J. Nanocrystalline MgO as a Dehydrogenation Catalyst. *J. Catal.* **2002**, *206*, 40–48. [CrossRef]
42. Kaichev, V.V.; Saraev, A.A.; Matveev, A.V.; Dubinin, Y.V.; Knop-Gericke, A.; Bukhtiyarov, V.I. In Situ NAP-XPS and Mass Spectrometry Study of the Oxidation of Propylene over Palladium. *J. Phys. Chem. C* **2018**, *122*, 4315–4323. [CrossRef]
43. Matveev, A.V.; Kaichev, V.V.; Saraev, A.A.; Gorodetskii, V.V.; Knop-Gericke, A.; Bukhtiyarov, V.I.; Nieuwenhuys, B.E. Oxidation of propylene over Pd(5 5 1): Temperature hysteresis induced by carbon deposition and oxygen adsorption. *Catal. Today* **2015**, *244*, 29–35. [CrossRef]
44. Medvedev, D.A.; Rybinskaya, A.A.; Kenzhin, R.M.; Volodin, A.M.; Bedilo, A.F. Characterization of electron donor sites on Al_2O_3 surface. *Phys. Chem. Chem. Phys.* **2012**, *14*, 2587–2598. [CrossRef] [PubMed]
45. Bedilo, A.F.; Shuvarakova, E.I.; Volodin, A.M.; Ilyina, E.V.; Mishakov, I.V.; Vedyagin, A.A.; Chesnokov, V.V.; Heroux, D.S.; Klabunde, K.J. Effect of Modification with Vanadium or Carbon on Destructive Sorption of Halocarbons over Nanocrystalline MgO: The Role of Active Sites in Initiation of the Solid-State Reaction. *J. Phys. Chem. C* **2014**, *118*, 13715–13725. [CrossRef]
46. Bedilo, A.F.; Shuvarakova, E.I.; Rybinskaya, A.A.; Medvedev, D.A. Characterization of Electron-Donor and Electron-Acceptor Sites on the Surface of Sulfated Alumina Using Spin Probes. *J. Phys. Chem. C* **2014**, *118*, 15779–15794. [CrossRef]
47. Koskin, A.P.; Andreev, R.V.; Primachenko, O.N.; Shuvarakova, E.I.; Bedilo, A.F. Perfluorosulfonic Acid Polymer Composites: Effect of the Support and Synthesis Method on the Acid and Catalytic Properties. *Mol. Catal.* **2020**, *492*, 111006. [CrossRef]
48. Shuvarakova, E.I.; Bedilo, A.F.; Kenzhin, R.M.; Ilyina, E.V.; Gerus, Y.Y. Synthesis and Investigation of Finely Dispersed Calcium Aluminates and Catalysts Based on Them. *Russ. J. Phys. Chem. B* **2022**, *16*, 411–420. [CrossRef]
49. Shuvarakova, E.I.; Ilyina, E.V.; Cherepanova, S.V.; Gerasimov, E.Y.; Bedilo, A.F.; Vedyagin, A.A. Synthesis of Vanadia-Mayenite Nanocomposites and Characterization of Their Structure, Morphology and Surface Sites. *J. Compos. Sci.* **2022**, *6*, 254. [CrossRef]
50. Vedyagin, A.A.; Volodin, A.M.; Kenzhin, R.M.; Stoyanovskii, V.O.; Rogov, V.A.; Kriventsov, V.V.; Mishakov, I.V. The role of chemisorbed water in formation and stabilization of active sites on Pd/Alumina oxidation catalysts. *Catal. Today* **2018**, *307*, 102–110. [CrossRef]
51. Ilyina, E.V.; Bedilo, A.F.; Cherepanova, S.V.; Gerus, Y.Y.; Shuvarakova, E.I.; Vedyagin, A.A. Aerogel synthesis of calcium aluminates with varied stoichiometry. *J. Sol-Gel Sci. Technol.* **2022**, *104*, 259–266. [CrossRef]
52. Ye, T.-N.; Li, J.; Kitano, M.; Hosono, H. Unique nanocages of $12CaO{\cdot}7Al_2O_3$ boost heterolytic hydrogen activation and selective hydrogenation of heteroarenes over ruthenium catalyst. *Green Chem.* **2017**, *19*, 749–756. [CrossRef]
53. Fang, M.; Machalaba, N.; Sánchez-Delgado, R.A. Hydrogenation of arenes and N-heteroaromatic compounds over ruthenium nanoparticles on poly(4-vinylpyridine): A versatile catalyst operating by a substrate-dependent dual site mechanism. *Dalton Trans.* **2011**, *40*, 10621–10632. [CrossRef]
54. Fang, M.; Sánchez-Delgado, R.A. Ruthenium nanoparticles supported on magnesium oxide: A versatile and recyclable dual-site catalyst for hydrogenation of mono- and poly-cyclic arenes, N-heteroaromatics, and S-heteroaromatics. *J. Catal.* **2014**, *311*, 357–368. [CrossRef]
55. Wei, Z.; Shao, F.; Wang, J. Recent advances in heterogeneous catalytic hydrogenation and dehydrogenation of N-heterocycles. *Chin. J. Catal.* **2019**, *40*, 980–1002. [CrossRef]
56. Rahi, R.; Fang, M.; Ahmed, A.; Sánchez-Delgado, R.A. Hydrogenation of quinolines, alkenes, and biodiesel by palladium nanoparticles supported on magnesium oxide. *Dalton Trans.* **2012**, *41*, 14490. [CrossRef]
57. Fujita, K.-I.; Tanaka, Y.; Kobayashi, M.; Yamaguchi, R. Homogeneous Perdehydrogenation and Perhydrogenation of Fused Bicyclic N-Heterocycles Catalyzed by Iridium Complexes Bearing a Functional Bipyridonate Ligand. *J. Am. Chem. Soc.* **2014**, *136*, 4829–4832. [CrossRef]
58. Oh, J.; Bathula, H.B.; Park, J.H.; Suh, Y.-W. A sustainable mesoporous palladium-alumina catalyst for efficient hydrogen release from N-heterocyclic liquid organic hydrogen carriers. *Commun. Chem.* **2019**, *2*, 68. [CrossRef]
59. Oh, J.; Jeong, K.; Kim, T.W.; Kwon, H.; Han, J.W.; Park, J.H.; Suh, Y.-W. 2-(N-Methylbenzyl)pyridine: A Potential Liquid Organic Hydrogen Carrier with Fast H_2 Release and Stable Activity in Consecutive Cycles. *ChemSusChem* **2018**, *11*, 661–665. [CrossRef]
60. Oh, J.; Kim, T.W.; Jeong, K.; Park, J.H.; Suh, Y.-W. Enhanced Activity and Stability of a Carbon-Coated Alumina-Supported Pd Catalyst in the Dehydrogenation of a Liquid Organic Hydrogen Carrier, Perhydro 2-(n-methylbenzyl)Pyridine. *ChemCatChem* **2018**, *10*, 3892–3900. [CrossRef]
61. Kim, Y.; Song, Y.; Choi, Y.; Jeong, K.; Park, J.H.; Ko, K.C.; Na, K. Catalytic Consequences of Supported Pd Catalysts on Dehydrogenative H2 Evolution from 2-[(n-Methylcyclohexyl)methyl]piperidine as the Liquid Organic Hydrogen Carrier. *ACS Sustain. Chem. Eng.* **2020**, *9*, 809–821. [CrossRef]
62. Bathula, H.B.; Oh, J.; Jo, Y.; Suh, Y.-W. Dehydrogenation of 2-[(n-Methylcyclohexyl)Methyl]Piperidine over Mesoporous Pd-Al_2O_3 Catalysts Prepared by Solvent Deficient Precipitation: Influence of Calcination Conditions. *Catalysts* **2019**, *9*, 719. [CrossRef]
63. Guo, Y.; Liu, X.; Duan, P.; Xu, D.; Luque, R. Catalytic Hydrodenitrogenation of Pyridine under Hydrothermal Conditions: A Comprehensive Study. *ACS Sustain. Chem. Eng.* **2020**, *9*, 362–374. [CrossRef]
64. Guo, Y.; He, H.; Liu, X.; Chen, Z.; Rioux, R.M.; Janik, M.J.; Savage, P.E. Ring-opening and hydrodenitrogenation of indole under hydrothermal conditions over Ni, Pt, Ru, and Ni-Ru bimetallic catalysts. *Chem. Eng. J.* **2021**, *406*, 126853. [CrossRef]
65. Klabunde, K.J.; Stark, J.; Koper, O.; Mohs, C.; Park, D.G.; Decker, S.; Jiang, Y.; Lagadic, I.; Zhang, D.J. Nanocrystals as Stoichiometric Reagents with Unique Surface Chemistry. *J. Phys. Chem.* **1996**, *100*, 12142–12153. [CrossRef]

66. Bedilo, A.F.; Ilyina, E.V.; Mishakov, I.V.; Vedyagin, A.A. Nanocrystalline Aerogels of Metal Oxides as Destructive Sorbents and Catalysts. *Chem. Sustain. Dev.* **2011**, *19*, 25–32.
67. Scofield, J.H. Hartree-Slater subshell photoionization cross-sections at 1254 and 1487 eV. *J. Electron Spectrosc. Relat. Phenom.* **1976**, *8*, 129–137. [CrossRef]
68. Shirley, D.A. High-Resolution X-Ray Photoemission Spectrum of the Valence Bands of Gold. *Phys. Rev. B* **1972**, *5*, 4709–4714. [CrossRef]
69. Fairley, N. CasaXPS VAMAS Processing Software. 2010. Available online: http://www.casaxps.com (accessed on 20 January 2022).

catalysts

MDPI

Article

Influence of Synthesis Methodology on the Properties and Catalytic Performance of Tin, Niobium, and Tin-Niobium Oxides in Fructose Conversion

Thatiane Veríssimo dos Santos Martins, Dhara Beatriz de Amorim Pryston, Simoni Margareti Plentz Meneghetti * and Mario Roberto Meneghetti *

Group of Catalysis and Chemical Reactivity, Institute of Chemistry and Biotechnology, Federal University of Alagoas, Av. Lourival Melo Mota, Tabuleiro do Martins, Maceió 57072-970, Brazil
* Correspondence: simoni.plentz@gmail.com (S.M.P.M.); mrm@qui.ufal.br (M.R.M.)

Abstract: Pure and mixed oxides were synthesized using three methods, namely, coprecipitation, hydrothermal treatment using CTAB and Pechini treatment using glycerol, and investigated for the transformation of fructose, aiming to determine the influence of textural, structural, and acid-base properties on conversion and selectivity. All systems led to fructose conversion in an aqueous medium, and the factors that influenced the transformation were the textural and structural properties, as well as the number of acid sites present in the catalysts. The best conversions were observed using mixed oxides, highlighting SnNb (CTAB) and SnNb (GLY), showing the importance of the modulation of properties using the synthesis method. All systems were selective mainly for 5-HMF (5-hydroxymethylfurfural) and, to a lesser extent, for the products of the retro-aldolic route, and this selectivity was preserved, regardless of the catalytic system used.

Keywords: biomass; biorefinery; fructose; mixed oxide; tin; niobium

1. Introduction

The dependence of society on fossil sources to produce chemicals and energy has generated several environmental and economic concerns, and research has been intensified to establish environmentally friendly alternatives [1,2]. Renewable biomass, in particular carbohydrates, is an excellent candidate for use as a raw material that facilitates the transition from the fossil industry to biorefineries, since it is abundant, presents excellent cost-effectiveness and constitutes a promising starting material for obtaining a wide range of products of great industrial interest, many of which are considered commodities [3,4].

Among carbohydrates, fructose is considered an important chemical platform to produce chemical inputs such as glucose, 5-hydroxymethylfurfural (5-HMF), glyceraldehyde, pyruvaldehyde, lactic acid, dihydroxyacetone, formic acid, acetic acid and levulinic acid [5–8]. Thus, the search for active catalytic systems in the conversion of fructose has been growing, and it is worth mentioning that the nature of the catalysts directs the routes in the transformation of sugars, thereby improving the conversion results of these pathways [9,10].

The reactions conducted in the presence of homogeneous catalytic systems are more selective and efficient than those conducted in heterogeneous systems [11]. However, the disadvantages related to equipment corrosion, difficulty in separating the final product and limitations during recycling and reuse of the catalyst [11,12] drive the quest for efficient heterogeneous catalysts that are easy to separate and reuse.

Among them, metal oxides have attracted great attention as attractive solid catalysts for biomass conversion, since they have unique properties such as thermal and chemical stability, typically having a dual character due to the presence of Lewis and Brønsted acid sites [11–13]. Despite some promising results with the use of monometallic oxides, the strategy of synthesizing mixed oxides aims to modulate the textural and chemical properties of the system by strengthening the thermal and chemical stability. One example is based on reports of the use of tin oxide and niobium in biorefinery processes [14–19].

The weak Lewis-acidity of the tin species was considered a determining factor in the isomerization reaction [20]. The use of sulfated tin oxide prepared by hydrothermal synthesis applied in the esterification of levulinic acid was also reported, and the results showed that the chemical structure and catalytic performance of these sulfated materials strongly depend on the treatment of nanoparticles before the sulfate procedure [21]. Studies of mesoporous tin oxides applied in the acetalization of glycerol under solvent-free conditions demonstrated that the activity is intrinsically related to the structure and acidity of these systems [22].

Niobium oxide has a wide range of applications in various industrial sectors [19], and studies have shown its catalytic potential, with high selectivity to 5-HMF (5-hydroxymethylfurfural) due to its high Bronsted-acidity [19]. Other work reported that its acidic properties are essential for the conversion of glucose into levulinic acid [23].

Considering all the properties of the pure oxides mentioned above, it is assumed that the mixed oxides of these materials can represent interesting systems in the field of biomass transformation, as they constitute new materials with physicochemical characteristics and particular catalytic properties compared to pure oxides. It is important to note that these properties can be modulated through different synthesis methods with controlled morphologies, particle sizes, crystalline structures and amounts/natures of acidic sites. In the case of a synthetic approach, it is necessary to apply a simple method that can result in materials with adequate catalytic efficiency.

In a recent study by our research group, pure and mixed oxides based on tin and niobium were synthesized using the Pechini method. In this case, the possibility of using glycerol instead of ethylene glycol in the esterification step was demonstrated, thus generating materials with excellent properties, which were obtained via a greener synthesis-process [24].

In this work, aiming to expand the information available on the influence of the synthesis methodology on the properties of this type of material, coprecipitation, surfactant-modeled-hydrothermal (CTAB) and Pechini (using glycerol) methods were used. Complementarily, a study was carried out to determine the influence of their textural, structural, and acid-base properties on the transformation of fructose. According to the research carried out, this is the first time that oxides based on tin and niobium have been synthesized by different methods and applied in this reaction.

2. Results and Discussion

Catalytic systems based on tin and niobium oxides were obtained by (co)precipitation, modeled by the CTAB template and by the Pechini method, using glycerol as the polyol [24]. Such systems were characterized and investigated in the conversion of fructose, aiming to determine how their properties drive this reaction.

2.1. Characterization of the Catalysts

Initially, for the mixed oxides, the amount of niobium present in the materials was determined, and values of 11.6, 15.8 and 16.9% were attained for SnNb (PPT), SnNb (CTAB) and SnNb (GLY), respectively [24].

The X-ray diffraction patterns of the various materials are shown in Figure 1. For the pure tin oxides, despite the synthesis method used, the signals present in the diffractograms (Figure 1a) confirm the polycrystalline and single-phase nature of the systems. In this case, tin oxides can be classified as tetragonal rutile (JCPDS no. 41-1445), as confirmed by the crystallographic planes (110), (101), (200), (211), (220), (002) (310), (112), (301), (202) and (321). Comparatively, Sn(PPT) showed a narrowing of the diffraction peaks, suggesting a gradual increase in crystallinity due to crystallite agglomeration. The tin oxides synthesized by the other methods showed a considerable decrease in intensity in the reflection plane (110), which suggests that there is great disorder generated in the material [22].

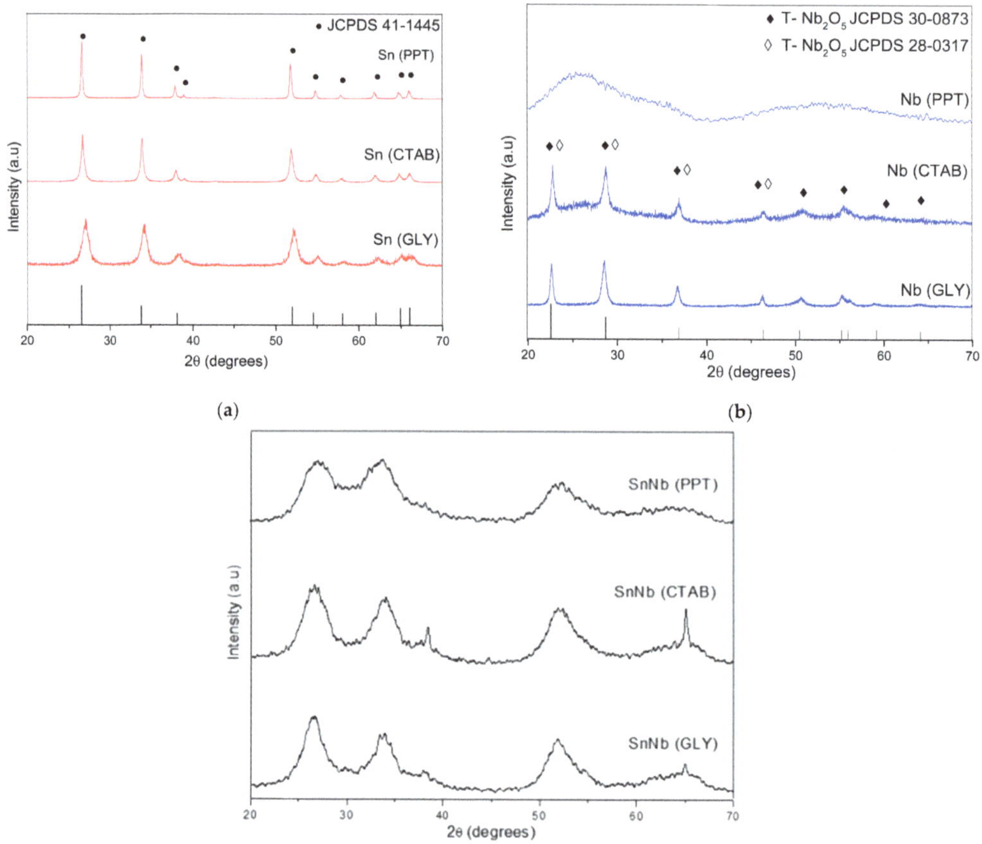

Figure 1. XRD patterns of the catalysts: (**a**) Sn(PPT), Sn(CTAB) and Sn(GLY); (**b**) Nb(PPT), Nb(CTAB) and Nb(GLY); and (**c**) SnNb(PPT), SnNb(CTAB) and SnNb(GLY).

XRD patterns for niobium oxide are shown in Figure 1b. Nb(PPT) showed an amorphous profile, but Nb(CTAB) and Nb(GLY) showed the crystallographic planes (001), (180), (181), (002), (321), (182), (215), (216) and (481), which indicate the predominance of the orthorhombic phase (T Nb_2O_5-JCPDS 30-0873) [25]. However, it is not possible to rule out the presence, to a lesser extent, of the pseudohexagonal phase with crystallographic planes in (001), (100), (101), (002), (110) and (102) (TT Nb_2O_5-JCPDS 280317). The most accentuated differences are noticed in the range between 25 and 40°, and it is important to note that the positions of the signals referring to the hexagonal phase and the orthorhombic phases are so close that it is difficult to differentiate between the two. This result is attributed to the presence of vacancies or impurities in the materials [26,27].

For all mixed oxides (Figure 1c), it is possible to observe the predominant presence of the tetragonal phase, with broader and less-intense signals compared to the precursors, suggesting that the difference in ionic radii due to the replacement of Sn^{4+} by Nb^{5+} caused the distortion of the cell lattice of SnO_2. Analyzing the diffractograms, it is possible to verify the fact that there was an enlargement of the reflection peaks, which may be related to the crystallite size and the short-range disorder.

The spectra in the mid-infrared region are shown in Figure 2. For SnO_2, the presence of a broad band in the range of 497 and 662 cm^{-1} was observed, which is attributed to the O-Sn-O and Sn-O stretching modes and SnO_2 flexion [28,29]. For niobium oxides, typical vibrations were observed in the range of 410 to 931 cm^{-1}, giving rise to a broad band attributed to the vibrations of the Nb-O-Nb bonds of the slightly distorted octahedral NbO_6 and the symmetrical stretching of the Nb-O species present in highly distorted environments [30]. The signal at 1625 cm^{-1} can be attributed to water molecules adsorbed on the surface of this material. There is also the presence of a band at 3412 cm^{-1}, probably due to the vibration of the Nb-OH bond [30].

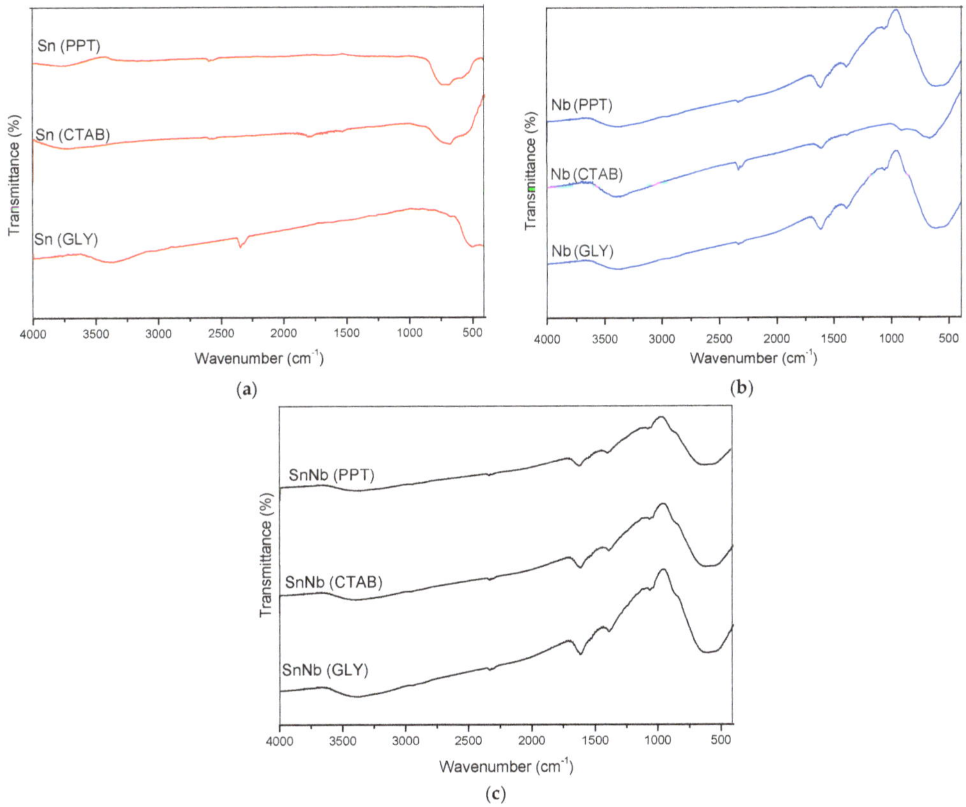

Figure 2. Infrared spectra of the catalysts: (**a**) Sn(PPT), Sn(CTAB) and Sn(GLY); (**b**) Nb(PPT), Nb(CTAB) and Nb(GLY); and (**c**) SnNb(PPT), SnNb(CTAB) and SnNb(GLY).

The mixed oxides presented vibrations similar to the pure ones. There is a broad band in the region between 410 and 931 cm^{-1} that is attributed to the symmetrical and asymmetrical stretching of metal-oxygen bonds, respectively. The signal at 3412 cm^{-1} and the bands at ~1100 cm^{-1} and 1350–1400 cm^{-1} can be attributed to bound metal-OH [31].

Some characteristics of the oxides were investigated using Raman spectroscopy (see Figure 3). It is important to remember that the SnO_2 synthesized by the various synthesis protocols has a rutile structure (more stable phase) and belongs to the space group D_{4h}^{14} (P42/mnm) [32]. Among these vibrations, the A_{1g}, B_{1g}, B_{2g} and E_g modes are active in Raman, while A_{2u}, E_u and B_{1u} are infrared active and optically inactive [33]. It has already been reported in the literature that the A_{1g}, B_{1g} and B_{2g} bands indicate the vibration modes of the Sn-O bonds, while the E_g band reveals the oxygen vacancies [32]. The SnO_2 Raman spectrum (Figure 3a) mainly exhibited two bands at approximately 630 and 770 cm^{-1}, which are attributed to vibration modes A_{1g} and B_{2g}, respectively. This fact is a strong indication that SnO_2 has a tetragonal structure. In addition to the two signals, there is another weak signal at approximately 694 cm^{-1}, which is attributed to the IR-active A_{2u} LO mode (the mode of longitudinal optical-phonons). Another mode was detected at 572 cm^{-1}, attributed to the smaller particle-size of SnO_2 [34].

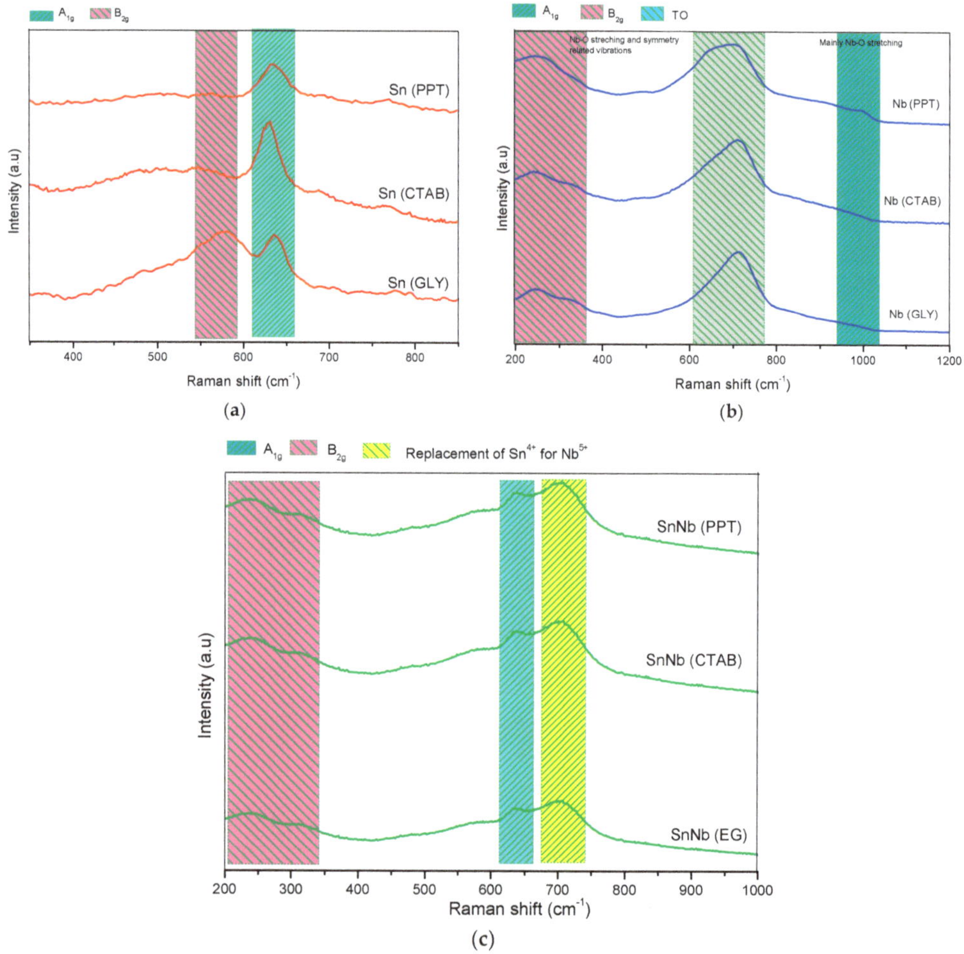

Figure 3. Raman spectra of the catalysts: (**a**) Sn(PPT), Sn(CTAB) and Sn(GLY); (**b**) Nb(PPT), Nb(CTAB) and Nb(GLY); and (**c**) SnNb(PPT), SnNb(CTAB) and SnNb(GLY).

Raman spectra of the niobium oxides are shown in Figure 3b, and the low-resolution signal located at 990 cm^{-1} is attributed to the vibration modes of $A_{1g}(\nu1)$ of the octahedral NbO_6 [35,36]. The signals that appear between 200 and 300 cm^{-1} are characteristic of the bending modes of the Nb–O–Nb bonds [37], and the signals present in the range of 400–800 cm^{-1} are attributed to symmetric and asymmetric stretching of the Nb-O-Nb bond [37]. A signal observed at approximately ~700 cm^{-1} is attributed to the orthorhombic Nb_2O_5 TO mode [35]. The Nb_2O_5 samples synthesized by the various methods belong to the space group ($P2_12_12_1$). In this crystalline structure, there are two Nb^{5+} sites, and the first Nb^{5+} is bonded to four O^{2-} atoms, to form corner-sharing NbO_4 tetrahedra. The second Nb^{5+} is bonded to six O^{2-} atoms, to form NbO_6 octahedrons [36].

Figure 3c shows the spectra of the mixed oxides. The band at approximately 632 cm^{-1} is attributed to the A_{1g} mode associated with symmetrical stretching of the Sn-O bond [37]. Considering that mixed oxides are formed by Sn, O and Nb atoms and that the crystalline structure of tin oxide belongs to the P42/mnm point group and tetragonal Bravais lattice, it can be inferred that the crystalline structure is distorted by the insertion of niobium atoms into the lattice, causing the appearance of oxygen vacancies during synthesis. This network-sharing promotes the appearance of defects, and this same phenomenon has already been reported in other studies, for metal oxides [38].

For all spectra of pure or mixed oxide (Figure 3) it is not possible to discard the presence of the band related to the E_g mode, which is a strong indicator of oxygen vacancies. The appearance of this band is expected at approximately 474 cm^{-1}, and just a low resolution signal is observed in some cases.

DRS UV–vis spectra were also produced (Figure 4), and provide important information about the nature of niobium and tin atoms in these systems. The UV–Vis DRS spectra for the SnO_2 samples (Figure 4a) showed an intense absorption band at approximately 300 nm, due to the high binding-energy, which is characteristic of cassiterite, and the strong interaction of the ligand-to-metal-charge-transfer (LMCT)-type [39]. For Nb_2O_5 samples (Figure 4b), bands at approximately 350 nm were observed, corresponding to orthorhombic and/or hexagonal geometries [40].

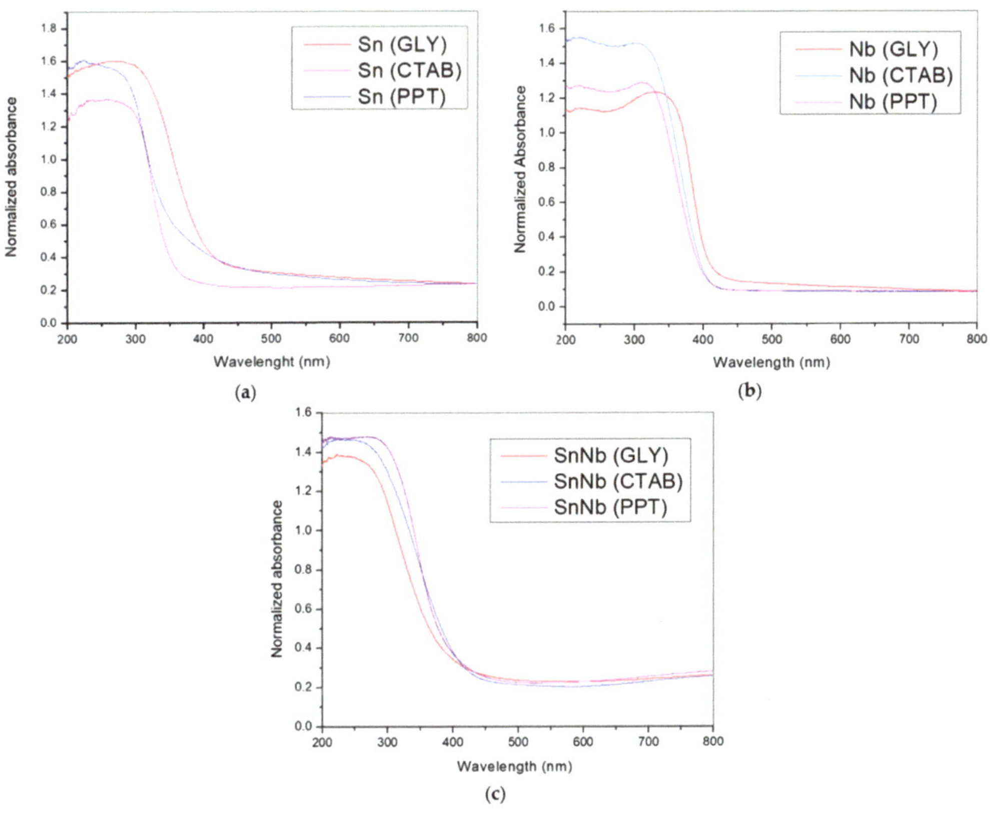

Figure 4. DRS-UV–vis spectra of the catalysts: (**a**) Sn(PPT), Sn(CTAB) and Sn(GLY); (**b**) Nb(PPT), Nb(CTAB) and Nb(GLY); and (**c**) SnNb(PPT), SnNb(CTAB) and SnNb(GLY).

The spectra of mixed oxides synthesized by different methods (Figure 4c) are different from those of pure materials, and such evidence is also corroborated by XRD and

Raman spectroscopy. Previous studies suggest that the tendency to change the absorption maximum to shorter wavelengths (200–250 nm) may be indicative of the superposition of tetrahedral Sn species and the strong interaction with octahedral niobium species (formation of Nb-O-Sn bonds), indicating charge-transfer transitions between atoms [41].

The textural properties and the nitrogen adsorption- and desorption-isotherms are shown in Figure 5. According to IUPAC, they are type IV with H1 hysteresis loops, suggesting that cylindrical channels are well defined. Sn(PPT) and Nb(PPT) are detected in areas of 30 and 115.5 m^2g^{-1}, respectively. However, SnNb(PPT) showed a decrease in surface area (78.51 m^2g^{-1}) when compared to Nb(PPT), which can be attributed to the agglomeration of metallic cations during the synthesis process.

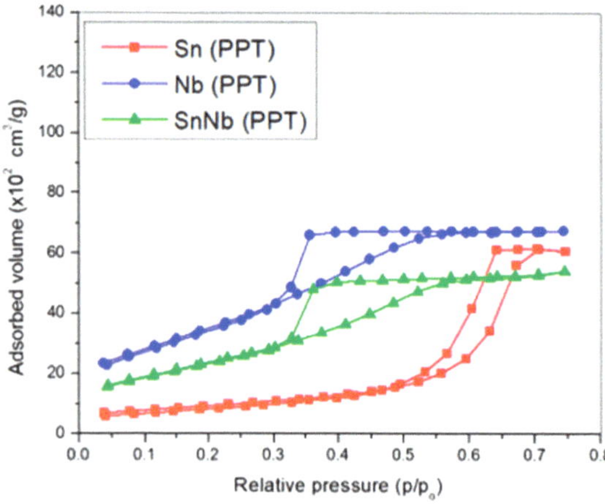

(**a**)

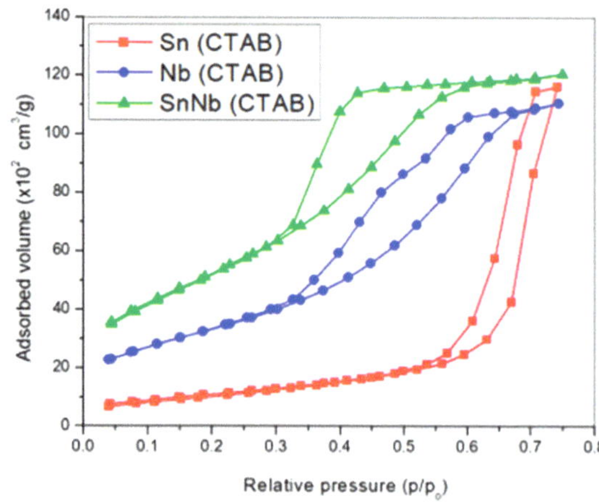

(**b**)

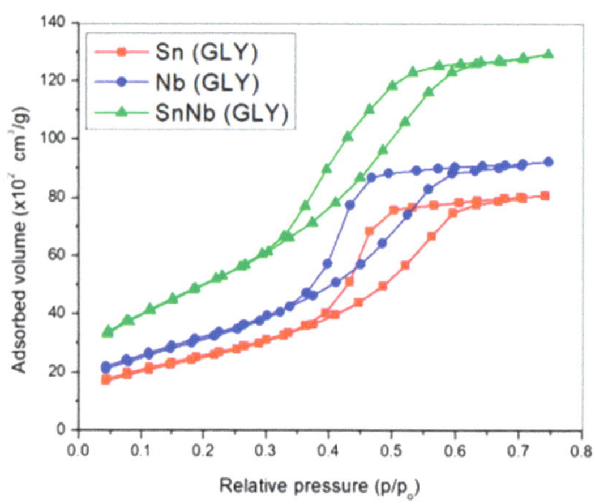

(**c**)

Sample	S_{BET} [a] ($m^2.g^{-1}$)	D_{BJH} [b] (Å)
Sn (PPT)	30.0	93.6
Nb (PPT)	115.5	36.0
SnNb (PPT)	78.51	36.4
Sn (CTAB)	46.0	121.8
Nb (CTAB)	130.0	45.2
SnNb (CTAB)	173.2	36.4
Sn (GLY)	69.6	51.0
Nb (GLY)	85.2	50.8
SnNb (GLY)	166.6	40.4

[a] S_{BET}, BET specific surface area.
[b] D_{BJH}, pore diameter.
The specific surface areas and pore diameters were calculated by the BET and BJH equations.

Figure 5. Nitrogen adsorption- and desorption-isotherms and textural properties of the catalysts: (**a**) Sn(PPT), Nb(PPT) and SnNb(PPT); (**b**) Sn(CTAB), Nb(CTAB) and Nb(GLY) and SnNb(CTAB); and (c) Sn(GLY), Nb(GLY) and SnNb(GLY).

For SnNb(CTAB), there was an increase in surface area, which suggests that the presence of the template influences this parameter, as there is a reduction in the aggregation of metallic cations, since the hydrogen atoms of the hydroxyl groups (-OH groups) related to the metallic species can interact with the main cationic groups $(C_{16}H_{33})(CH_3)_3N^+$ (hydrophilic groups) of CTAB micelles, disfavoring the crystallization of individual cations. This same increasing effect is observed in the case of SnNb(GLY), and there are indications that the higher surface-area values of mixed oxides can be attributed to the strong interaction between species, which inhibits individual crystallization, reducing the number of clusters and the particle growth [42,43].

The thermal analyses of the materials were evaluated, and the presence of organic matter was not verified through the TG curves (Figure S1, Supplementary Materials). All oxides suffer a mass loss (between 0.8 and 12%) in just one step in the temperature range of 100 to 150 °C, corresponding to the release of physiosorbed water, suggesting that the materials exhibit thermal stability in the evaluated range.

The catalytic activity of materials such as oxides can be driven by the presence of acidic sites. Therefore, it is important to know the nature, quantity and strength of the acid sites present in their structures. Thus, the acid properties of the solids were studied by monitoring the pyridine adsorption using infrared spectroscopy, to assess the presence of Lewis and Brønsted acid-sites (Figure S2, Supplementary Information). The bands at 1445, 1540 and 1489 cm^{-1} were attributed to Lewis acid sites, Brønsted acids, and the overlapping of these sites, respectively [11]. For all SnO_2 samples, weak vibrations were observed, corresponding to the presence of Lewis and/or Brønsted acid-sites.

For all Nb_2O_5 samples, signs similar to those found in the literature were observed, since this system consists mainly of NbO_6 distorted octahedra and NbO_4 tetrahedra, and the surface hydroxyl groups (-OH) function as Brønsted acid sites, while the tetrahedra function as Lewis acid sites [44,45]. From this perspective, structural variation in the oxides leads to changes in the nature of acidic sites and in their strength [46]. In addition, the use of Nb_2O_5 in aqueous reactions suggests a "water-tolerant solid-acid catalysts" system, and this type of system proves to be resistant to water, in contrast to the classic Lewis acids such as $AlCl_3$, which suffer severe poisoning by water, undergoing deactivation due to the strong coordination of water in the metallic center (Lewis acids) or hydrolytic decomposition of the system [47]. In this work, Nb_2O_5 showed intense absorption-bands for Brønsted acid sites and weak and/or moderate bands for Lewis acid sites.

Some studies indicate that the properties of mixed oxides strongly depend on the preparation method employed, and it was observed in the present study that with the presence of niobium in the lattice, there was also an increase in the Lewis–Brønsted ratio (I_A/I_B), mainly for SnNb(CTAB) and SnNb(GLY), resulting in catalytic systems with high synergy and an improved balance in acid strength (Figure 6). It should be noted that SnNb(PPT) did not have such a sharp increase in (A_L/A_B) as was observed in the others, clearly showing the influence of the synthesis method on its properties.

To evaluate the resistance and stability of the acid sites, pyridine was used, and infrared spectra were acquired at 100, 200 and 300 °C. When pyridine is chemically absorbed at Lewis or Brønsted acid-sites, according to the temperature required to break such interactions, it can be considered weak (below 200 °C), moderate (between 200 and 300 °C) or strong (above 300 °C). For SnO_2, weak vibrations related to Lewis and/or Brønsted acid-sites were observed. For Nb_2O_5, strong vibrations were observed for the Brønsted acid sites, and weak and/or moderate interactions were also detected for the Lewis acid sites. On the other hand, for the mixed oxides, a better-resolved profile of the spectra was observed, with more prominent signals related to the Lewis acids, and Brønsted-related bands also detected, in addition to the overlap of both species (Figure S2, Supplementary Information). This can be better observed by examining the A_L/A_B results, where the significant increase is notorious for all temperatures (Figure 6). The modulation of these characteristics in mixed oxides is essential to make this system more versatile for application in various biorefinery platforms [38].

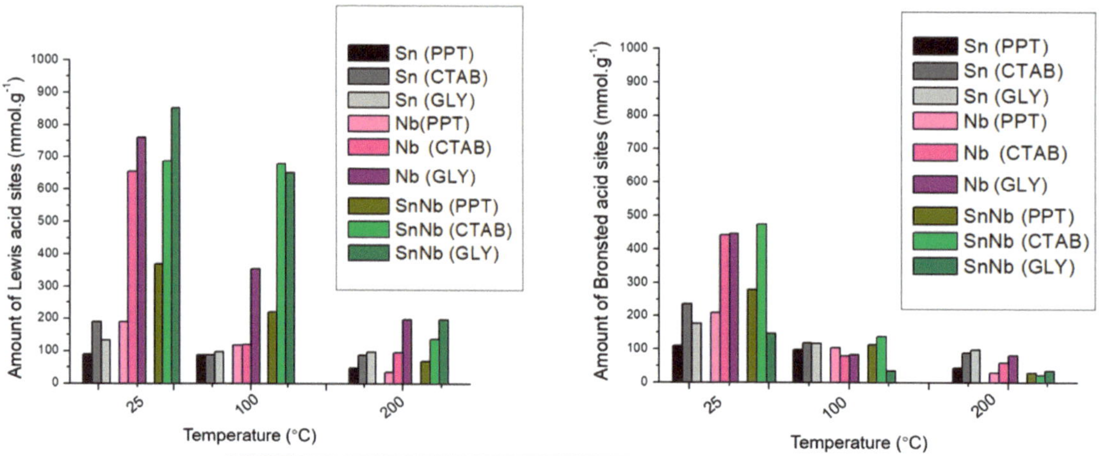

	A_L/A_B		
Catalysts↓	**25 °C**	**100 °C**	**200 °C**
Sn(PPT)	0.8	0.9	1.1
Sn(CTAB)	0.8	0.7	1.0
Sn(GLY)	0.7	0.8	1.1
Nb(PPT)	0.9	1.1	1.2
Nb(CTAB)	1.5	1.5	1.6
Nb(GLY)	0.9	0.9	1.3
SnNb(PPT)	1.3	2.0	2.5
SnNb(CTAB)	1.5	4.9	6.5
SnNb(GLY)	1.4	4.4	6.0

Figure 6. Amount of Lewis and Bronsted acid-sites and ratio of the number of Lewis and Bronsted acid-sites (A_L/A_B) of the catalysts.

In this context, it is evident that the synthesis of SnNb was fundamental for a balance between the Lewis and Brønsted acid-sites, which allows these systems to become more versatile for application in different biorefinery platforms. In previous studies, it was found that the number, nature and relative strength of acid sites in some materials can be adjusted using different synthetic strategies, including the use of templates and different calcination temperatures [22].

SEM micrographs of the materials are shown in Figure 7, and the images for SnNb(CTAB) and SnNb(GLY) showed great similarity, indicating the formation of materials with morphological disorder due to the greater interaction among the species during synthesis, resulting in the formation of smaller particles. Similar behavior was observed in the synthesis of porous, spherical materials from different oxides (Ce, Fe and Mn), since the functionality of the reagent species played an important role in the structure, and using glycerol instead of ethylene glycol resulted in smaller channels [48].

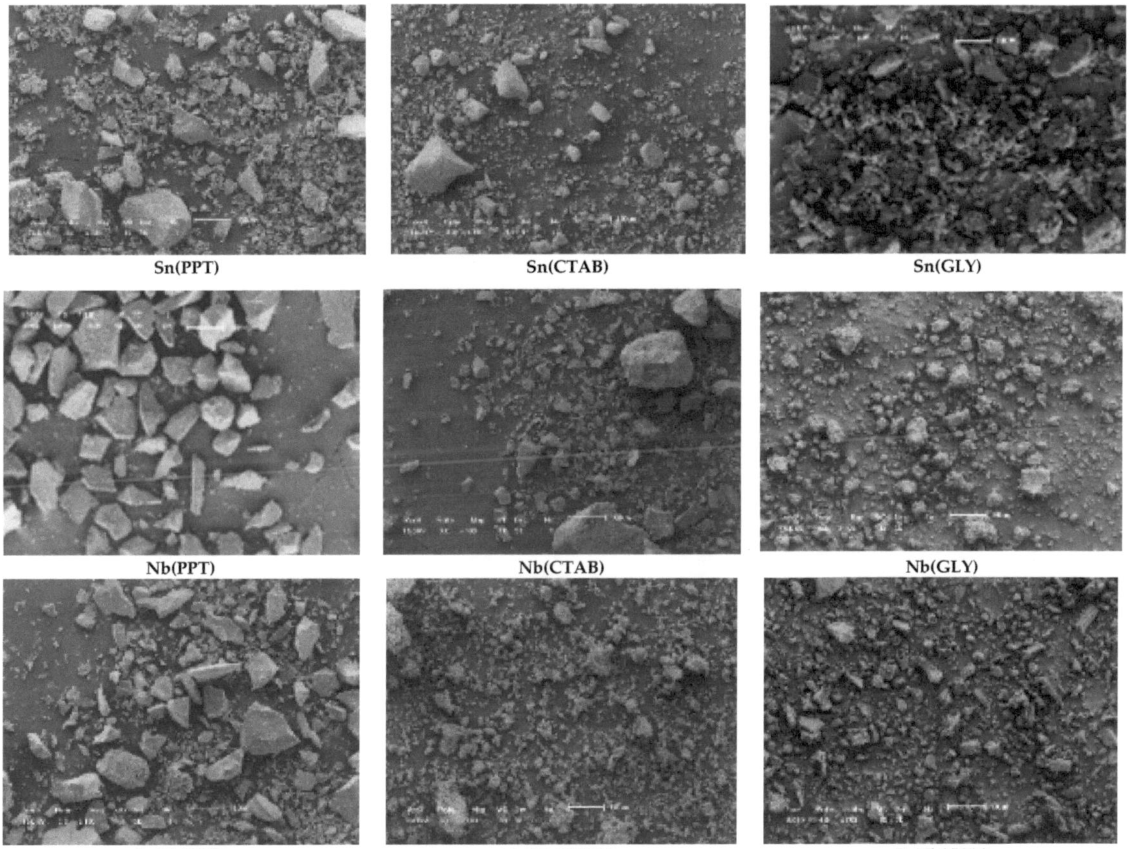

Figure 7. Scanning-electron-microscopy (SEM) images of the catalysts. The scale bar of all images is 100 μm (see Figures S3–S5, Supplementary Information).

2.2. Catalytic Assays in Fructose Transformation

The series of oxides based on tin and niobium obtained by various synthesis methods were investigated as catalysts in the conversion of fructose in an aqueous medium, at 150 °C (see Figure 8 and Figure S6, Supplementary Information). Using the mixed oxides, significantly higher values of fructose conversion were achieved than those observed for the pure oxides. For example, the conversion was 75.7, 76.9 and 69.8% in 2 h of reaction when using SnNb (G), SnNb (CTAB) and SnNb (PPT), respectively, and this tendency was maintained during the investigated reaction time.

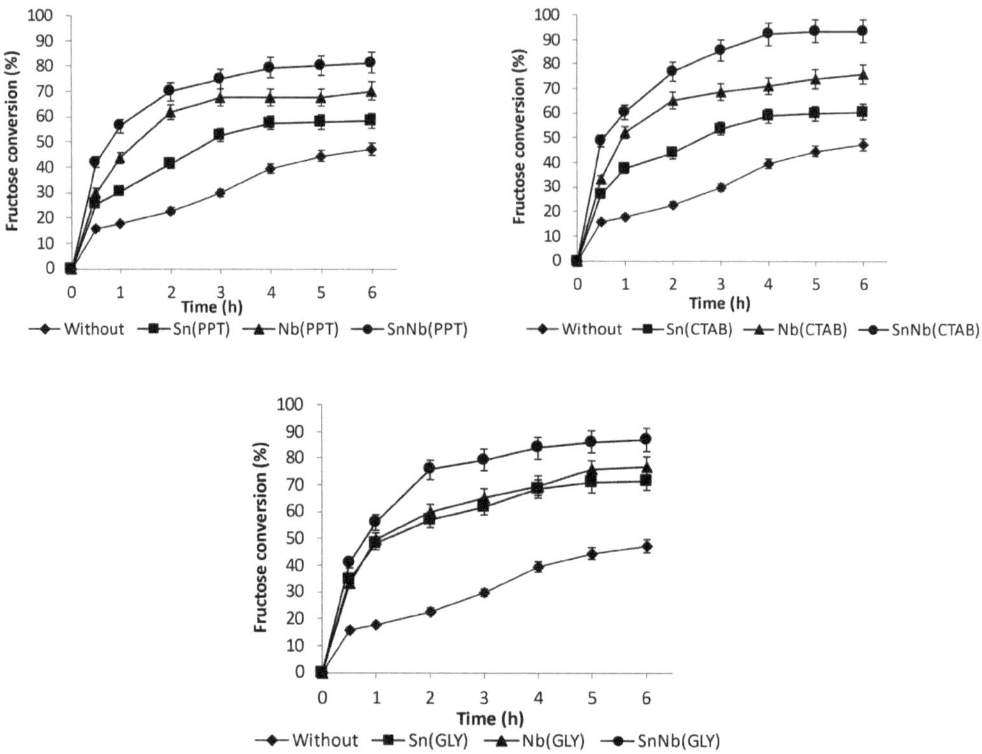

Figure 8. Fructose conversion at 150 °C using Sn(PPT), Nb(PPT) and SnNb(PPT); Sn(CTAB), Nb(CTAB) and Nb(GLY) and SnNb(CTAB); and Sn(GLY), Nb(GLY) and SnNb(GLY).

The co-precipitation method, often referred to as analogous to the sol-gel process, is an efficient chemical route for obtaining metal oxides, and has the advantages of obtaining high yields and purity, being easily reproducible and having a low cost [49,50]. The method that uses surfactant is known to provide a greater interaction between the reagents, minimizing possible diffusion-restrictions during the synthesis. In this case, the presence of the template directly influences the modulation of the specific surface area and pore size, as there is a reduction in the aggregation of metallic cations, due to their interaction with the cationic groups, for example with CTAB, disfavoring the individual crystallization of the cations [22,42,43]. The methodology developed by Pechini, in which the metal cations are chelated and, with the help of polyalcohol, are cross-linked to create a gel through the esterification, allows for a wide distribution of metallic cations throughout the polymer structure, promoting an increase in the surface area and improving the distribution of metals in the polymeric network, and being quite suitable for the manufacture of highly pure oxides [15,51].

Among all systems, the SnNb(CTAB) and SnNb(GLY) systems provided the highest fructose conversions, and in addition to containing the highest numbers of acidic sites, they have the highest surface areas, of 173.2 and 166.6 m^2g^{-1}, respectively. SnNb(PPT) has practically half of the surface area of these materials (78.51 m^2g^{-1}). The larger surface area promotes better contact between substrate and catalyst, and, as already mentioned, when using template (CTAB) or the Pechini method a greater interaction between species occurs during synthesis, providing greater homogeneity of the material. As evidence of this, using these two methods it was possible to achieve a higher niobium content in the materials (15.8% in SnNb(CTAB), 16.9% in SnNb(GLY), and only 11.6% in SnNb(PPT)). In addition, it is important to note that the results obtained using Raman spectroscopy and DRS analysis for the mixed oxides suggest the occurrence of a structural distortion, which can cause the appearance of oxygen vacancies during synthesis, resulting in defects which constitute active sites for fructose conversion. [17].

The soluble products formed during the transformation of fructose included glucose, 5-HMF (5-hydroxymethylfufural), levulinic acid, formic acid, glyceraldehyde, dihydroxyacetone, pyruvaldehyde, lactic acid and acetic acid. The selectivity profiles are shown in Figure 9 for mixed oxides (for pure oxides, in Figure S7, Supplementary Information), and all systems exhibit great selectivity for 5-HMF and, to a lesser extent, for products of the retro-aldolic pathway (lactic acid and its intermediates glyceraldehyde, dihydroxyacetone and pyruvaldehyde). In general, a trend toward a decrease in selectivity to 5-HMF (5-hydroxymethylfurfural) is observed with an increase in A_L/A_B (Figure 6), confirming that the Brønsted acid sites present in the material are mainly responsible for the dehydration of fructose. The Lewis acid sites play a fundamental role in the isomerization of glucose into fructose and in the retro-aldolic pathway.

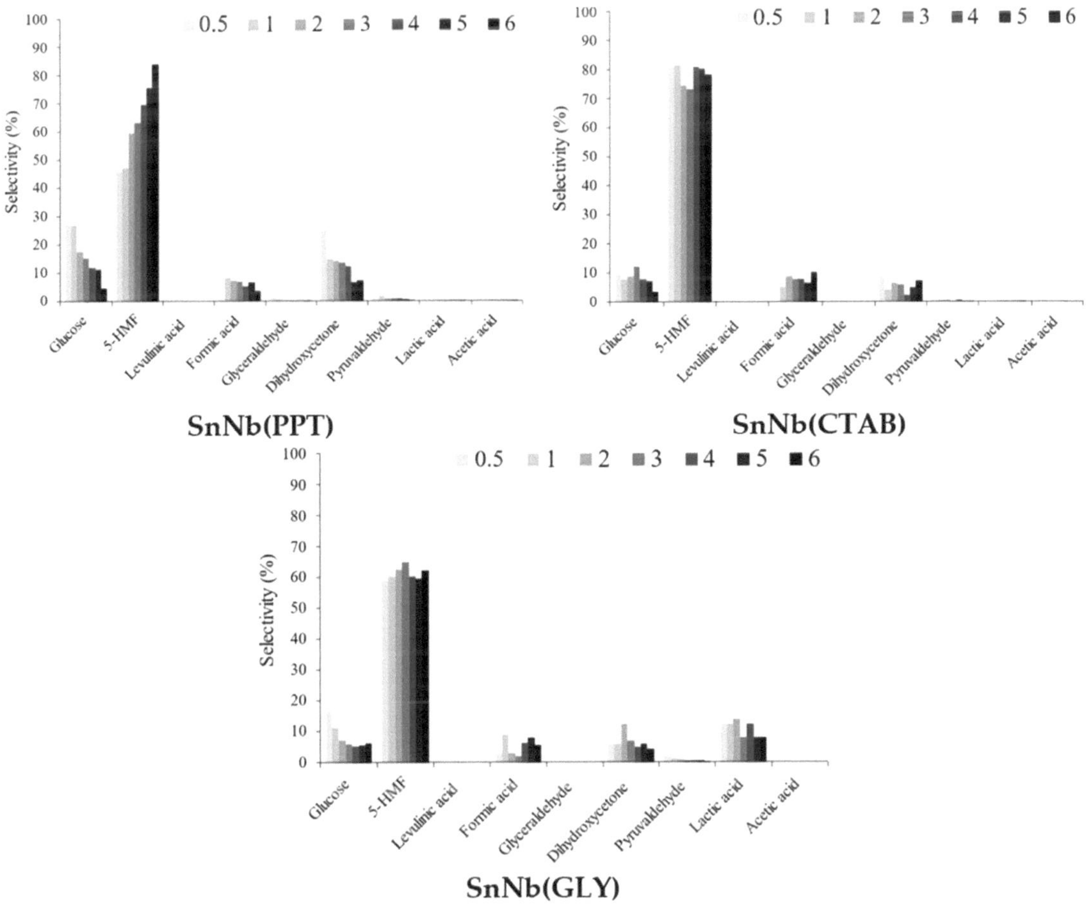

Figure 9. Selectivity for soluble products identified in the fructose conversion at 150 °C, using 1.5×10^{-3} g of SnNb(PPT), SnNb(CTAB) and SnNb(GLY).

During these reactions, some insoluble or partially soluble products may be formed, causing turbidity and brown coloration in the samples. Such materials, known as humines, are formed mainly by the polymerization of fructose or the decomposition of 5-HMF (5-hydroxymethylfufural), due to high temperatures, long reaction-times or the nature of the catalyst [52,53]. Regardless of the synthesis method, the qualitative analysis of the reactional samples (Figure 10) allows us to infer that using niobium oxides considerably increases the coloring and turbidity, while using tin oxides produces the smallest change in these properties. By using mixed oxides, the formation of low-solubility products was significantly reduced, in comparison to the use of pure niobium-oxides. Therefore, the structural and textural changes and the number of acid sites observed in mixed oxides resulted in better conversions, maintaining the advantage of reduced humine-formation.

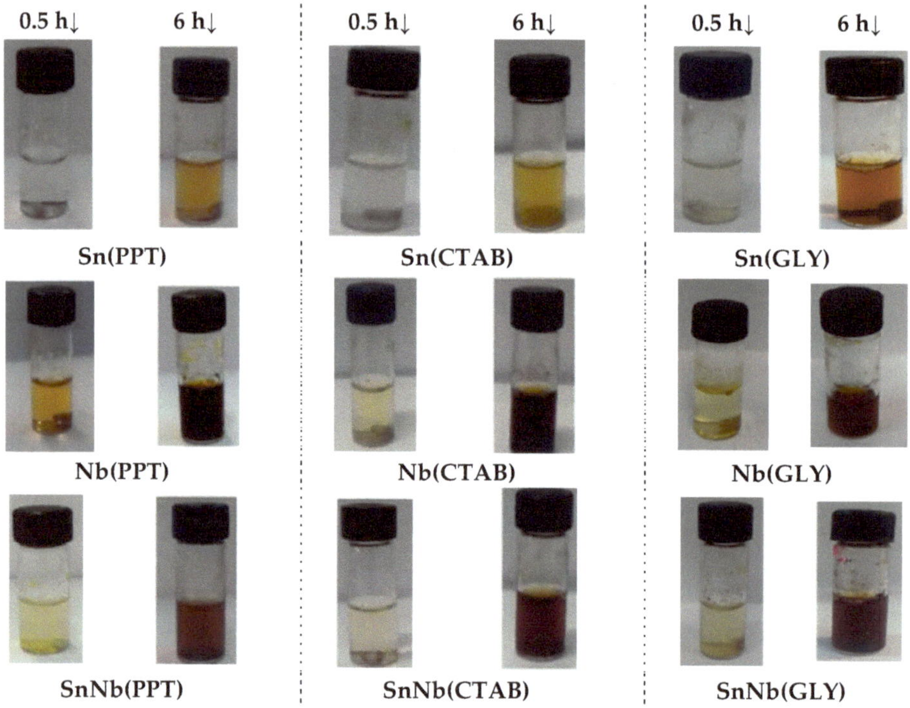

Figure 10. Visual aspects of the samples during the reaction using catalysts (1.5×10^{-3} g) at 150 °C (0.5 and 6 h).

The possibility of recovery and reuse of the catalyst is of great importance in the processes of biorefineries, and in previous studies the chemical stability of some of these mixed oxides was evidenced, as the catalyst can be recovered and reused for four cycles, and no leaching of species for the reaction medium was observed [24,54].

Finally, a comparison with studies involving bifunctional catalysts applied in fructose conversion in aqueous medium reported in the literature, allows us to indicate that the systems discussed here have great potential, as conversions of 75.7, 76.9 and 69.8% were observed at 2 h of reaction when using SnNb (G), SnNb (CTAB) and SnNb (PPT), respectively. In addition, it is important to highlight the fact that all systems exhibit high selectivity for 5-HMF and, to a lesser extent, for products from the retro-aldol pathway. To exemplify, studies using mixed oxides of cerium and niobium (25, 50 and 75% of Nb_2O_5) obtained using the coprecipitation method, led to fructose conversions lower than 75%, and the maximum selectivity for 5-HMF was around 45%. The reuse of the 25% CeO_2/75% Nb_2O_5 catalyst showed a reduction in the catalytic activity after the first reaction [55]. In the same way, the use of mixed tin-oxides containing different proportions of molybdenum (SnMoX, X = 5, 25, 50 and 75%), taken by the precipitation method, were evaluated in the conversion of fructose compared to pure oxides. Regarding the mixed oxides, using SnMo25 at 150 °C a conversion of 73.2% was observed. However, a change in selectivity was observed as the Mo content in the materials increased. Thus, the SnMo5 system showed high selectivity for 5-HMF (~75%), while for SnMo25 the selectivity for this product was very low (approximately 20%), with a preferential tendency for products from the retro-aldol pathway. In terms of stability, SnMo25 was used in six cycles, and no significant decrease in yield, selectivity or substantial leaching were detected [13].

3. Materials and Methods

3.1. Materials

Tin(IV) chloride (99.995% trace-metals basis, Sigma–Aldrich, St. Louis, MI, USA)), niobium(V) chloride (>99.995%, Sigma–Aldrich), citric acid (>99.0%, Sigma–Aldrich), ammonium hydroxide (Sigma–Aldrich), nitric acid (>65.0%, Sigma–Aldrich), glycerol (>99.0%, Sigma–Aldrich), ethylene glycol (>99.0%, Sigma–Aldrich), cetyltrimethylammonium bromide (CTAB) (>99.0%, Sigma–Aldrich) and fructose (>99.0%, Sigma–Aldrich) were obtained commercially, and used as received. Sulfuric acid (>95%) was purchased from Dynamics.

3.2. Preparation of Catalysts Using the Coprecipitation Method

Pure and mixed catalysts were obtained using aqueous solutions of tin chloride (IV) and/or niobium chloride (V) ($SnCl_4$ and/or $NbCl_5$) at a concentration of 0.1 mol/L, and coprecipitation was achieved by the slow addition of ammonium hydroxide. Soon after, the precipitated material was left to rest for 4 h and then vacuum filtered, washed with deionized water, and dried at 115 °C for 8 h. The material obtained was sieved with a 100-mesh granulometry sieve, and then macerated for further analysis. Finally, the solids were calcined at 500 °C, for 4 h.

3.3. Preparation of Catalysts Using the CTAB-Surfactant-Template Method

Initially, 9 g of the surfactant cetyltrimethylammonium bromide (CTAB) was dissolved in 75 mL of deionized water under constant stirring, to ensure the homogeneity of the system. Then, 12 mL of NH_4OH (25% wt.) dissolved in 48 mL of distilled water was added under stirring. The metal precursors (tin chloride (IV) and/or niobium chloride (V)) were dissolved in 150 mL of deionized water and added dropwise to this first solution under constant stirring, until a white paste was formed. This suspension was maintained under stirring for 3 h, followed by aging at 25 °C for 48 h. The obtained product was filtered, washed, dried, macerated and calcined at 500 °C, for 4 h.

3.4. Preparation of Catalysts Using the Modified Pechini Method

Initially, solutions of tin chloride (IV) and/or niobium chloride (V) ($SnCl_4$ and/or $NbCl_5$) and citric acid, with concentrations of 0.479, 0.462 and 1.66 mol/L were prepared using a mass ratio of 3:1 (citric acid: metal). Then, this material was stirred and heated to 70 °C. Upon reaching the desired temperature, the material was removed from the system and immersed in an ice bath, to promote the onset of precipitation. Subsequently, the solution was returned to the stirring system, and ammonium hydroxide was gradually added for 4 h, until complete precipitation. Then, the citrate was filtered, washed, and dried at 120 °C. The formed citrates were dissolved in distilled water and nitric acid. Then, glycerol was added in a molar proportion of 60:40, in relation to the amount of citric acid. The solution was heated to 70 °C under stirring to promote the polymerization reaction. After the elimination of nitrous oxides and water, the resin was obtained, and finally, the solid was crushed and heated at 350 °C for 3 h to eliminate the polymeric matrix, and calcinated at 500 °C for 4 h.

3.5. Characterization of the Catalysts

All materials were macerated in a mortar and sieved (100 mesh), and the description of the catalysts and the codes used for their identification are shown in Table 1.

Table 1. Code of the catalysts and the respective descriptions.

Code	Description
Sn(PPT) Nb(PPT)	Pure oxide obtained using the coprecipitation method
SnNb(PPT)	Mixed oxide obtained using the coprecipitation method
Sn(CTAB) Nb(CTAB)	Pure oxide obtained using CTAB as the template method
SnNb(CTAB)	Mixed oxide obtained using CTAB as the template method
Sn(GLY) Nb(GLY)	Pure oxide obtained using the Pechini method, using glycerol
SnNb(GLY)	Mixed oxide obtained using the Pechini method, using glycerol

For catalyst characterization, X-ray-diffraction (XRD) measurements were performed using a Shimadzu diffractometer, model XRD-6000, and a Cu Kα radiation (1.5418 Å) source with a nickel filter to obtain wide-angle diffraction patterns in the 2θ = 3–40° range. Crystallite size (nm) was calculated using the Debye–Scherrer equation. Nitrogen-adsorption measurements were performed at 77.15 K, using a gas-adsorption analyzer (Micromeritics, model ASAP-2020). The textural properties were estimated from N_2 adsorption isotherms, using the Brunauer–Emmett–Teller (BET) equation and Barrett–Joyner–Halenda (BJH) method. Before the analysis, all the samples were pretreated under vacuum at 423.15 K for 24 h. Thermogravimetric (TG/DTG) analyses were performed using a Shimadzu analyzer, model DTG-60H, with an N_2 flow-rate of 40 mL·min^{-1} in the temperature range of 30–800 °C at a heating rate of 10 °C·min^{-1}. Fourier-transform-infrared (FTIR) spectra were obtained with a Shimadzu IR Prestige 21 infrared spectrophotometer, using tablets of potassium bromide (KBr). Eighty scans were performed in transmittance mode in the spectral range of 4000–400 cm^{-1} at a resolution of 4.0 cm^{-1}. Raman spectroscopy was performed at ~30 °C, using a 532 nm laser, 10 mW power, 100× objective, and 16 data acquisitions in 120 s with respect to silica on HORIBA scientific equipment, model XploRA. The UV–Vis-DRS spectra were obtained on a Shimadzu model UV-2600 at a resolution of 8 cm^{-1}. Scanning-electron-microscopy analyses were performed on a Shimadzu instrument, model SSX-550 Superscan, with a tungsten filament. Initially, the samples were metallized in a Sanyu Electron instrument, quick coater SC-701, for 8 min with a 10 mA gold current target.

3.6. *Fructose Conversion*

All experiments were performed in 4 mL closed reactors under magnetic stirring and heating, at several reaction times. The fructose solution (0.016 g of fructose in 2 mL of deionized water) and 1.5×10^{-3} g of catalysts were used. Conversion, yield, and selectivity were calculated from the results of quantification using HPLC. For this, the reaction product was filtered using 0.45 μm porosity membrane-filters (MilliporeTM), and then injected into a chromatograph (HPLC) with refractive-index (IR) detection. The analyses were performed on a SHIMADZU (Japan) model CTO-20A liquid chromatograph with an isocratic pump, RID-10A refractive index detector and manual injection system, with a sampling handle of 20 μL. The chromatographic column Varian MetaCarb H Plus 300 × 7.8 mm and precolumn Varian MetaCarb H Plus were obtained from Varian (Palo Alto, CA, USA), operating with a column temperature of 55 °C, and the eluent was a sulfuric acid solution (0.005 mol·L^{-1}) with a flow of 0.70 mL·min^{-1}. For the quantification of the reaction products, calibration curves of standard solutions were prepared. The chromatograms obtained were treated using LabSolutions software (Shimadzu Corporation 2010, Kyoto, Japan). The fructose conversion and the yield and selectivity of each product formed were calculated as presented in the Supplementary Information. For reuse studies, the catalyst was removed from the reaction medium by centrifugation, and calcined at 550 °C for 4 h.

4. Conclusions

Using the catalysts presented in this study, it was possible to efficiently convert fructose in an aqueous medium, and the factors that influenced the transformation were the textural and structural properties, as well as the number of acid sites. The materials that presented the best conversions were the mixed oxides, with a prominence of SnNb(CTAB) and SnNb(GLY), showing the importance of the modulation of properties using the synthesis method. All systems were selective mainly for 5-HMF (5-hydroxymethylfurfural) and, to a lesser extent, for the products of the retro-aldolic route, and this selectivity was preserved, regardless of the catalytic system used. The reuse test for SnNb(GLY), which showed stability for four cycles with no significant reduction in terms of conversion, suggested that the hydrophobic characteristics of the rutile structure of SnO_2 in contrast to Nb_2O_5 resulted in water-tolerant acid sites.

Supplementary Materials: The following supporting information can be downloaded at: https://www.mdpi.com/article/10.3390/catal13020285/s1: Information concerning fructose conversion, yield and selectivity of each product formed, Figure S1: Thermal profiles (TG) of the catalysts: (a) Sn(PPT), Sn(CTAB) and Sn(GLY); (b) Nb(PPT), Nb(CTAB) and Nb(GLY); and (c) SnNb(PPT), SnNb(CTAB) and SnNb(GLY); Figure S2: FTIR spectra using pyridine as probe molecule: (a) Sn(PPT), Sn(CTAB) and Sn(GLY); (b) Nb(PPT), Nb(CTAB) and Nb(GLY); and (c) SnNb(PPT), SnNb(CTAB) and SnNb(GLY); Figure S3. Scanning electron microscopy (SEM) images of the catalysts Sn(PPT), Sn(CTAB) and Sn(GLY).; Figure S4. Scanning electron microscopy (SEM) images of the catalysts Nb(PPT), Nb(CTAB) and Nb(GLY).; Figure S5. Scanning electron microscopy (SEM) images of the catalysts SnNb(PPT), SnNb(CTAB) and SnNb(GLY).; Figure S6. Fructose conversion at 150 °C using (a) Sn(PPT), Sn(CTAB) and Sn(GLY), (b) Nb(PPT), Nb(CTAB) and Nb(GLY) and (c) Nb(GLY) SnNb(PPT) SnNb(CTAB) and SnNb(GLY).; Figure S7. Selectivity for soluble products identified in the fructose conversion at 150 °C using 1.5×10^{-3} g of catalyst (a) Sn(PPT) and Nb(PPT); (b) Sn(CTAB) and Nb(CTAB); (c) Sn(GLY) and Nb(GLY).; Table S1: Yield (%) for soluble products identified in the fructose conversion at 150 °C in absence of catalyst or using 1.5×10^{-3} g of catalyst (a) Sn(PPT) and Nb(PPT); (b) Sn(CTAB) and Nb(CTAB); (c) Sn(GLY) and Nb(GLY).

Author Contributions: Conceptualization, T.V.d.S.M., D.B.d.A.P., S.M.P.M. and M.R.M.; methodology, T.V.d.S.M., D.B.d.A.P., S.M.P.M. and M.R.M..; investigation, T.V.d.S.M., D.B.d.A.P., S.M.P.M. and M.R.M.; resources, S.M.P.M. and M.R.M.; writing—original draft preparation, T.V.d.S.M., D.B.d.A.P., S.M.P.M. and M.R.M.; writing—review and editing, T.V.d.S.M., D.B.d.A.P., S.M.P.M. and M.R.M.; visualization, T.V.d.S.M., D.B.d.A.P., S.M.P.M. and M.R.M.; supervision, S.M.P.M. and M.R.M.; project administration, S.M.P.M. and M.R.M.; funding acquisition, S.M.P.M. and M.R.M. All authors have read and agreed to the published version of the manuscript.

Funding: This research was supported by the National Council for Scientific and Technological Development (CNPq), the Brazilian Federal Agency for the Improvement of Higher Education (CAPES), the Brazilian Innovation Agency (FINEP), and the Alagoas Research Foundation (FAPEAL). T.V.d.S.M. and D.B.d.A.P. express their appreciation for fellowships granted by CAPES and CNPq. S.M.P.M. and M.R.M. thank CNPq for research fellowships.

Data Availability Statement: Not applicable.

Acknowledgments: The authors also thank the LSCAT/CTEC/UFAL, TecNano/ICF/UFAL, and GON/IF/UFAL teams for their contributions.

Conflicts of Interest: The authors declare no conflict of interest.

References

1. Binder, J.B.; Raines, R.T. Simple Chemical Transformation of Lignocellulosic Biomass into Furans for Fuels and Chemicals. *J. Am. Chem. Soc.* **2009**, *131*, 1979–1985. [CrossRef] [PubMed]
2. Wang, W.; Gu, Y.; Zhou, C.; Hu, C. Current Challenges and Perspectives for the Catalytic Pyrolysis of Lignocellulosic Biomass to High-Value Products. *Catalysts* **2022**, *12*, 1524. [CrossRef]
3. Corma, A.; Iborra, S.; Velty, A. Chemical Routes for the Transformation of Biomass into Chemicals. *Chem. Rev.* **2007**, *107*, 2411–2502. [CrossRef]

4. Wu, X.; Fu, J.; Lu, X. Hydrothermal Decomposition of Glucose and Fructose with Inorganic and Organic Potassium Salts. *Bioresour. Technol.* **2012**, *119*, 48–54. [CrossRef] [PubMed]
5. Qu, Y.; Huang, C.; Zhang, J.; Chen, B. Efficient Dehydration of Fructose to 5-Hydroxymethylfurfural Catalyzed by a Recyclable Sulfonated Organic Heteropolyacid Salt. *Bioresour. Technol.* **2012**, *106*, 170–172. [CrossRef]
6. Lee, Y.-Y.; Wu, K.C.-W. Conversion and Kinetics Study of Fructose-to-5-Hydroxymethylfurfural (HMF) Using Sulfonic and Ionic Liquid Groups Bi-Functionalized Mesoporous Silica Nanoparticles as Recyclable Solid Catalysts in DMSO Systems. *Phys. Chem. Chem. Phys.* **2012**, *14*, 13914–13917. [CrossRef]
7. Mascal, M.; Nikitin, E.B. High-Yield Conversion of Plant Biomass into the Key Value-Added Feedstocks 5-(Hydroxymethyl)-Furfural, Levulinic Acid, and Levulinic Esters Via5-(Chloromethyl)Furfural. *Green Chem.* **2010**, *12*, 370–373. [CrossRef]
8. Upare, P.P.; Yoon, J.-W.; Kim, M.Y.; Kang, H.-Y.; Hwang, D.W.; Hwang, Y.K.; Kung, H.H.; Chang, J.-S. Chemical Conversion of Biomass-Derived Hexose Sugars to Levulinic Acid over Sulfonic Acid-Functionalized Graphene Oxide Catalysts. *Green Chem.* **2013**, *15*, 2935. [CrossRef]
9. Zaera, F. Nanostructured Materials for Applications in Heterogeneous Catalysis. *Chem. Soc. Rev.* **2013**, *42*, 2746–2762. [CrossRef]
10. Armor, J.N. A History of Industrial Catalysis. *Catalysts* **2011**, *163*, 3–9. [CrossRef]
11. Fernandes Barbosa, F.; Pinheiro Braga, T. Catalytic Conversion of Glycerol to Acetol and Acrolein Using Metal Oxides: Surface Reactions, Prospects and Challenges. *ChemCatChem* **2022**, *15*, e202200950. [CrossRef]
12. dos Santos, T.V.; da Silva Avelino, D.O.; Meneghetti, M.R.; Meneghetti, S.M.P. Mixed Oxides Based on SnO_2 Impregnated with MoO_3: A Robust System to Apply in Fructose Conversion. *Catal. Commun.* **2018**, *114*, 120–123. [CrossRef]
13. dos Santos, T.V.; dos Santos Brainer, N.; de Amorin Pryston, D.B.; da Silva Avelino, D.O.; Dornelas, C.B.; Meneghetti, M.R.; Meneghetti, S.M.P. Study of Neat and Mixed Sn(IV) and Mo(VI) Oxides for Transesterification and Esterification: Influence of the Substrate on Leaching. *Catal. Lett.* **2019**, *149*, 3132–3137. [CrossRef]
14. Bayu, A.; Abudula, A.; Guan, G. Reaction Pathways and Selectivity in Chemo-Catalytic Conversion of Biomass-Derived Carbohydrates to High-Value Chemicals: A Review. *FPT* **2019**, *196*, 106162. [CrossRef]
15. Grigorev, M.E.; Mikhailov, S.P.; Bykov, A.V.; Tiamina, I.Y.; Nikoshvili, L.Z.; Sulman, M.G.; Vasiliev, A.L.; Sidorov, A.I.; dos Santos, T.V.; Meneghetti, M.R.; et al. Surface Interactions with the Metal Oxide Surface Control Ru Nanoparticle Formation and Catalytic Performance. *Colloids Surf. A Physicochem. Eng. Asp.* **2021**, *610*, 125722. [CrossRef]
16. Almerindo, G.I.; Probst, L.F.D.; Campos, C.E.M.; de Almeida, R.M.; Meneghetti, S.M.P.; Meneghetti, M.R.; Clacens, J.-M.; Fajardo, H.V. Magnesium Oxide Prepared via Metal–Chitosan Complexation Method: Application as Catalyst for Transesterification of Soybean Oil and Catalyst Deactivation Studies. *J. Power Sources* **2011**, *196*, 8057–8063. [CrossRef]
17. Yu, Z.; Lu, X.; Liu, C.; Han, Y.; Ji, N. Synthesis of γ-Valerolactone from Different Biomass-Derived Feedstocks: Recent Advances on Reaction Mechanisms and Catalytic Systems. *Renew. Sustain. Energy Rev.* **2019**, *112*, 140–157. [CrossRef]
18. Yang, F.; Tang, J.; Ou, R.; Guo, Z.; Gao, S.; Wang, Y.; Wang, X.; Chen, L.; Yuan, A. Fully Catalytic Upgrading Synthesis of 5-Ethoxymethylfurfural from Biomass-Derived 5-Hydroxymethylfurfural over Recyclable Layered-Niobium-Molybdate Solid Acid. *Appl. Catal. B Environ.* **2019**, *256*, 117786. [CrossRef]
19. Kreissl, H.T.; Nakagawa, K.; Peng, Y.-K.; Koito, Y.; Zheng, J.; Tsang, S.C.E. Niobium Oxides: Correlation of Acidity with Structure and Catalytic Performance in Sucrose Conversion to 5-Hydroxymethylfurfural. *J. Catal.* **2016**, *338*, 329–339. [CrossRef]
20. Xia, M.; Shen, Z.; Xiao, S.; Peng, B.; Gu, M.; Dong, W.; Zhang, Y. Synergistic Effects and Kinetic Evidence of a Transition Metal-Tin Modified Beta Zeolite on Conversion of Miscanthus to Lactic Acid. *Appl. Catal. A-Gen* **2019**, *583*, 117126. [CrossRef]
21. Popova, M.; Shestakova, P.; Lazarova, H.; Dimitrov, M.; Kovacheva, D.; Szegedi, A.; Mali, G.; Dasireddy, V.; Likozar, B.; Wilde, N.; et al. Efficient Solid Acid Catalysts Based on Sulfated Tin Oxides for Liquid Phase Esterification of Levulinic Acid with Ethanol. *Appl. Catal. A-Gen.* **2018**, *560*, 119–131. [CrossRef]
22. Manjunathan, P.; Marakatti, V.S.; Chandra, P.; Kulal, A.B.; Umbarkar, S.B.; Ravishankar, R.; Shanbhag, G.V. Mesoporous Tin Oxide: An Efficient Catalyst with Versatile Applications in Acid and Oxidation Catalysis. *Catal. Today* **2018**, *309*, 61–76. [CrossRef]
23. Wei, W.; Yang, H.; Wu, S. Efficient Conversion of Carbohydrates into Levulinic Acid over Chromium Modified Niobium Phosphate Catalyst. *Fuel* **2019**, *256*, 115940. [CrossRef]
24. dos Santos, T.V.; Pryston, D.B.A.; Assis, G.C.; Meneghetti, M.R.; Meneghetti, S.M.P. Tin, Niobium and Tin-Niobium Oxides Obtained by the Pechini Method Using Glycerol as a Polyol: Synthesis, Characterization and Use as a Catalyst in Fructose Conversion. *Catal. Today* **2021**, *379*, 62–69. [CrossRef]
25. Kato, K.; Tamura, S. Die Kristallstruktur von T -Nb_2O_5. *Acta Crystallogr. Sect. B Struct. Crystallogr. Cryst. Chem.* **1975**, *31*, 673–677. [CrossRef]
26. Quinelato, A.L.; Longo, E.; Leite, E.R.; Varela, J.A. Synthesis of Nanocrystalline Tetragonal Zirconia by a Polymeric Organometallic Method. *Appl. Organomet. Chem.* **1999**, *13*, 501–507. [CrossRef]
27. Prado, A.G.S.; Bolzon, L.B.; Pedroso, C.P.; Moura, A.O.; Costa, L.L. Nb_2O_5 as Efficient and Recyclable Photocatalyst for Indigo Carmine Degradation. *Appl. Catal. B Environ.* **2008**, *82*, 219–224. [CrossRef]
28. Sagadevan, S.; Podder, J. Investigation on Structural, Surface Morphological and Dielectric Properties of Zn-Doped SnO_2 Nanoparticles. *Mater. Res.* **2016**, *19*, 420–425. [CrossRef]
29. Elci, A.; Demirtas, O.; Ozturk, I.M.; Bek, A.; Nalbant Esenturk, E. Synthesis of Tin Oxide-Coated Gold Nanostars and Evaluation of Their Surface-Enhanced Raman Scattering Activities. *J. Mater. Sci.* **2018**, *53*, 16345–16356. [CrossRef]

30. Castro, D.C.; Cavalcante, R.P.; Jorge, J.; Martines, M.A.U.; Oliveira, L.C.S.; Casagrande, G.A.; Machulek Jr., A. Synthesis and Characterization of Mesoporous Nb_2O_5 and Its Application for Photocatalytic Degradation of the Herbicide Methylviologen. *J. Braz. Chem. Soc.* **2015**, *27*, 303–313.
31. da Conceição, L.R.V.; Carneiro, L.M.; Rivaldi, J.D.; de Castro, H.F. Solid Acid as Catalyst for Biodiesel Production via Simultaneous Esterification and Transesterification of Macaw Palm Oil. *Ind. Crops Prod.* **2016**, *89*, 416–421. [CrossRef]
32. Azam, A.; Habib, S.; Salah, N.; Ahmed, F. Microwave-Assisted Synthesis of SnO_2 Nanorods for Oxygen Gas Sensing at Room Temperature. *IJN* **2013**, *8*, 3875–3882. [CrossRef] [PubMed]
33. Xu, G.; Zhang, Y.-W.; Sun, X.; Xu, C.-L.; Yan, C.-H. Synthesis, Structure, Texture, and CO Sensing Behavior of Nanocrystalline Tin Oxide Doped with Scandia. *J. Phys. Chem. B* **2005**, *109*, 3269–3278. [CrossRef]
34. Mariammal, R.N.; Ramachandran, K.; Renganathan, B.; Sastikumar, D. On the Enhancement of Ethanol Sensing by CuO Modified SnO_2 Nanoparticles Using Fiber-Optic Sensor. *Sens. Actuators B Chem.* **2012**, *169*, 199–207. [CrossRef]
35. Varghese, B.; Haur, S.C.; Lim, C.-T. Nb_2O_5 Nanowires as Efficient Electron Field Emitters. *J. Phys. Chem. C* **2008**, *112*, 10008–10012. [CrossRef]
36. Zhao, Y.; Zhou, X.; Ye, L.; Chi Edman Tsang, S. Nanostructured Nb_2O_5 Catalysts. *Nanotechnol. Rev.* **2012**, *3*, 17631.
37. Yu, A.; Frech, R. Mesoporous Tin Oxides as Lithium Intercalation Anode Materials. *J. Power Sources* **2002**, *104*, 97–100. [CrossRef]
38. Zuo, J.; Xu, C.; Liu, X.; Wang, C.; Wang, C.; Hu, Y.; Qian, Y. Study of the Raman spectrum of nanometer SnO_2. *J. Appl. Phys.* **1994**, *75*, 1835–1836. [CrossRef]
39. Soltan, W.B.; Lassoued, M.S.; Ammar, S.; Toupance, T. Vanadium Doped SnO2 Nanoparticles for Photocatalytic Degradation of Methylene Blue. *J. Mater. Sci. Mater. Electron.* **2017**, *28*, 15826–15834. [CrossRef]
40. Morais, L.A.; Adán, C.; Araujo, A.S.; Guedes, A.P.M.A.; Marugán, J. Synthesis, Characterization, and Photonic Efficiency of Novel Photocatalytic Niobium Oxide Materials. *Glob. Chall.* **2017**, *1*, 1700066. [CrossRef] [PubMed]
41. Scotti, N.; Ravasio, N.; Evangelisti, C.; Psaro, R.; Penso, M.; Niphadkar, P.; Bokade, V.; Guidotti, M. Epoxidation of Karanja (Millettia Pinnata) Oil Methyl Esters in the Presence of Hydrogen Peroxide over a Simple Niobium-Containing Catalyst. *Catalysts* **2019**, *9*, 344. [CrossRef]
42. Lee, D.-U.; Jang, S.-R.; Vittal, R.; Lee, J.; Kim, K.-J. CTAB Facilitated Spherical Rutile TiO_2 Particles and Their Advantage in a Dye-Sensitized Solar Cell. *Sol. Energy* **2008**, *82*, 1042–1048. [CrossRef]
43. Casino, S.; Di Lupo, F.; Francia, C.; Tuel, A.; Bodoardo, S.; Gerbaldi, C. Surfactant-Assisted Sol Gel Preparation of High-Surface Area Mesoporous TiO_2 Nanocrystalline Li-Ion Battery Anodes. *J. Alloys Compd.* **2014**, *594*, 114–121. [CrossRef]
44. Nakajima, K.; Baba, Y.; Noma, R.; Kitano, M.; Kondo, J.N.; Hayashi, S.; Hara, M. $Nb_2O_5 \cdot nH_2O$ as a Heterogeneous Catalyst with Water-Tolerant Lewis Acid Sites. *J. Am. Chem. Soc.* **2011**, *133*, 4224–4227. [CrossRef]
45. Nowak, I.; Ziolek, M. Niobium Compounds: Preparation, Characterization, and Application in Heterogeneous Catalysis. *Chem. Rev.* **1999**, *99*, 3603–3624. [CrossRef]
46. Skrodczky, K.; Antunes, M.M.; Han, X.; Santangelo, S.; Scholz, G.; Valente, A.A.; Pinna, N.; Russo, P.A. Niobium Pentoxide Nanomaterials with Distorted Structures as Efficient Acid Catalysts. *Commun. Chem.* **2019**, *2*, 129. [CrossRef]
47. Siddiki, S.H.; Rashed, M.N.; Ali, M.A.; Toyao, T.; Hirunsit, P.; Ehara, M.; Shimizu, K.I. Lewis Acid Catalysis of Nb_2O_5 for Reactions of Carboxylic Acid Derivatives in the Presence of Basic Inhibitors. *ChemCatChem* **2019**, *11*, 383–396. [CrossRef]
48. Rudisill, S.G.; Hein, N.M.; Terzic, D.; Stein, A. Controlling Microstructural Evolution in Pechini Gels through the Interplay between Precursor Complexation, Step-Growth Polymerization, and Template Confinement. *Chem. Mater.* **2013**, *25*, 745–753. [CrossRef]
49. Babooram, K. Novel solution routes to ferroelectrics and relaxors. In *Handbook of Advanced Dielectric, Piezoelectric and Ferroelectric Materials*; Elsevier: Amsterdam, The Netherlands, 2008; pp. 852–883.
50. Wu, W.; Jiang, C.Z.; Roy, V.A.L. Designed synthesis and surface engineering strategies of magnetic iron oxide nanoparticles for biomedical applications. *Nanoscale* **2016**, *8*, 19421–19474. [CrossRef]
51. Dimesso, L. Pechini Processes: An Alternate Approach of the Sol–Gel Method, Preparation, Properties, and Applications. *Handb. Sol-Gel Sci. Technol.* **2016**, 1067–1088.
52. Tsilomelekis, G.; Orella, M.J.; Lin, Z.; Cheng, Z.; Zheng, W.; Nikolakis, V.; Vlachos, D.G. Molecular Structure, Morphology and Growth Mechanisms and Rates of 5-Hydroxymethyl Furfural (HMF) Derived Humins. *Green Chem.* **2016**, *18*, 1983–1993. [CrossRef]
53. Hao, H.; Shen, F.; Yang, J.; Qiu, M.; Guo, H.; Qi, X. Synthesis of Sulfonated Carbon from Discarded Masks for Effective Production of 5-Hydroxymethylfurfural. *Catalysts* **2022**, *12*, 1567. [CrossRef]
54. dos Santos, T.V.; da Silva Avelino, D.O.; Pryston, D.B.A.; Meneghetti, M.R.; Meneghetti, S.M.P. Tin, Molybdenum and Tin-Molybdenum Oxides: Influence of Lewis and Bronsted Acid Sites on Xylose Conversion. *Catal. Today* **2022**, *394–396*, 125–132. [CrossRef]
55. Stošić, D.; Bennici, S.; Rakić, V.; Auroux, A. CeO_2–Nb_2O_5 mixed oxide catalysts: Preparation, characterization and catalytic activity in fructose dehydration reaction. *Catal. Today* **2012**, *192*, 160–168. [CrossRef]

catalysts

MDPI

Article

Guaiacol to Aromatics: Efficient Transformation over In Situ-Generated Molybdenum and Tungsten Oxides

Mariyam Mukhtarova, Maria Golubeva *, Alexey Sadovnikov and Anton Maximov *

A.V. Topchiev Institute of Petrochemical Synthesis, Russian Academy of Sciences (TIPS RAS), Moscow 119991, Russia

* Correspondence: vinnikova@ips.ac.ru (M.G.); max@ips.ac.ru (A.M.)

Abstract: The development of catalysts for the hydrodeoxygenation of bio-based feedstocks is an important step towards the production of fuels and chemicals from biomass. This paper describes in situ-generated bulk molybdenum and tungsten oxides in the hydrodeoxygenation of the lignin-derived compound guaiacol. The catalysts obtained were studied using powder X-ray diffraction, X-ray photoelectron spectroscopy, scanning electron microscopy, high-resolution transition electron microscopy, diffuse reflectance infrared Fourier transform spectroscopy, and Raman spectroscopy. The use of metal carbonyls as precursors was shown to promote the formation of amorphous molybdenum oxide and crystalline tungsten phosphide under hydrodeoxygenation conditions. The catalysts' activity was investigated under various reaction conditions (temperature, H_2 pressure, solvent). MoO_x was more active in the partial and full hydrodeoxygenation of guaiacol at temperatures of 200–380 °C (5 MPa H_2, 6 h). However, cyclohexane, which is an undesirable product, was formed in significant amounts using MoO_x (5 MPa H_2, 6 h), while WO_x was more selective to aromatics. When using dodecane as a solvent (380 °C, 5 MPa H_2, 6 h), the benzene-toluene-xylenes fraction was obtained with a 96% yield over the WO_x catalyst.

Keywords: MoO_x; WO_x; guaiacol; hydrodeoxygenation; vacancy sites; dispersed catalysts; transition metal oxides; aromatic hydrocarbons; BTX

1. Introduction

Lignocellulosic biomass is a promising and widespread source of renewable energy [1,2]. One of the components of lignocellulosic biomass is lignin, which consists of oxygen-containing derivatives of phenylpropane. The structure of lignin allows it to be used as a source for the production of phenols and various aromatic compounds [3].

The main method for the processing of low-molecular-weight lignin derivatives is catalytic hydrodeoxygenation (HDO). Supported noble metals are actively involved in the HDO of different oxygenated compounds and show high catalytic activity. Nevertheless, the high cost of noble metals is a significant disadvantage in their application [4]. Transition metal phosphides, carbides, and nitrides are reported to be potential candidates to replace expensive catalysts (Table 1) [4].

Table 1. Some results from the literature of the lignin-derived compound HDO over Mo- and W-containing catalysts.

Catalyst	Feedstock	Reaction Conditions [a]	Conversion, %	Selectivity, %	Ref.
Mo_xC/CNF	0.68 g cinnamaldehyde in 50 mL toluene	200 °C, 2 MPa H_2, ~2.75 h [a]	90	40.9 for hydrocinnamaldehyde 25.8 for β-methylstyrene	[5]
W_xC/CNF		200 °C, 2 MPa H_2, ~5.75 h [a]	90	43 for hydrocinnamaldehyde 34 for β-methylstyrene	
Pd/α-MoC	anisole	200 °C [b]	21.5	94 for benzene	[6]
Mo_2C@ BMZIF-700 °C (4 h)	0.12 g guaiacol in 20 mL n-decane	330 °C, 4 MPa H_2, 4 h [a]	97	70 for phenol	[7]
Mo_2N@ NC500/SBA-15	3 wt. % guaiacol in n-decane	380 °C, 2 MPa H_2 [c]	100	92 to benzene + toluene	[8]
in situ formed MoP	10 wt. % guaiacol in n-dodecane	360 °C, 5 MPa H_2, 6 h [a]	90	80 for phenol	[9]
in situ formed WP		340 °C, 5 MPa H_2, 1 h [a]	53	78 for phenol	
MoP		340 °C, 5 MPa H_2, 6 h [a]	90	66 for phenol	[10]
WP		380 °C, 5 MPa H_2, 6 h [a]	89	84 for phenol	
in situ formed MoO_x		380 °C, 1 MPa H_2, 6 h [a]	100	89 for BTX	This work
in situ formed WO_x		380 °C, 5 MPa H_2, 6 h [a]	100	96 for BTX	

[a] a batch reactor was used; [b] a fixed bed quartz tubular reactor was used; [c] a micro-hydrogenation reactor was used.

However, the activity of such catalysts is partially related to the presence of oxygen on their surface, including in the metal oxide form. Therefore, transition metal oxides can also be considered as alternative HDO catalysts. Oxygen vacancies (coordinatively unsaturated metal sites) in oxides are reported to be active sites in HDO reactions [11,12]. Vacancies are formed as a result of H_2 adsorption on the oxide surface and the subsequent removal of oxygen as a part of water [13]. The electron-lone-pair oxygen in a substrate is adsorbed on the active site which leads to HDO as a result of the cleavage of the C–O bond [11].

Mo- and W-containing compounds are well-known in the field of catalysis. It has been reported that Mo and W are resistant to oxygen, acids, and alkalis and, therefore, they are used for oxygen removal [14]. Efficient hydrodeoxygenation of various substrates over molybdenum oxides have been previously reported in a number of works [11,13], while pure tungsten oxides have not provided high selectivity for hydrodeoxygenation products [15]. Prasomsri et al. found that MoO_3 can be used for the effective HDO of linear and cyclic ketones, phenolics, and cyclic ethers at 250–400 °C under low H_2 pressure with the formation of olefins and aromatics [13]. Selectivity for hydrocarbons was reportedly more than 97% at 400 °C. Zhang et al. carried out the HDO of guaiacol, o-cresol, eugenol, trans-anethole, vanillin, and diphenyl ether over MoO_3 under mixed H_2 + N_2 pressure at 300–360 °C [11]. Aromatics were obtained with the selectivity of 45.3–93.8% at 340 °C. Whiffen et al. tested MoO_3 and MoO_2 in the HDO of p-cresol at 325–375 °C, under 4.14–4.83 MPa H_2 in a batch reactor [16]. Toluene was the main reaction product over both catalysts. MoO_2 was noted to be more active in the HDO than MoO_3. Several studies were devoted to m-cresol HDO to toluene over WO_x promoted by other metals [17–19]. Wang et al. demonstrated that Pt-WO_x/C catalyst can convert m-cresol into toluene with 94% selectivity (350 °C, 0.1 MPa H_2, H_2 flow rate: 12 mL/min, WHSV = 60 h^{-1}) [17]. In another work, they used n-hexane which produced H_2 by dehydration [18]. Direct bonding of Pt and WO_x contributed to oxide stabilization and oxygen vacancies formation, which are responsible for direct hydrogenolysis of the C–O bond in m-cresol [17]. Chen et al. reported similar regularities using a Ni-WO_x/C catalyst [19]. Toluene was obtained from m-cresol with 89% selectivity at 350 °C, under 0.1 MPa H_2, WHSV of 1 h^{-1}.

Previously, the in situ formation of transition metal catalysts during hydroprocessing of various model compounds and real feedstocks were studied [9,20–24]. In our recent works, the in situ formation of phosphide catalysts, including molybdenum and tungsten phosphides in the HDO of guaiacol, was investigated for the first time [9,20,21]. The absence of a catalyst synthesis stage, and therefore a simplification of the process, is one of the advantages of catalyst in situ formation. The use of metal carbonyls in combination with an oil-soluble phosphorus precursor made it possible to obtain the corresponding phosphides. It was found that, in addition to phosphides, the catalyst surface also contains metal oxides.

In the present work, molybdenum and tungsten oxides were obtained in situ from carbonyls during the HDO of guaiacol (a low-molecular-weight lignin derivative) and investigated in this process for the first time. We report that these bulk catalysts are highly active in the HDO of guaiacol to aromatics, in particular, to a benzene-toluene-xylenes (BTX) fraction. MoO_x converted guaiacol into BTX with the highest selectivity of 89% (380 °C, 1 MPa H_2, 6 h) at full conversion, while using WO_x the highest selectivity for BTX was 96% (380 °C, 5 MPa H_2, 6 h). In comparison with oxides obtained in the present work, phosphides and carbides promoted the obtaining of phenol as a main reaction product of guaiacol conversion (Table 1). In situ-generated MoP and WP from the same metal precursors in the HDO of guaiacol allows the obtaining of phenol as a main reaction product [9]. The highest phenol selectivity rates were 80% and 78% using MoP and WP, respectively. Ex situ-generated bulk MoP and WP also made a contribution to the phenol obtained from guaiacol (with 66% and 84% selectivity over MoP and WP, respectively) [10]. Kurisingal et al. made a report about selective guaiacol HDO to phenol using bimetallic Zn-Co zeolitic imidazolate framework (BMZIF)-decorated molybdenum carbide [7]. The distribution of active species, such as Mo_2C, MoO_3, ZnO, and CoO was found to have a high influence on guaiacol conversion and phenol selectivity. The most active Mo_2C@BMZIF- 700 °C (4 h) catalyst promoted a 97% guaiacol conversion and 70% phenol selectivity. Wang et al. demonstrated that an SBA-15-supported ultrastable Mo_2N@CN catalyst showed similar catalytic activity in the guaiacol HDO to the catalysts reported in the present work [8]. A benzene and toluene mixture was obtained with 92% selectivity at 380 °C, while at 320 °C phenol was the main product. However, the synthesis of nitrides is a laborious procedure, while we report an easy method for catalyst synthesis.

2. Results and Discussion

2.1. Catalyst Characterization

In the present work, thein situ-formation of molybdenum and tungsten oxides from molybdenum and tungsten carbonyls during guaiacol hydrodeoxygenation was investigated. Metal carbonyl complexes are known to react with water, other sources of OH-groups, and oxygen with the formation of oxides [25–31]. A substrate used in this work has an OH-group, and the HDO process involves the formation of water. The catalysts obtained were called MoO_x and WO_x, and were studied using powder X-ray diffraction (XRD), X-ray photoelectron spectroscopy (XPS), high-resolution transmission electron microscopy (HRTEM) with energy-dispersive X-ray spectroscopy (EDX), scanning electron microscopy (SEM), diffuse reflectance infrared Fourier transform spectroscopy (DRIFTS), and Raman spectroscopy.

The XRD patterns of catalyst precursors are shown in Figure S1 and correspond to crystalline $Mo(CO)_6$ (PDF №75-1336) and $W(CO)_6$ (PDF № 84-2314). Figure 1a shows the XRD of MoO_x obtained at temperatures of 120–360 °C. MoO_x is shown to be amorphous over the whole temperature range. The evolution of the WO_x catalyst with temperature can be followed using XRD (Figure 1b). At 120 °C the WO_x diffractogram has characteristic peaks at 15.52° and 16.68°, which are assigned to $W(CO)_6$. When temperature rises to 180 °C, carbonyl partially decomposes with the formation of a WO_3 phase (PDF № 5-363), of which characteristic peaks are observed at 70.64° and 72.88°. With a further increase in temperature, the peaks related to carbonyl disappear, which characterizes its complete decomposition and oxidation. In addition, at 240 °C, WO_3 is converted to $W_{18}O_{49}$ (PDF № 71-2450). This phase was identified in the sample obtained at 240–360 °C. The characteristic peaks at 23.52° and 48.12° correspond to the (010) and (020) planes in $W_{18}O_{49}$ [32]. The average crystallite sizes of $W_{18}O_{49}$ were 2.0 ± 0.6 nm, 13.3 ± 0.9 nm and 14.6 ± 0.1 nm at

240, 300, and 360 °C, respectively. Therefore, a temperature increase affects the growth of $W_{18}O_{49}$ crystallites. Thus, the XRD of WO_x catalyst shows $W(CO)_6$ decomposition with the formation of WO_3 at 180 °C, and further WO_3 reduction at 240 °C to $W_{18}O_{49}$.

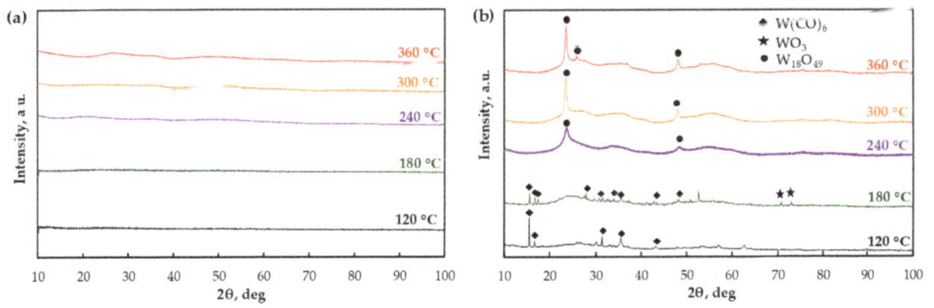

Figure 1. Powder X-ray diffraction of (**a**) MoO_x, (**b**) WO_x formed in situ during guaiacol HDO (10 wt. % solution in toluene). Reaction conditions: 120–360 °C, 5 MPa H_2, 6 h.

To identify electronic states on the surface of the catalysts, the XPS technique was applied. For MoO_x obtained at 120 °C (Figure 2a), the doublet at 232.5 ($3d_{5/2}$) and 235.7 ($3d_{3/2}$) eV in the Mo 3d region can be assigned to Mo^{6+} in MoO_3 [33]. The peaks at 231.8 ($3d_{5/2}$) and 234.8 ($3d_{3/2}$) eV are ascribed to Mo^{5+} [34]. The presence of Mo^{5+} species indicates the existence of the oxygen vacancies in MoO_3 [35,36]. Choi and Thompson noted that the reduction of Mo^{6+} to Mo^{4+} can proceed through a +5 intermediate oxidation state [37]. Figure 2b shows the XP-spectrum of WO_x in the W 4f region. The peaks with binding energies of 34.7 ($4f_{7/2}$) and 35.6 ($4f_{5/2}$) eV correspond to W^{4+} in WO_2 [38]. The other two peaks observed at 35.9 ($4f_{7/2}$) and 38.0 ($4f_{5/2}$) eV are attributed to W^{6+} in WO_3 [39].

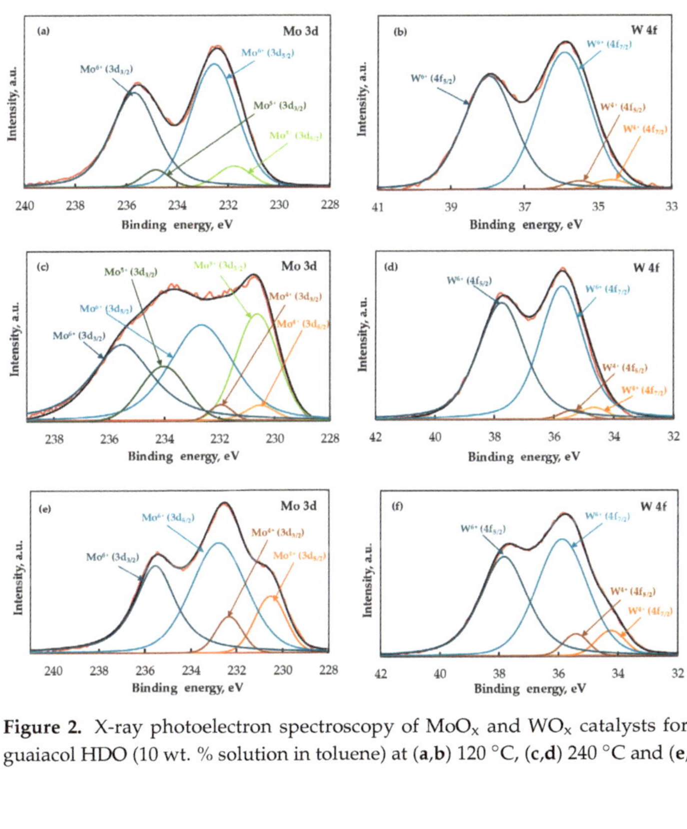

Figure 2. X-ray photoelectron spectroscopy of MoO_x and WO_x catalysts formed in situ during guaiacol HDO (10 wt. % solution in toluene) at (**a**,**b**) 120 °C, (**c**,**d**) 240 °C and (**e**,**f**) 360 °C.

The XP-spectrum of MoO_x obtained at 240 °C is presented in Figure 2c. The peaks located at 230.6 ($3d_{5/2}$) and 234.0 ($3d_{3/2}$) eV correspond to Mo^{5+} species [34]. The doublet at 232.7 ($3d_{5/2}$) and 235.5 ($3d_{3/2}$) eV is attributed to Mo^{6+} [33]. The doublet at 230.5 ($3d_{5/2}$) and 231.9 ($3d_{3/2}$) eV is assigned to Mo^{4+} in MoO_2 [33]. Therefore, with a temperature increase from 120 to 240 °C, Mo^{6+} and Mo^{5+} begin to reduce to Mo^{4+}. The XP-spectrum of WO_x obtained at 240 °C is presented in Figure 2d. The binding energies of 34.7 and 35.3 eV are defined as W^{4+} in the $4f_{7/2}$ and $4f_{5/2}$ regions, respectively. The doublet observed at 35.7 ($4f_{7/2}$) and 37.8 ($4f_{5/2}$) eV is attributed to W^{6+} in WO_3 [39].

The XP-spectrum of MoO_x (Figure 2e) obtained at 360 °C shows two peaks located at 230.5 ($3d_{5/2}$) and 232.3 ($3d_{3/2}$) eV and assigned to Mo^{4+} in MoO_2 [33]. The predominant doublet with peaks at 232.8 ($3d_{5/2}$) and 235.5 ($3d_{3/2}$) eV corresponds to Mo^{6+} in MoO_3 [33]. It is important to note that the presence of Mo^{5+} is not observed at 360 °C. Thus, at this temperature Mo^{5+} is completely reduced to Mo^{4+}. WO_x XP-spectrum (Figure 2f) shows the peaks with binding energies of 34.3 and 35.4 eV corresponding to W^{4+} in WO_2 in $4f_{7/2}$ and $4f_{5/2}$ regions, respectively [38]. The peaks observed at 35.9 ($4f_{7/2}$) and 37.8 ($4f_{5/2}$) eV are attributed to W^{6+} in WO_3 [39].

According to the XPS data, the total content of W^{4+} in the WO_x catalyst increases with a temperature rise from 240 to 360 °C, which is associated with the reduction of W^{6+} to W^{4+} (Table S1). In the case of MoO_x, Mo^{6+} and Mo^{5+} species are also reduced to Mo^{4+} with a temperature increase.

The XP-spectra in C 1 s and O 1 s regions of the catalysts obtained at various temperatures are shown in Figures S2 and S3. The same peaks are shown to be observed in the spectra of both catalysts. For example, there are two peaks located at 283.7 and 284.3 eV in the spectrum of MoO_x (C 1 s region) obtained at 120 °C (Figure S2). These peaks correspond to the C-C bond [40,41]. The other peaks observed at 285.2 and 288.8 eV can be assigned to the C-O and O = C-O bonds, respectively [42,43]. The presence of these peaks is associated with the adsorption of reaction products on the catalyst surface. In addition, the peaks corresponding to C-O can also be associated with the carbonyl groups in $Mo(CO)_6$ and $W(CO)_6$.

As shown in Figure S3, there are three peaks in the XP-spectra of MoO_x and WO_x (O 1 s region) obtained at 120 and 240 °C. In the case of MoO_x obtained at 120 °C, the binding energies of 530.5 and 531.7 eV are assigned to the Mo-O and Mo-OH bonds [44]. The XP-spectrum of WO_x obtained at 120 °C has two peaks at 530.7 and 532.0 eV, which can be attributed to W-O and W-OH, respectively [45]. Thus, at 120 and 240 °C, the presence of molybdenum and tungsten oxides and OH-groups on the surface of the catalysts is observed. The peaks located at 533.0 and 533.7 eV correspond to the C = O/C-OH bonds [41]. These peaks can be associated with the presence of water and oxygenated compounds on the surface of catalysts [46]. There are only two peaks corresponding to Me-O and Me-OH in XP-spectra of MoO_x and WO_x obtained at 360 °C. It can be concluded that at 360 °C carbonyls are completely converted into the corresponding molybdenum and tungsten oxides.

Figure 3 presents the SEM images of MoO_x and WO_x obtained at different reaction temperatures (120, 240, 360 °C). The MoO_x catalyst is represented by close-to-spherical particles of different sizes. The morphology of MoO_x (Figure 3a,c,e) is shown to not undergo significant changes with a temperature increase. The various morphology of the WO_x catalyst (Figure 3b,d,f) is observed at different temperatures. At 120 °C, the catalyst particles are close to spherical; at 240 °C, the particles do not have a definite morphology; at 360 °C, the particles are urchin-like spheres.

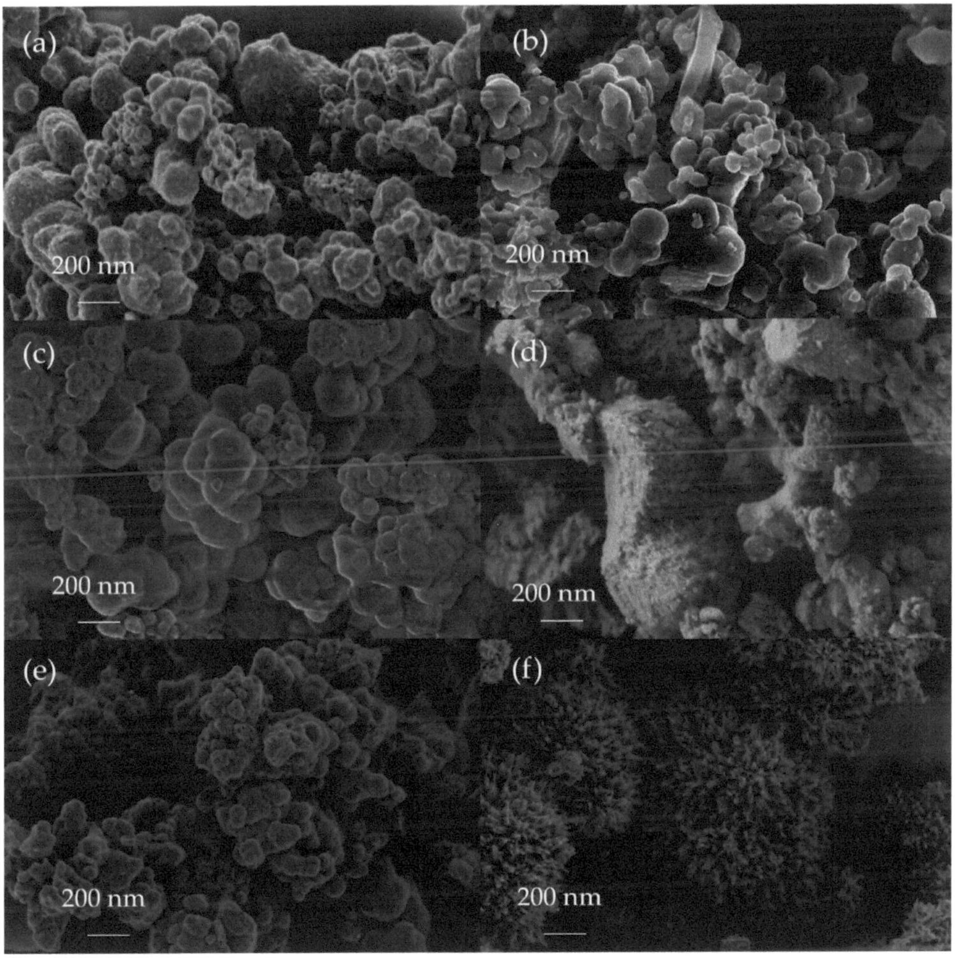

Figure 3. The SEM images of MoO_x and WO_x catalysts formed in situ during guaiacol HDO (10 wt. % solution in toluene) at (**a**,**b**) 120 °C, (**c**,**d**) 240 °C and (**e**,**f**) 360 °C.

HRTEM, DRIFTS and Raman spectroscopy was used to study catalysts formed at 360 °C. HRTEM was used to observe the morphology of the catalysts (Figure 4). As shown in Figure 4a, MoO_x represents agglomerates of particles. Moreover, the sample of MoO_x is amorphous (Figure 4b), which confirms the powder XRD results. Particle sizes are difficult to calculate due to agglomeration, since the edges of individual particles are not clearly visible. WO_x consists of urchin-like spheres, which are made up of nanorods with an average length of ~70–80 nm and width of ~8–10 nm (Figure 4c,d). According to the HRTEM, WO_x has a crystal lattice with the interplanar spacing of 0.38 nm, which corresponds to the (010) plane in $W_{18}O_{49}$ [47,48]. Figures S4 and S5 show the elemental maps of MoO_x and WO_x, respectively. Uniform distribution of metal and oxygen is observed for both samples. Energy-dispersive X-ray spectroscopy was used to study the elemental content of the catalysts. The ratio of O/Me calculated from the EDX spectra was ~1.5 for MoO_x and ~2.3 for WO_x (Figure S6).

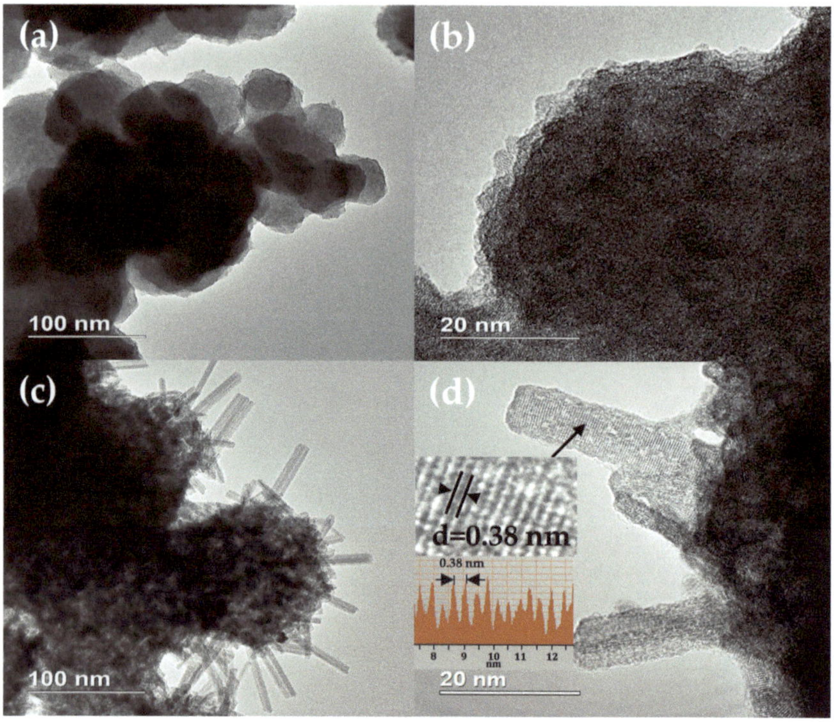

Figure 4. The microphotographs of the catalysts obtained in situ during guaiacol HDO (10 wt. % solution in toluene) (**a**,**b**) MoO_x, (**c**,**d**) WO_x.

The DRIFT spectra of MoO_x are shown in Figure S7. The intense bands of Me-O at 1000–900 cm^{-1} (at room temperature) are attributed to molybdenum oxide (Figure S7) [49]. There is also a broad band at 3400–3200 cm^{-1} from the stretching vibration, and at 1630 cm^{-1} from the bending vibration of OH-groups in the spectra. This is caused by the presence of water on the catalyst surface. The band growth from 3300 to 1630 cm^{-1} occurs due to water removal from a surface layer of the sample with an increase in temperature from 100 to 300 °C [50]. The presence of the bands at 3080 and 2950–2850 cm^{-1} is assigned to the C-H bonds of aryl and alkyl groups, respectively. In addition, the bands at 1392 cm^{-1} and 1425 cm^{-1} are attributed to the bending vibration of C-H bonds of the alkyl group. A further increase and decrease in temperature also shows the presence of aromatic compounds (the bands at 3080 and 1510 cm^{-1}) [51] and alkyl groups (the bands at 2950–2850 cm^{-1}) [52]. The band at 3740 cm^{-1} is attributed to the vibrations of OH-groups in reaction products.

The DRIFT spectra of WO_x are shown in Figure S8. The intense band at 990 cm^{-1} at a room temperature is attributed to the Me-O bond, which indicates the presence of tungsten oxide [52]. Water (the bands at 3300 and 1620 cm^{-1} from the stretching and bending vibrations of O-H, respectively) [50] and organic compounds are also present in small amounts (the bands at 3080 cm^{-1} from the bending vibrations of C-H in aromatic ring and at 1510 cm^{-1} from bending vibrations of aromatic ring) on the catalyst surface [51]. When the sample is calcined from room temperature to 300 °C, water is actively released from the sample layer to the surface, as evidenced by the growth of the bands at 3400 and 1620 cm^{-1}. The bands from the vibrations of the C-H bonds in aromatic compounds (3080 and 1510 cm^{-1}) and the C-H bonds in alkyl groups (CH_2, CH_3), and the bands at 2950–2850 cm^{-1} from stretching vibrations and 1460 cm^{-1} from bending vibrations also become more intense [53]. At 450 °C, the band at 3150 cm^{-1}, characteristic of condensed aromatic rings, appears on the surface. The band at 3750 cm^{-1} exists over a whole temperature range and is attributed to OH-groups in different aromatic products of guaiacol HDO (cresols and phenol) [51]. Thus, according to the DRIFTS data, the reaction products are absorbed on the surface of the catalysts, which is also confirmed by the XPS in the C 1 s and O 1 s regions.

The Raman spectra of MoO_x and WO_x is presented in Figure S9. According to the results of Raman spectroscopy, MoO_x represents crystalline MoO_3. The Raman bands at 995, 820, 670, 289 cm^{-1} are attributed to the Mo = O bond and the band at 210 cm^{-1} corresponds to the Mo–O–Mo bond. According to the XRD, the sample of MoO_x obtained at 360 °C is amorphous, which does not match with the results of Raman spectroscopy. Ajito et al. demonstrated that amorphous MoO_3 can be crystallized by the laser used to obtain the Raman spectrum [54]. There are two intense peaks in the Raman spectra of WO_x. The G peak at 1500–1600 cm^{-1} refers to the C–C bond, which is assigned to sp^2 hybridized carbon atoms, and the D peak at 1330–1390 cm^{-1}, which refers to the C–C bond (sp^3 hybridized carbon atoms). The presence of these peaks is associated with the adsorption of reaction products on the WO_x surface [55]. The peaks at 790, 700 and 250 cm^{-1} are broad, blurry and non-intense and refer to amorphous WO_3. However, according to the XRD, this catalyst has the crystal structure of $W_{18}O_{49}$. These differences can be explained by the low stability of $W_{18}O_{49}$. Lu et al. noted that $W_{18}O_{49}$ can be transformed into WO_3 by the laser used to carry out the Raman spectra [38].

2.2. Catalytic Activity

2.2.1. The Effect of Temperature in Toluene

The activity of catalysts obtained in situ was investigated in the HDO of guaiacol. The effect of temperature on the catalytic activity was evaluated over a wide temperature range of 120–380 °C, under an initial H_2 pressure of 5 MPa. The reaction time was 6 h, and toluene was used as a solvent. The conversion of guaiacol over MoO_x was noted to be extremely low and reached 1–4% at 120–200 °C (Figure S10a). A rise in temperature from 200 to 240 °C promoted a sharp increase in guaiacol conversion from 4 to 18%—which may be related to the full precursor transformation to oxides—the partial reduction of the catalyst, and, as a result, an increase in the number of oxygen vacancies. As shown in Figure 5a, the full conversion of guaiacol was reached at 320 °C. At 240–300 °C, phenol was observed as a main reaction product (Figure 5b). Anisole, cresols, and cyclohexene were also obtained. Since the bond dissociation energy of C_{AR}-OCH_3 is 45 kJ/mol and lower than that of the C_{Ar}–OH bond, the conversion of guaiacol to phenol was more favored than conversion to anisole [56]. The highest selectivity for phenol was 66% at 280 °C. A following temperature increase led to direct deoxygenation (DDO) of phenol to benzene. Therefore, the cleavage of the C_{Ar}–OH bond was favored at higher temperatures [11]. In this way, the selectivity for benzene gradually increased with a rise in temperature from 280 to 380 °C. Cyclohexane and xylenes were also detected in large amounts at 320–380 °C.

At 120–220 °C, guaiacol conversion over WO_x was very low and reached 1–5% (Figure S10b). The full conversion of guaiacol was reached at 360 °C (Figure 5c). Phenol was a major product at 240–300 °C. The decrease in phenol selectivity with a temperature increase from 240 °C to 380 °C is associated with its conversion to benzene (Figure 5d). Compared to MoO_x, WO_x demonstrated low catalytic activity in the hydrogenation of benzene to cyclohexene and cyclohexane. Oxygen vacancies in metal oxides are noted to act as active sites in hydrogenation reactions under typical HDO conditions [16,57]. With the temperature increase from 240 to 360 °C the selectivity for xylenes increased from 11 to 52%. As is the case with MoO_x, anisole was not detected among the reaction products at 320–380 °C. Most likely, it is caused by the conversion of anisole to phenol or cresols.

Phenol was the major product of guaiacol HDO at 240–300 °C over both catalysts. With a temperature increase to 320–340 °C, phenol selectivity over MoO_x decreased, and cyclohexane became the main product. In the case of WO_x, phenol was also obtained in significant amounts at 320–340 °C. However, xylenes formed with higher selectivity. With a further increase in temperature to 360–380 °C, aromatics, such as benzene and xylenes, were obtained in large amounts over both catalysts. At this temperature, WO_x is more selective to aromatics than MoO_x.

The comparison of catalyst activities in the formation of HDO products (benzene, xylenes, cyclohexene, cyclohexane) is presented in Figure S11. The highest yield of HDO products (96%) was shown to be reached at 360 °C using MoO_x. Compared to MoO_x, the highest yield of HDO products over WO_x was 94% and was reached at 380 °C.

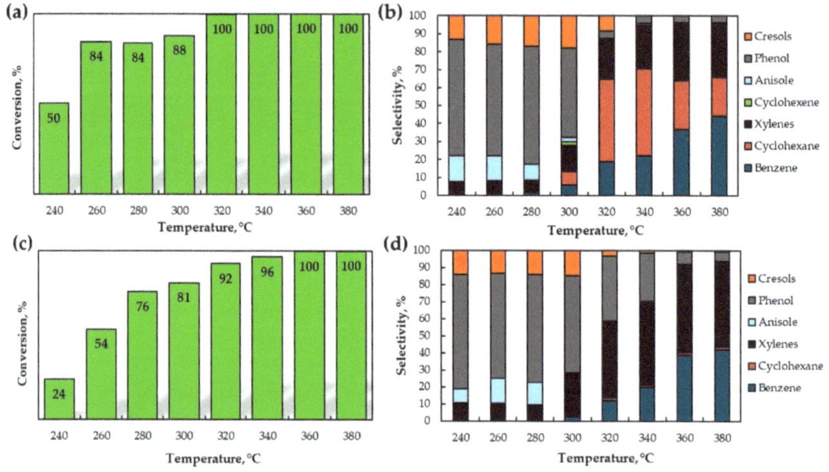

Figure 5. The results of guaiacol HDO over in situ-formed (**a**,**b**) MoO_x (**c**,**d**) WO_x. Reaction conditions: 120–380 °C, 5 MPa H_2, 6 h; 10 wt. % guaiacol in toluene solution.

2.2.2. The Effect of Pressure in Toluene

Because cyclohexane was formed in significant amounts when MoO_x was used as a catalyst, the effect of H_2 pressure (1–5 MPa) on product formation was studied (Figure 6). Guaiacol conversion was 100% in all cases. In the case of MoO_x, benzene, xylenes, cyclohexane and phenol were formed as products (Figure 6a). An increase in pressure led to a decrease in selectivity for benzene and an increase in selectivity for cyclohexane. Thus, carrying out the HDO of guaiacol over MoO_x under low H_2 pressure contributes to higher benzene selectivity. At the same time, H_2 pressure is shown to have almost no effect on benzene selectivity when WO_x is used as a catalyst (Figure 6b).

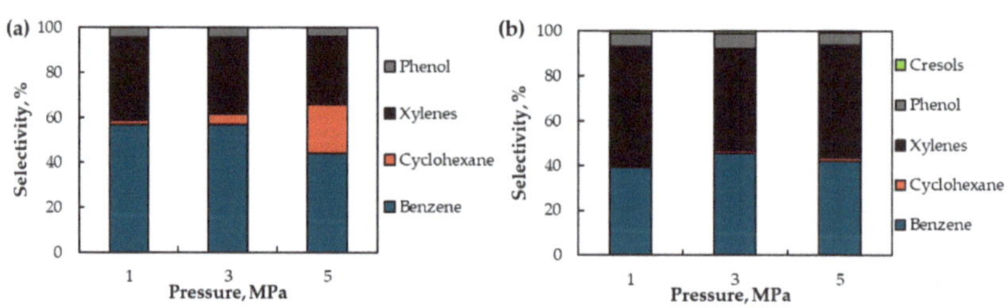

Figure 6. The results of guaiacol HDO over in situ formed (**a**) MoO_x (**b**) WO_x. Reaction conditions: 380 °C, 1–5 MPa H_2, 6 h; 10 wt. % guaiacol in toluene solution.

2.2.3. Reaction Pathways

The HDO of guaiacol over MoO_x and WO_x can occur via three routes, i.e., direct deoxygenation to anisole (DDO), demethoxylation to phenol (DMO), and DMO followed by methylation to cresols (Figure 7). The formation of cresols also occurs through transalkylation (TRA) of anisole [58]. Cresols can then convert to xylenes via two pathways: methylation (ME) to xylenols followed by DDO, or DDO to toluene followed by ME. However, xylenols were not detected, and toluene was used as a solvent. Phenol is formed not only by DMO of guaiacol but also by demethylation (DME) of anisole. Phenol then converts to benzene through DDO. Benzene is transformed into cyclohexene and cyclohexane by hydrogenation (HYD). Cyclohexene may also be formed by dehydratation of cyclohexanol, however, cyclohexanol was not identified in the reaction medium. Cyclohexane is also obtained by the HYD of cyclohexene.

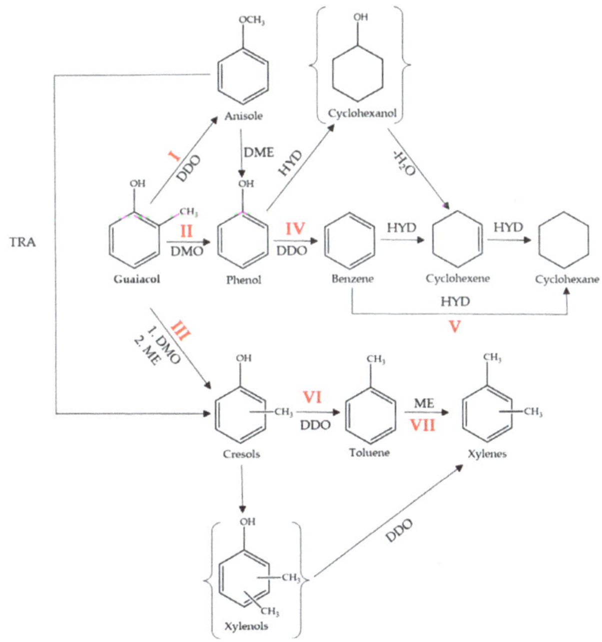

Figure 7. Proposed reaction pathways of guaiacol HDO over in situ formed MoO_x and WO_x. DDO—direct deoxygenation, HYD—hydrogenation, DMO—demethoxylation, DME—demethylation, ME—methylation, TRA—transalkylation. The following conditions favored each path: **I**—MoO_x or WO_x, 240–280 °C, 5 MPa (anisole is a minor product); **II**—MoO_x or WO_x, 240–300 °C, 5 MPa; **III**—MoO_x or WO_x, 240–300 °C, 5 MPa (cresols are minor products); **IV**—MoO_x, 320–360 °C, 5 MPa; 380 °C, 1–5 MPa; WO_x, 360 °C, 5 MPa; 380 °C, 1–5 MPa; **V**—MoO_x, 320–380 °C, 5 MPa; MoO_x, 340 °C, 5 MPa, 6 h; **VI, VII**—MoO_x or WO_x, 320–360 °C, 5 MPa; 380 °C, 1–5 MPa.

2.2.4. The Effect of Temperature and Pressure in Dodecane

Since xylenes were one of the reaction products, toluene was also probably formed. Thereby, the solvent was changed from toluene to dodecane. The catalytic tests were carried out at 300–380 °C, 5 MPa H_2, and 6 h of reaction. The yield of HDO products (benzene, toluene, xylenes, cyclohexene, cyclohexane) and the BTX fraction for both catalysts are shown in Figure 8. For both catalysts, the rise in the reaction temperature contributes to an increase in the yield of HDO products and the BTX fraction. At the same time, WO_x is more selective in the formation of BTX fraction, while in the presence of MoO_x, cyclohexane is also one of the HDO products. In addition, the catalytic tests were carried out at 380 °C, 1–5 MPa H_2 and 6 h of reaction. As shown in Figure S12, the selectivity for BTX products changes insignificantly, however, higher pressure results in more cyclohexane selectivity, especially when MoO_x is used.

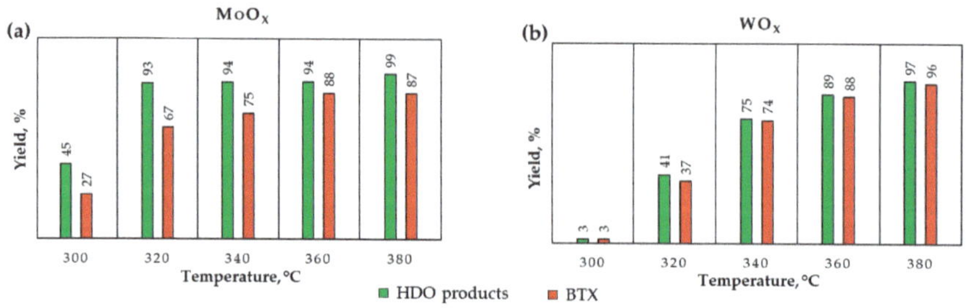

Figure 8. The yield of HDO products and BTX over in situ-formed (**a**) MoO_x (**b**) WO_x. Reaction conditions: 300–380 °C, 5 MPa H_2, 6 h; 10 wt. % guaiacol in dodecane solution.

2.2.5. The Recycling Tests

Five runs of recycling tests were carried out to estimate the stability of MoO_x and WO_x catalysts (Figure S13). The stability tests were carried out at average conversions (320 °C, 5 MPa H_2, 1 h). The conversion of guaiacol over MoO_x was 73% at the first run. It gradually decreased and was 56% after the fifth run. Such a decrease in conversion indicates insignificant deactivation of the MoO_x catalyst during the recycling test runs. Phenol and HDO product selectivity remained almost unchanged. At the same conditions, the conversion of guaiacol over WO_x changed significantly from 43% at the first run to 8% at the fifth run. The reason for this may be the adsorption on the catalyst surface of significant amounts of reaction products (confirmed by DRIFTS and Raman spectroscopy), and, as a consequence, the blocking of active sites. Phenol and HDO product selectivity changed insignificantly.

3. Materials and Methods

3.1. Materials

All reagents were purchased from commercial suppliers and used as received. Molybdenum hexacarbonyl (LLC "Redkino experimental plant," Redkino, Russia, wt. % of Mo 36.5–37.5%), and tungsten hexacarbonyl (LLC "Redkino experimental plant," Redkino, Russia, wt. % of W 51–52%) were used as precursors for catalyst preparation. Guaiacol (Acros organics, Geel, Belgium, >99%) was used as a substrate for hydrodeoxygenation. Toluene (Component-reaktiv, Moscow, Russia, >98.5%) or dodecane (Sigma-Aldrich, Burlington, VT, USA, >99%) were used as solvents. Acetone (Component-reaktiv, Moscow, Russia, >99.5%,) and petroleum ether 40/70 (Component-reaktiv, Moscow, Russia, tech.) were used for catalyst washing. H_2 (≥98%, Air Liquide, Paris, France) and Ar (≥98%, Air Liquide, Paris, France) gases were also used.

3.2. Catalyst Preparation and Catalytic Tests

The catalysts were prepared in a 45 mL stainless-steel batch autoclave reactor during the HDO of guaiacol. A mixture of 84.4 mg $Mo(CO)_6$ (or 113.7 mg $W(CO)_6$) and guaiacol (2 g of 10 wt. % solution in toluene or dodecane) was added to the reactor. The HDO of guaiacol was investigated at 120–400 °C for 6 h while undergoing constant magnetic stirring (7000 rpm). The initial pressure of H_2 was 5 MPa. After the catalytic test, the reactor was cooled to a room temperature. Liquid products were isolated from the reactor and separated from the catalysts by centrifugation (5000× g rpm). Catalysts obtained during the reaction were washed by petroleum ether and acetone until the solution over the catalysts became colorless. Then, the catalysts were dried in an argon atmosphere at room temperature.

3.3. Characterization

Powder X-ray diffraction was used to determine the phase composition of the catalysts. The X-ray diffractograms were obtained for a range of 5−100° 2θ by using a Rigaku Rotaflex

RU-200 diffractometer (CuK_α radiation) equipped with a Rigaku D/Max-RC goniometer (a rotation speed of 1°/min; a step 0.04°). Qualitative phase analysis of the samples was carried out using the PDF-2 ICDD database of powder diffraction patterns. The average crystallite sizes were estimated using the Scherrer equation.

X-ray photoelectron spectra were obtained using a PREVAC EA15 electronic spectrometer (AlK_α radiation, hν = 1486.74 eV, 150 W). Spectra were recorded at a pressure not exceeding 5×10^{-9} mbar. Spectra deconvolution was carried out using a PeakFit software.

The morphology of the catalysts obtained was observed using scanning electron microscopy and high-resolution transmission electron microscopy. Scanning electron microscopy with energy-dispersive X-ray spectroscopy was performed using a Carl Zeiss NVision 40 microscope equipped with an Oxford Instruments X-Max EDX detector operated at 20 kV. High-resolution transmission electron microscopy with energy-dispersive X-ray spectroscopy were performed using a FEI Tecnai Osiris transmission electron microscope equipped with a field emission electron gun operated at 200 kV.

Diffuse reflectance infrared Fourier transform spectroscopy was carried out in a high-temperature PIKE Diffus IR cell coupled with a VERTEX-70 Bruker Fourier transform IR spectrometer. DRIFT spectra of catalysts (194 scans at a resolution of 2 cm^{-1} in a range of 600–4000 cm^{-1}) were investigated in an argon flow in a temperature range of 25–450 °C. Spectra processing was carried out using OPUS-7 software. Raman spectra were obtained using a Senterra II spectrometer (laser power is 0.25 mW, a wavelength is 532 nm; a magnification is 50×).

Liquid reaction products were identified by gas chromatography–mass spectrometry using a Thermo Scientific ISQ 7000 GC-MS system equipped with a Restek 5XI-17SIL MS CAP capillary column (30 m × 0.25 mm × 0.25 μm), with helium being used as a carrier gas. The quantitative analysis of liquid products was carried out using a Crystallux 4000 M gas chromatograph equipped with a flame ionization detector, an Optima-1 capillary column (25 m × 0.32 mm × 0.35 μm), with helium used as a carrier gas. Chromatograms were obtained and analyzed using a NetChromWin software. Guaiacol conversion (%), product selectivity (%), and yield of full HDO products (%) were calculated using the following equations:

$$Conversion\ of\ guaiacol\ (\%) = \frac{mole\ of\ substrate\ consumed}{initial\ mole\ of\ substrate} \times 100\%$$

$$Selectivity\ (\%) = \frac{mole\ of\ product\ formed}{\sum moles\ of\ all\ products} \times 100\%$$

$$Yield\ of\ full\ HDO\ products\ (\%) = \frac{\sum moles\ of\ full\ HDO\ products}{initial\ mole\ of\ substrate} \times 100\%$$

$$Yield\ of\ BTX\ fraction\ (\%) = \frac{\sum moles\ of\ BTX}{initial\ mole\ of\ substrate} \times 100\%$$

4. Conclusions

Th possibility of obtaining in situ MoO_x and WO_x from the corresponding carbonyls during the HDO of guaiacol is shown in this study. According to XRD, MoO_x is amorphous, while WO_x displayed the crystalline patterns of WO_3 at 180 °C and $W_{18}O_{49}$ at 240–360 °C. The HRTEM of the catalysts obtained at 360 °C confirms the results of the XRD. It was shown via XPS that an increase in reaction temperature promoted the reduction of the surface M^{6+} to M^{4+}. In the case of MoO_x, the reduction proceeded through the formation of an intermediate M^{5+}. The presence of M^{5+} and M^{4+} may indicate vacancy sites on the surface of the catalysts.

The catalysts obtained were investigated in the HDO of guaiacol. Low guaiacol conversions at temperatures of 120–180 °C were associated with incomplete decomposition of the precursors. The active phase of the catalysts began to form at temperatures above 200 °C. An increase in the reaction temperature promoted the formation of vacancy sites on which hydrodeoxygenation could proceed. The full conversion of guaiacol was reached at 320 °C (5 MPa, 6 h) and 360 °C (5 MPa, 6 h) using MoO_x and WO_x, respectively. MoO_x, compared to WO_x, is shown to demonstrate higher activity in the full HDO of guaiacol. However, WO_x is more selective in the production of aromatics, in particular the BTX fraction. The highest yield of the BTX was 96% (380 °C, 5 MPa, 6 h) using the WO_x catalyst.

Supplementary Materials: The following supporting information can be downloaded at: https://www.mdpi.com/article/10.3390/catal13020263/s1, Figure S1: X-ray diffractograms of (**a**) molybdenum and (**b**) tungsten carbonyls—the precursors of the catalysts; Table S1: Oxidation states (Mo 3d, W 4f regions) on the surface of the catalysts, identified by the XPS; Figure S2: The XPS spectra in C 1 s region of the catalysts obtained at different temperatures; Figure S3: The XPS spectra in O 1 s region of the catalysts obtained at different temperatures; Table S2: Oxidation states (C 1 s region) on the surface of the catalysts, identified by the XPS; Table S3: Oxidation states (O 1 s region) on the surface of the catalysts, identified by the XPS; Figure S4: The elemental mapping of MoO_x. Maps are assigned to (**a**) molybdenum, (**b**) oxygen, (**c**) molybdenum and oxygen, and (**d**) the HAADF image; Figure S5: The elemental mapping of WO_x. Maps are assigned to (**a**) tungsten, (**b**) oxygen, (**c**) tungsten and oxygen, and (**d**) the HAADF image; Figure S6: The EDX spectra of the catalysts: (**a**) MoO_x (**b**) WO_x; Figure S7: The DRIFT spectra of MoO_x in a range of 4000–600 cm^{-1}; Figure S8: The DRIFT spectra of WO_x in a range of 4000–600 cm^{-1}; Figure S9: The Raman spectra of MoO_x and WO_x; Figure S10: The results of the guaiacol HDO over in situ formed (**a**) MoO_x (**b**) WO_x. Reaction conditions: 120–220 °C, 5 MPa H_2, 6 h; 10 wt. % guaiacol in toluene solution; Figure S11: The yield of HDO products over in situ-formed MoO_x and WO_x. Reaction conditions: 240–380 °C, 5 MPa H_2, 6 h; 10 wt. % guaiacol in toluene solution.; Figure S12: The results of guaiacol HDO over in situ formed (**a**) MoO_x (**b**) WO_x. Reaction conditions: 380 °C, 1–5 MPa H_2, 6 h; 10 wt. % guaiacol in dodecane solution.; Figure S13: The recycling test runs of (**a**) MoO_x (**b**) WO_x. Reaction conditions: 320 °C, 5 MPa H_2, 1 h; 10 wt. % guaiacol in dodecane solution.

Author Contributions: Conceptualization, M.M., M.G. and A.M.; validation, M.M. and M.G.; formal analysis, M.M. and M.G.; investigation, M.M. and A.S.; resources, A.S.; writing—original draft preparation, M.M. and M.G.; writing—review and editing, M.G. and A.M.; visualization, M.M. and M.G.; supervision, M.G. and A.M.; project administration, M.G. and A.M. All authors have read and agreed to the published version of the manuscript.

Funding: This work was carried out within the State Program of TIPS RAS.

Data Availability Statement: Not applicable.

Acknowledgments: The authors thank Ivan Levin from TIPS RAS for the XRD analysis, Olga Arapova from TIPS RAS for the Raman spectroscopy and DRIFTs analysis, and Yury Grigoriev from A.V. Shubnikov Institute of Crystallography RAS for the HRTEM analysis. This work was performed using the equipment of the Shared Research Center Analytical center of deep oil processing and petrochemistry of the A.V. Topchiev Institute of Petrochemical Synthesis RAS and the JRC PMR IGIC RAS.

Conflicts of Interest: The authors declare no conflict of interest.

References

1. Nanda, S.; Mohammad, J.; Reddy, S.N.; Kozinski, J.A.; Dalai, A.K. Pathways of lignocellulosic biomass conversion to renewable fuels. *Biomass Convers. Biorefin.* **2014**, *4*, 157–191. [CrossRef]
2. Cherubini, F.; Strømman, A.H. Production of Biofuels and Biochemicals from Lignocellulosic Biomass: Estimation of Maximum Theoretical Yields and Efficiencies Using Matrix Algebra. *Energy Fuels* **2010**, *24*, 2657–2666. [CrossRef]
3. Tribot, A.; Amer, G.; Abdou, A.M.; de Baynast, H.; Delattre, C.; Pons, A.; Mathias, J.; Callois, J.; Vial, C.; Michaud, P.; et al. Wood-lignin: Supply, extraction processes and use as bio-based material. *Eur. Polym. J.* **2019**, *112*, 228–240.
4. Zhang, J.; Sun, J.; Wang, Y. Recent Advances in Selectively Catalytic Hydrodeoxygenation of Lignin-derived Oxygenates to Arenes. *Green Chem.* **2020**, *22*, 1072–1098. [CrossRef]
5. Führer, M.; Van Haasterechta, T.; Bitter, J.H. Cinnamaldehyde hydrogenation over carbon supported molybdenum and tungsten carbide catalysts. *Chem. Commun.* **2022**, *58*, 13608–13611. [CrossRef]
6. Yang, Y.; Liu, X.; Xu, Y.; Gao, X.; Dai, Y.; Tang, Y. Palladium-Incorporated α-MoC Mesoporous Composites for Enhanced Direct Hydrodeoxygenation of Anisole. *Catalysts* **2021**, *11*, 370. [CrossRef]

7. Kurisingal, J.F.; Lee, S.; Lee, J.G.; An, K. Zeolitic Imidazolate Framework Decorated Molybdenum Carbide Catalysts for Hydrodeoxygenation of Guaiacol to Phenol. *Catalysts* **2022**, *12*, 1605. [CrossRef]
8. Wang, F.; Wen, C.; Lu, M.; Zhang, P.; Zhu, J.; Li, M.; Shan, Y.; Song, C. SBA-15-supported ultrastable Mo_2N@CN catalysts for hydrodeoxygenation of guaiacol. *Biomass Bioenergy* **2023**, *168*, 106680. [CrossRef]
9. Golubeva, M.A.; Mukhtarova, M.; Bugaev, A.L.; Naranov, E.R. In Situ Generated Dispersed Catalysts Based on Molybdenum and Tungsten Phosphides in Hydroprocessing of Guaiacol. *Pet. Chem.* **2023**. [CrossRef]
10. Golubeva, M.A.; Mukhtarova, M.; Sadovnikov, A.A.; Maximov, A.L. Bulk Molybdenum and Tungsten Phosphides for Selective Phenol Production from Guaiacol. *ACS Omega* **2022**, *7*, 40586–40595. [CrossRef]
11. Zhang, X.; Tang, J.; Zhang, Q.; Liu, Q.; Li, Y.; Chen, L.; Wang, C.; Ma, L. Hydrodeoxygenation of lignin-derived phenolic compounds into aromatic hydrocarbons under low hydrogen pressure using Molybdenum oxide as catalyst. *Catal. Today* **2019**, *319*, 41–47. [CrossRef]
12. Jiang, S.; Ji, N.; Diao, X.; Li, H.; Rong, Y.; Lei, Y.; Yu, Z. Vacancy Engineering in Transition Metal Sulfide and Oxide Catalysts for Hydrodeoxygenation of Lignin-Derived Oxygenates. *ChemSusChem* **2021**, *14*, 4377–4396. [CrossRef] [PubMed]
13. Prasomsri, T.; Nimmanwudipong, T.; Román-Leshkov, Y. Effective hydrodeoxygenation of biomass-derived oxygenates into unsaturated hydrocarbons by MoO_3 using low H_2 pressures. *Energy Environ. Sci.* **2013**, *6*, 1732. [CrossRef]
14. Bagnato, G.; Sanna, A.; Paone, E.; Catizzone, E. Recent catalytic advances in hydrotreatment processes of pyrolysis bio-oil. *Catalysts* **2021**, *11*, 157. [CrossRef]
15. Thibodeau, T.J.; Canney, A.S.; DeSisto, W.J.; Wheeler, M.C.; Amar, F.G.; Frederick, B.G. Composition of tungsten oxide bronzes active for Hydrodeoxygenation. *Appl. Catal. A* **2010**, *388*, 86–95. [CrossRef]
16. Whiffen, V.M.L.; Smith, K.J. Hydrodeoxygenation of 4-Methylphenol over Unsupported MoP, MoS_2 and MoO_x Catalysts. *Energy Fuels* **2010**, *24*, 4728–4737. [CrossRef]
17. Wang, C.; Mironenko, A.V.; Raizada, A.; Chen, T.; Mao, X.; Padmanabhan, A.; Vlachos, D.G.; Gorte, R.J.; Vohs, J.M. Mechanistic Study of the Direct Hydrodeoxygenation of m-Cresol over WO_x-Decorated Pt/C Catalysts. *ACS Catal.* **2018**, *8*, 7749–7759. [CrossRef]
18. Wang, C.; Wittreich, G.R.; Lin, C.; Huang, R.; Vlachos, D.G.; Gorte, R.J. Hydrodeoxygenation of m-Cresol Over Pt-WO_x/C Using H_2 Generated In Situ by n-Hexane Dehydrogenation. *Catal. Lett.* **2020**, *150*, 913–921. [CrossRef]
19. Chen, T.; Kwon, O.; Huang, R.; Lin, C.; Vohs, J.M. WO_x promoted nickel catalyst for hydrodeoxygenation of m-cresol. *J. Catal.* **2021**, *400*, 294–300. [CrossRef]
20. Golubeva, M.A.; Maximov, A.L. Hydroprocessing of furfural over in situ generated nickel phosphide based catalysts in different solvents. *Appl. Catal. A* **2020**, *608*, 117890. [CrossRef]
21. Golubeva, M.A.; Maximov, A.L. Investigations on the Formation of Transition Metal Phosphides during the Hydrotreating of Light Cycle Oil Russ. *J. Appl. Chem.* **2021**, *94*, 1536–1545.
22. Vagvala, T.C.; Pandey, S.S.; Ogomi, Y.; Ma, T.; Hayase, S. Investigation of metal xanthates as latent curing catalysts for epoxy resin via formation of in-situ metal sulfides. *Inorg. Chim. Acta* **2015**, *435*, 292–298. [CrossRef]
23. Vutolkina, A.V.; Baygildin, I.G.; Glotov, A.P.; Cherednichenko, K.A.; Maksimov, A.L.; Karakhanov, E.A. Dispersed Ni-Mo sulfide catalysts from water-soluble precursors for HDS of BT and DBT via in situ produced H_2 under Water gas shift conditions. *Appl. Catal. B* **2021**, *282*, 119616. [CrossRef]
24. Kuchinskaya, T.; Kniazeva, M.; Samoilov, V.; Maximov, A. In Situ Generated Nanosized Sulfide Ni-W Catalysts Based on Zeolite for the Hydrocracking of the Pyrolysis Fuel Oil into the BTX Fraction. *Catalysts* **2020**, *10*, 1152. [CrossRef]
25. Zhang, K.; McCleese, C.; Lin, P.; Chen, X.; Morales, M.; Cao, W.; Seo, F.J.; Burda, C.; Baumgart, H. Synthesis of ALD Tungsten Trioxide Thin Films from $W(CO)_6$ and H_2O Precursors. *ECS Trans.* **2015**, *69*, 199–209. [CrossRef]
26. Davazoglou, D.; Moutsakis, A.; Valamontes, V.; Psychorish, V. Tungsten Oxide Thin Films Chemically Vapor Deposited at Low Pressure by $W(CO)_6$ Pyrolysis. *J. Electrochem. Soc.* **1997**, *144*, 595–599. [CrossRef]
27. Suvanto, M.; Pakkanen, T. Deposition of tungsten hexacarbonyl on alumina: A diffuse reflectance infrared Fourier transform spectroscopy study. *J. Mol. Catal. A Chem.* **1999**, *138*, 211–220. [CrossRef]
28. Ou, N.C.; Su, X.; Bock, D.C.; McElwee-White, L. Precursors for chemical vapor deposition of tungsten oxide and molybdenum oxide. *Coord. Chem. Rev.* **2020**, *421*, 213459. [CrossRef]
29. Dhas, N.A.; Gedanken, A. Characterization of Sonochemically Prepared Unsupported and Silica-Supported Nanostructured Pentavalent Molybdenum Oxide. *J. Phys. Chem. B* **1997**, *101*, 9495–9503. [CrossRef]
30. Dimitrova, Z.; Gogova, D. On the structure, stress and optical properties of CVD tungsten oxide films. *Mater. Res. Bull.* **2005**, *40*, 333–340. [CrossRef]
31. Fu, J.; Liu, S.; Zheng, W.; Huang, R.; Wang, C.; Lawal, A.; Alexopoulos, K.; Liu, S.; Wang, Y.; Yu, K.; et al. Modulating the dynamics of Brønsted acid sites on $PtWO_x$ inverse catalyst. *Nat. Catal.* **2022**, *5*, 144–153. [CrossRef]
32. Fang, Z.; Jiao, S.; Kang, Y.; Pang, G.; Feng, S. Photothermal Conversion of $W_{18}O_{49}$ with a Tunable Oxidation State. *ChemistryOpen* **2017**, *6*, 261–265. [CrossRef] [PubMed]
33. Baltrusaitis, J.; Mendoza-Sanchez, B.; Fernandez, V.; Veenstra, R.; Dukstiene, N.; Roberts, A.; Fairley, N. Generalized molybdenum oxide surface chemical state XPS determination via informed amorphous sample model. *Appl. Surf. Sci.* **2015**, *326*, 151–161. [CrossRef]

34. Xie, F.; Choy, W.C.H.; Wang, C.; Li, X.; Zhang, S.; Hou, J. Low-Temperature Solution-Processed Hydrogen Molybdenum and Vanadium Bronzes for an Efficient Hole-Transport Layer in Organic Electronics. *Adv. Mater.* **2013**, *25*, 2051–2055. [CrossRef] [PubMed]
35. Kim, H.; Cook, J.B.; Lin, H.; Ko, J.S.; Tolbert, S.H.; Ozolins, V.; Dunn, B. Oxygen vacancies enhance pseudocapacitive charge storage properties of MoO_{3-x}. *Nat. Mater.* **2017**, *16*, 454–460. [CrossRef] [PubMed]
36. Apergi, S.; Koch, C.; Brocks, G.; Olthof, S.; Tao, S. Decomposition of Organic Perovskite Precursors on MoO_3: Role of Halogen and Surface Defects. *Appl. Mater. Interfaces* **2022**, *14*, 34208–34219. [CrossRef] [PubMed]
37. Choi, J.G.; Thompson, L.T. XPS study of as-prepared and reduced molybdenum oxides. *Appl. Surf. Sci.* **1996**, *93*, 143–149. [CrossRef]
38. Lu, D.Y.; Chen, J.; Zhou, J.; Deng, S.Z.; Xu, N.S.; Xu, J.B. Raman spectroscopic study of oxidation and phase transition in $W_{18}O_{49}$ nanowires. *J. Raman Spectrosc.* **2007**, *38*, 176–180. [CrossRef]
39. Katoh, M.; Takeda, Y. Chemical State Analysis of Tungsten and Tungsten Oxides Using an Electron Probe Microanalyzer. *Jpn. J. Appl. Phys.* **2004**, *43*, 7292–7295. [CrossRef]
40. Lin, Y.-S.; Lai, J.-Y.; Tsai, T.-H.; Chuang, P.-Y.; Chen, Y.-C. Effects of oxygen addition on electrochromic properties in low temperature plasma-enhanced chemical vapor deposition-synthesized MoO_xC_y thin films for flexible electrochromic devices. *Thin Solid Films* **2011**, *519*, 3875–3882. [CrossRef]
41. Światowska-Mrowiecka, J.; de Diesbach, S.; Maurice, V.; Zanna, S.; Klein, L.; Briand, E.; Vickridge, I.; Marcus, P. Li-Ion Intercalation in Thermal Oxide Thin Films of MoO_3 as Studied by XPS, RBS, and NRA. *J. Phys. Chem. C* **2008**, *112*, 11050–11058. [CrossRef]
42. Ghasempour, R.; Iraji, A. Hybrid multiwalled carbon nanotubes and trioxide tungsten nanoparticles for hydrogen gas sensing. *J. Phys. D Appl. Phys.* **2009**, *42*, 165105. [CrossRef]
43. Li, J.; Zhou, C.; Mu, J.; Yang, E.; Zhao, X. In situ synthesis of molybdenum carbide/N-doped carbon hybrids as an efficient hydrogen-evolution electrocatalyst. *RSC Adv.* **2018**, *8*, 17202–17208. [CrossRef] [PubMed]
44. Sajadi, M.; Ranjbar, M.; Rasuli, R. Two-step synthesis of Ag-decorated MoO_3 nanotubes, and the effect of hydrogen doping. *Appl. Surf. Sci.* **2020**, *527*, 146675. [CrossRef]
45. Vasilopoulou, M.; Soultati, A.; Georgiadou, D.G.; Stergiopoulos, T.; Palilis, L.C.; Kennou, S.; Stathopoulos, N.A.; Davazoglou, D.; Argitis, P. Hydrogenated under-stoichiometric tungsten oxide anode interlayers for efficient and stable organic photovoltaics. *J. Mater. Chem. A* **2014**, *2*, 1738. [CrossRef]
46. Wang, J.; Gao, Y.; You, Z.; Fan, J.; Zhang, J.; Wang, S.; Xu, J. Laser Induced Nano and Micro Structures of Molybdenum Surface Applied in Multistage Depressed Collector for Secondary Electron Suppression. *Appl. Sci.* **2019**, *9*, 4374. [CrossRef]
47. Feng, C.; Tang, L.; Deng, Y.; Wang, J.; Tang, W.; Liu, Y.; Chen, Z.; Yu, J.; Wang, J.; Liang, Q. Synthesis of branched $WO_3@W_{18}O_{49}$ homojunction with enhanced interfacial charge separation and full-spectrum photocatalytic performance. *Chem. Eng. J.* **2020**, *389*, 124474. [CrossRef]
48. Chen, C.L.; Mori, H. In situ TEM observation of the growth and decomposition of monoclinic $W_{18}O_{49}$ nanowires. *Nanotechnology* **2009**, *20*, 285604. [CrossRef]
49. Hahn, T.; Bentrup, U.; Armbrüster, M.; Kondratenko, E.V.; Linke, D. The Enhancing Effect of Brønsted Acidity of Supported MoO_x Species on their Activity and Selectivity in Ethylene/trans-2-Butene Metathesis. *ChemCatChem* **2014**, *6*, 1664–1672. [CrossRef]
50. Zafeiratos, S.; Papakonstantinou, G.; Jacksic, M.M.; Neophytides, S.G. The effect of Mo oxides and TiO_2 support on the chemisorption features of linearly adsorbed CO on Pt crystallites: An infrared and photoelectron spectroscopy study. *J. Catal.* **2005**, *232*, 127–136. [CrossRef]
51. Baes, A.U.; Bloom, P.R. Diffuse Reflectance and Transmission Fourier Transform Infrared (DRIFT) Spectroscopy of Humic and Fulvic Acids Soil. *Sci. Soc. Am. J.* **1989**, *53*, 695–700. [CrossRef]
52. Barreto, M.S.C.; Reis, J.V.; Muraoka, T.; Jemo, M.; Vergutz, L.; Alleoni, L.R.F. Diffuse reflectance infrared Fourier transform spectroscopy for a qualitative evaluation of plant leaf pigment extraction. *Analyst* **2021**, *146*, 3440. [CrossRef] [PubMed]
53. Li, C.; Zhang, C.; Zhang, L.; Gholizadeh, M.; Hu, X. Biochar catalyzing polymerization of the volatiles from pyrolysis of poplar wood. *Int. J. Energy. Res.* **2021**, *45*, 13936–13951. [CrossRef]
54. Ajito, K.; Nagahara, L.A.; Tryk, D.A.; Hashimoto, K.; Fujishima, A. Study of the Photochromic Properties of Amorphous MoO_3 Films Using Raman Microscopy. *J. Phys. Chem.* **1995**, *99*, 16383–16388. [CrossRef]
55. Casiraghi, C.; Piazza, F.; Ferrari, A.C.; Grambole, D.; Robertson, J. Bonding in hydrogenated diamond-like carbon by Raman spectroscopy. *Diamond Relat. Mater.* **2005**, *14*, 1098–1102. [CrossRef]
56. Tran, N.T.T.; Uemura, Y.; Chowdhury, S.; Ramli, A. Vapor-phase Hydrodeoxygenation of Guaiacol on Al−MCM−41 Supported Ni and Co Catalysts. *Appl. Catal. A* **2016**, *512*, 93–100. [CrossRef]
57. Bui, V.N.; Laurenti, D.; Afanasiev, P.; Geantet, C. Hydrodeoxygenation of guaiacol with CoMo catalysts. Part I: Promoting effect of cobalt on HDO selectivity and activity. *Appl. Catal. B* **2011**, *101*, 239–245. [CrossRef]
58. Venkatesan, K.; Krishna, J.V.J.; Anjana, S.; Selvam, P.; Vinu, R. Hydrodeoxygenation kinetics of syringol, guaiacol and phenol over H-ZSM-5. *Catal. Commun.* **2021**, *148*, 106164. [CrossRef]

catalysts

MDPI

Article

Cr-Zn/Ni-Containing Nanocomposites as Effective Magnetically Recoverable Catalysts for CO_2 Hydrogenation to Methanol: The Role of Metal Doping and Polymer Co-Support

Svetlana A. Sorokina [1], Nina V. Kuchkina [1], Maxim E. Grigoriev [2], Alexey V. Bykov [2], Andrey K. Ratnikov [1], Valentin Yu. Doluda [2], Mikhail G. Sulman [2] and Zinaida B. Shifrina [1,*]

[1] A.N. Nesmeyanov Institute of Organoelement Compounds, Russian Academy of Sciences, 28 Vavilov St., Moscow 119991, Russia
[2] Department of Biotechnology and Chemistry, Tver State Technical University, 22 A. Nikitina St., Tver 170026, Russia
* Correspondence: shifrina@ineos.ac.ru

Abstract: CO_2 hydrogenation to methanol is an important process that could solve the problem of emitted CO_2 that contributes to environmental concern. Here we developed Cr-, Cr-Zn-, and Cr-Ni-containing nanocomposites based on a solid support (SiO_2 or Al_2O_3) with embedded magnetic nanoparticles (NPs) and covered by a cross-linked pyridylphenylene polymer layer. The decomposition of Cr, Zn, and Ni precursors in the presence of supports containing magnetic oxide led to formation of amorphous metal oxides evenly distributed over the support-polymer space, together with the partial diffusion of metal species into magnetic NPs. We demonstrated the catalytic activity of Cr_2O_3 in the hydrogenation reaction of CO_2 to methanol, which was further increased by 50% and 204% by incorporation of Ni and Zn species, respectively. The fine intermixing of metal species ensures an enhanced methanol productivity. Careful adjustment of constituent elements, e.g., catalytic metal, type of support, presence of magnetic NPs, and deposition of hydrophobic polymer layer contributes to the synergetic promotional effect required for activation of CO_2 molecules as well. The results of catalytic recycle experiments revealed excellent stability of the catalysts due to protective role of hydrophobic polymer.

Keywords: methanol synthesis; CO_2 hydrogenation; magnetic nanoparticles; chromium oxide; nanocomposites; supported catalyst; polymer layer

1. Introduction

Human impact on the environment has become a critical issue for the last 20 years. Emissions of greenhouse gases are now believed to be responsible for global warming and have gained worldwide concern. CO_2 is one of the most emitted gases with level in the atmosphere raised from 315 ppm in 1958 when continuous observations began to 420 ppm in 2021 [1,2]. In this regard, CO_2 utilization and transformation into valuable chemicals are of particular importance.

CO_2 can be considered as a carbon feedstock alternative to hydrocarbons whose combustion depletes the carbon sources. The possible transformation routes for CO_2 molecule via catalytic hydrogenation include the formation of formic acid, methane, methanol, and DME [3,4]. The production of methanol from CO_2 is considered as one of the most effective ways for decreasing the CO_2 concentration [5,6]. Methanol is an important platform molecule which can be converted into a variety of reagents, such as formaldehyde, acetic acids, and others [7–9]. It is a component of fuel blends, antifreezes, resins, and plastics [10]. Currently methanol is almost exclusively synthesized from syngas via Fischer-Tropsch process over the Cu-ZnO-Al_2O_3 catalyst [11]. However, recent reports have defined CO_2 methanolization as one of the most exciting commercial applications for CO_2 capture [4,10,12,13].

The heterogeneous catalysts employed for CO_2 to methanol hydrogenation are usually ternary catalytic systems comprised of bimetallic species deposited on a solid support. While Cu-ZnO-based materials are the most investigated systems due to high activity and ability to operate under moderate reaction conditions, recent studies have been focused on expanding the scope of catalytic systems and chemical element combinations [13,14]. Thus, Cu-In [15], Pd-Zn [16–18], Rh-In [19], Pd-Cu [20], and Ni-Ga [21,22] catalytic systems have been explored. The key role of support is also highlighted in numerous papers [14,23]. A strong metal-support interaction may direct the reaction pathway, enhancing the selectivity and activity. For example, Ni-based catalysts without proper modification deliver only methane as a main product of CO_2 hydrogenation, while Ni deposited on Ga_2O_3 provides methanol with high selectivity [22]. The effect of metal-support interaction is attributed to the electronic and structural modification due to formation of oxygen vacancies, electronic doping, formation of alloy interface, and charged metal moieties, etc. [14,24–26]. Al_2O_3, CeO_2, ZrO_2, SiO_2, and TiO_2 are the widely examined solid materials for this purpose.

However, the explored catalysts usually suffer from relatively poor activity and low selectivity due to the presence of reverse water-gas shift (RWGS) reaction that proceeds along with the target hydrogenation reaction. RWGS reaction leads to the formation of CO and H_2O vapors lowering the methanol yield [14]. Moreover, water vapors also cause a notable decrease in catalyst stability and reusability [27,28]. Another major shortcoming of the nanosized catalysts used in methanol synthesis is the NP agglomeration that occurs due to the sintering process induced by high-reaction temperature and pressure [29,30]. Strong and reliable stabilization of NPs may suppress the sintering. This could be achieved by appropriate interactions with a support and/or incorporation of additional ligand molecules creating a barrier between metal species [23].

Therefore, novel catalysts developed for methanol synthesis from CO_2 should withstand harsh reaction conditions, preserve stability under high water partial pressure, possess high thermostability, and have easily adoptable microstructures. The synthetic procedures employed in catalyst production should be simple and scalable due to potential industrial implementation of the process.

In our preceding work, we developed the catalysts based on metal NPs stabilized by aromatic polymers for different hydrogenation reactions, such as synthesis of methanol from syngas [31–33], synthesis of gamma-valerolactone from levulinic acid [34], and furfuryl alcohol from furfural [35]. These catalysts demonstrated exceptional activity and stability which have been preserved in several consecutive catalytic cycles. The synthesis of these nanocomposites has been carried out by thermal decomposition of metal precursors in the presence of a polymer acting as a stabilizing media. Despite the precious control over the NP size and morphology provided by this technique, its employment in CO_2 methanolization is not fully reasonable because of low catalyst yield. Indeed, the approaches to nanocomposite formation providing several grams of catalyst during simple synthetic procedures are preferable. This could be achieved by addition of a solid support to the catalyst formulation. More recently, we developed an approach to the formation of a thin polymer layer on a mesoporous silica gel containing magnetic nanoparticles [36]. After incorporation of palladium species by coordination with pyridine groups of the polymer, the resulting nanocomposites were tested in Suzuki-Miyaura reaction. The results of recycle experiments revealed the important role of the branched polymer layer in NPs stabilization and superior catalytic activity under repeated use [36].

In this work, we applied this approach to fabricate the novel nanostructured catalytic systems for methanol synthesis from $CO_2 + H_2$. Deposition of a thin polymer layer which is hydrophobic in nature on a solid support (SiO_2 or Al_2O_3) containing metal NPs could be one of the possible strategies for attenuating the negative influence of H_2O vapors. Moreover, the polymer adsorption contributes to NP segregation and stabilization, thus preventing the possible sintering and leaching of NPs. Both processes are known to be responsible for significant drops in the catalyst activity [29,30].

Here we synthesized and explored Cr-, Cr-Zn-, and Cr-Ni-containing magnetically recoverable catalysts based on two different inorganic supports—SiO_2 and Al_2O_3—covered by a polymer layer. Recently, Cr_2O_3 has been shown to possess a potential activity in CO_2 methanolization [37] while having an established productivity in the methanol synthesis from syngas [33,38,39]. The oxygen vacancies presented on the surface of Cr_2O_3 make it suitable for activation of thermodynamically stable and inert CO_2 molecule [37,40]. The activity of Cr_2O_3 can be adopted by doping with different metals like Ni, La, etc. [41]. In this work we expanded the studies of Cr_2O_3 catalytic activity in methanol synthesis via CO_2 hydrogenation. To elucidate the main factors underlying the high-catalyst activity, we fabricated both monometallic and bimetallic catalytic systems. Magnetic NPs included in the catalyst structure do not possess the catalytic activity in this reaction; however, they may boost methanol productivity while ensuring the effective magnetic separation for repeated uses and more sustainable processes.

2. Results and Discussion

To meet requirements for effective CO_2 methanolization the designed catalytic system should be (1) stable under harsh reaction conditions and high partial pressure of water vapors, (2) easily separable from reaction mixture to allow the repeated uses, and (3) synthesized via robust procedure. To develop such catalyst, we used a combination of synthetic methods.

At the first step, Fe_3O_4 NPs were formed in the pores of commercially available mesoporous solid support according to the procedure described previously [42]. Here we used two different supports: Al_2O_3 and SiO_2 to further assess their influence on the catalytic activity. Thermal decomposition of iron nitrate in the presence of the support and mild reducing agent led to the formation of magnetic nanoparticles of 13.2 and 15.6 nm for SiO_2- and Al_2O_3-based composites, respectively, oriented alongside the pores [42,43]. Magnetic NPs ensure the easy and simple separation of catalyst from the reaction mixture by external magnet. After formation of magnetic support, a thin layer of cross-linked pyridylphenylene polymer (PPP) has been deposited by Diels Alder polycondensation of two branched monomers in the presence of magnetic silica/aluminia. Structure of PPP together with the scheme of formation of a solid support -Fe_3O_4 –PPP composite are given in Figure S1 in SI. PPP is a thermally stable polymer whose thermal decomposition starts at 500 °C which is important for practical application [32,34]. It is worth noting that deposition of PPP on the surface of support does not decrease the porosity of material due to cross-linked hyperbranched structure, as was established earlier [36]. This means that a polymer layer does not hinder the active sites of the future catalyst and does not prevent the diffusion of substrate molecules. The presence of pyridine moieties in PPP structure may contribute to more effective chemisorption of CO_2 molecules being the first step of catalysis because of electropositivity of CO_2 carbon atom.

The metal species, namely Cr, Cr-Zn, and Cr-Ni, have been incorporated into the structure of SiO_2/Al_2O_3-Fe_3O_4-PPP composites by wet impregnation method of corresponding metal acetyl acetonates followed by a thermal decomposition in a furnace tube at 350 °C. To elucidate the possible promoting effect of Fe_3O_4 molecules, the nanocomposites without magnetic NPs have been synthesized as well.

2.1. Structure and Morphology of the Nanocomposites

To identify the position of all species in the composites, STEM EDS mapping was performed. Figure 1 shows the STEM dark field image and EDS maps of SiO_2-Fe_3O_4-PPP-Cr-Zn nanocomposite. The carbon map demonstrates a full coverage of silica by polymer layer. Fe map totally repeats the shape of Si map indicating the formation of Fe_3O_4 NPs in silica pores. Cr and Zn maps are similar to Fe map which may correspond to the incorporation of Cr and Zn species in magnetite structure. It is worth noting that deposition of catalytic species on the top of magnetic oxide with formation of heterogeneous hybrid nanocomposite is more preferable for effective catalysis rather than homogeneous nucleation of separately being Cr_2O_3 and ZnO_2 nanoparticles [25,44].

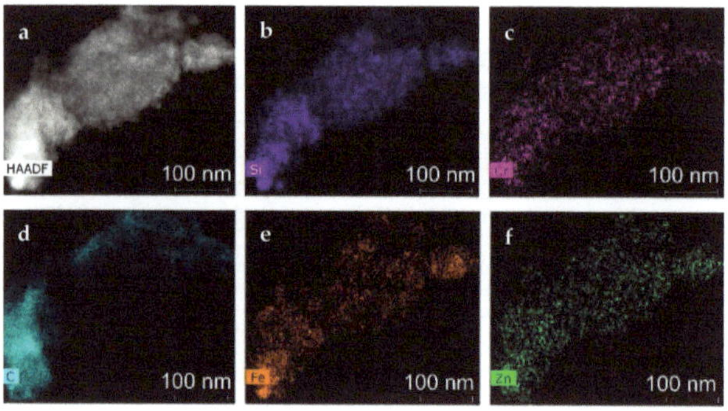

Figure 1. STEM dark field image (**a**) and EDS maps of Si (**b**), Cr (**c**), C (**d**), Fe (**e**) and Zn (**f**) of SiO_2-Fe_3O_4-PPP-Cr-Zn nanocomposite.

STEM EDS images for SiO_2-Fe_3O_4-PPP-Cr-Ni nanocomposite are given in Figure 2. Again, C, Si, Fe, Cr, and Ni maps show a good fit revealing the formation of magnetic and catalytic species in the silica covered by polymer. However, Ni species are distributed across the sample with some aggregation of NPs. This may be due to a poor intermixing of Ni species with a magnetite phase.

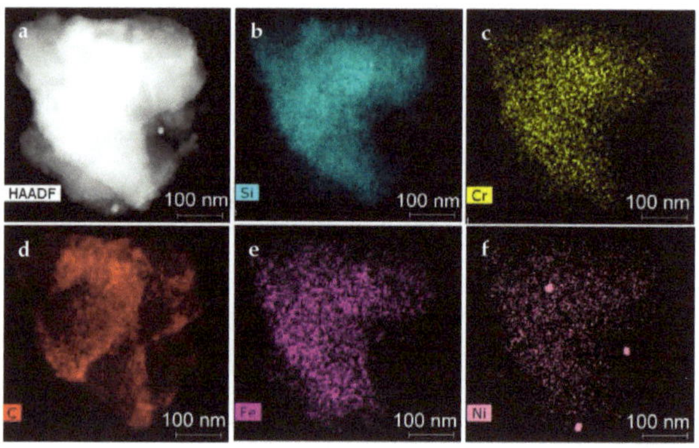

Figure 2. STEM dark field image (**a**) and EDS maps of Si (**b**), Cr (**c**), C (**d**), Fe (**e**) and Ni (**f**) of SiO_2-Fe_3O_4-PPP-Cr-Ni nanocomposite.

STEM EDS maps of Al_2O_3-Fe_3O_4-PPP-Cr-Zn, SiO_2-PPP-Cr-Zn, and SiO_2-Fe_3O_4-PPP-Cr are presented in Figures S2–S4, respectively. Replacement of SiO_2 with Al_2O_3 did not influence the deposition of PPP on the support. The positions of metal species for all composites demonstrated their perfect distribution over the support-polymer area and a good correlation with Fe map.

To assess the chemical composition of the materials obtained, EDS spectra were recorded (Figure 3a,b for SiO_2-Fe_3O_4-PPP-Cr-Zn, and SiO_2-Fe_3O_4-PPP-Cr-Ni, respectively, and Figure S5a,b for SiO_2-PPP-Cr-Zn and SiO_2-Fe_3O_4-PPP-Cr, respectively). The spectra confirmed the presence of constituent elements together with successful deposition of a polymer layer which is evidenced by the presence of carbon in the composite. The elemental compositions for all samples are given in Table S1.

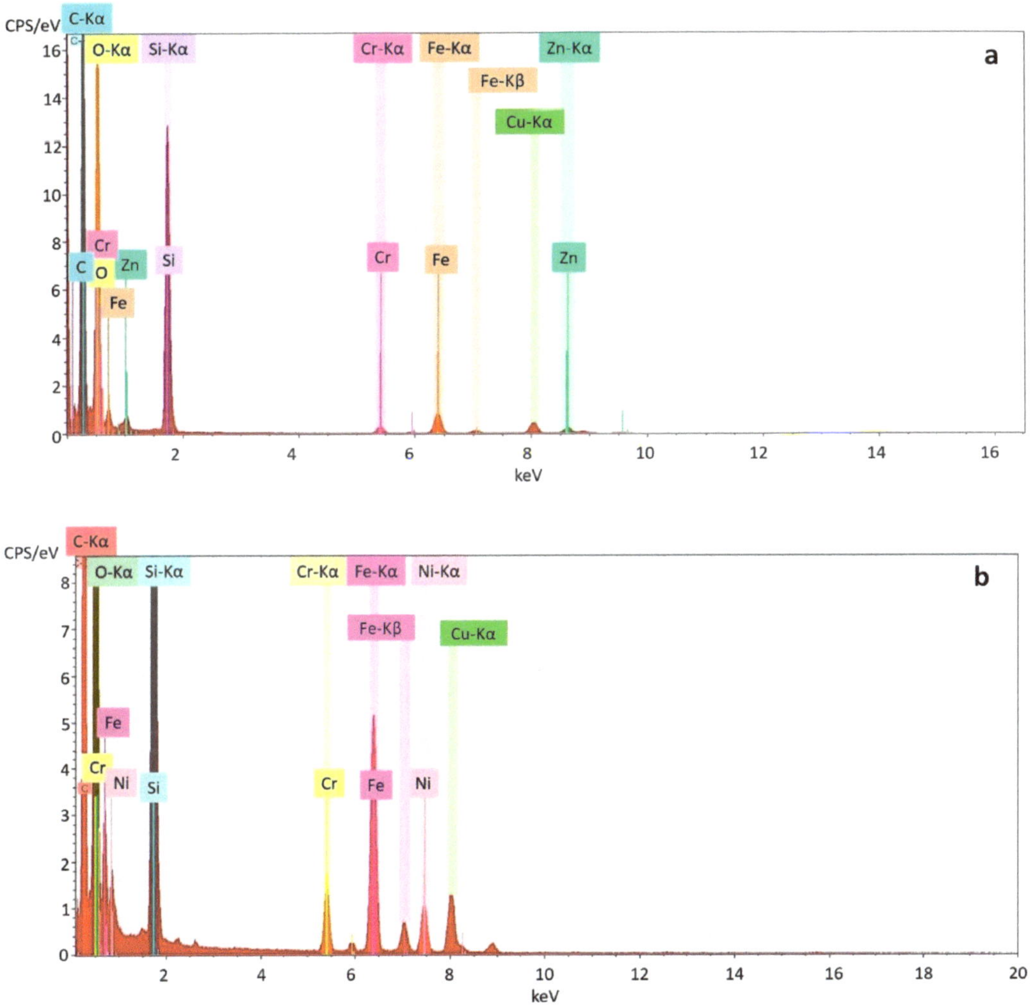

Figure 3. EDS spectra of SiO_2-Fe_3O_4-PPP-Cr-Zn (**a**) and SiO_2-Fe_3O_4-PPP-Cr-Ni (**b**) nanocomposites. Cu-Kα signals are presented due to Cu-grid used for recording of the spectra.

Figure 4 displays the XRD spectra of SiO_2-Fe_3O_4-PPP (a), SiO_2-Fe_3O_4-PPP-Cr-Zn (b), and SiO_2-Fe_3O_4-PPP-Cr-Ni (c) nanocomposites, which are very similar to each other and show a polymer halo at ~10° 2θ and a broad signal at 22° corresponding to amorphous silica. A set of reflections characteristic for a spinel structure of magnetite is also observed for all samples. Surprisingly, no additional reflections are detected, which could be attributed to the formation of zinc or chromium oxides. This could be due to the formation of amorphous oxide structures or different ferrites whose reflections are overlapped with those of Fe_3O_4 or due to intermixing of Zn, Cr, and Ni species with a magnetite phase that do not disturb the Fe_3O_4 crystalline structure. The more careful inspection of XRD-spectra revealed a slight shift of (311) peak of Fe_3O_4 to smaller angles for Cr-Zn-containing composites. The enlarged depiction of this region is presented in Figure S6. This could indicate a partial penetration of Cr and Zn species in a magnetite structure as it was reported earlier [42,45,46]. The lattice constants calculated from XRD data showed an increase from 8.3801 for SiO_2-Fe_3O_4-PPP to 8.3982 for SiO_2-Fe_3O_4-PPP-Cr-Zn. The effect could be attributed to the substitution of some Fe ions with Zn^{2+} and Cr^{3+}, and formation of ferrites. Another possible explanation is the penetration of Zn^{2+} and Cr^{3+} possessing the higher ionic radii into Fe_3O_4 NP.

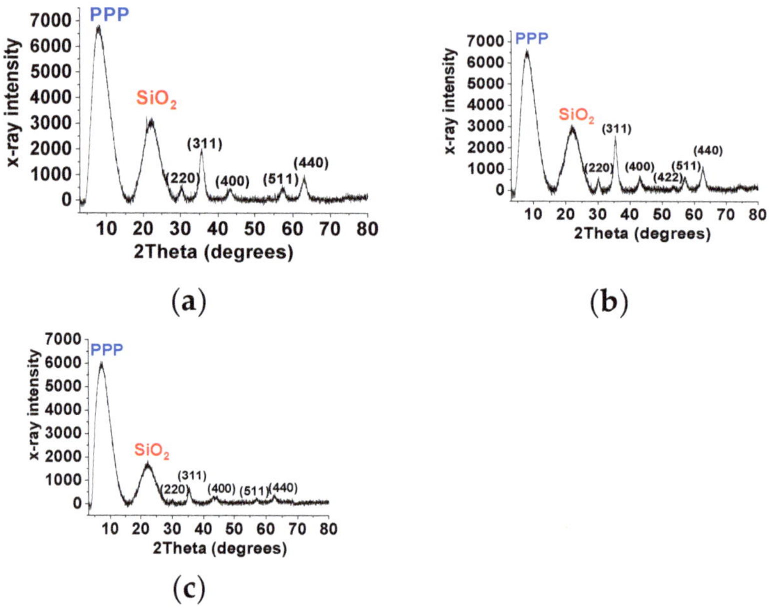

Figure 4. XRD patterns of SiO_2-Fe_3O_4-PPP (**a**); SiO_2-Fe_3O_4-PPP-Cr-Zn (**b**); SiO_2-Fe_3O_4-PPP-Cr-Ni (**c**).

In case of composite without Fe_3O_4 only the signals corresponding to PPP and SiO_2 are presented in diffractogram (Figure S7a). This suggests the formation of amorphous Cr- and Zn-containing species. At the same time, for Al_2O_3-based composites containing Fe_3O_4, the diffraction pattern contains a set of peaks corresponding to Al_2O_3 and reflections attributed to Fe_3O_4 structure (Figure S7b,c). Similar to composites based on SiO_2-, the reflections characteristic for Cr and Zn oxides are absent.

To achieve further insight into the structure of the composites obtained, XPS spectra were recorded. The survey spectra of the samples are given in Figure S8. The oxidation states of the metal species were analyzed by high resolution (HR) XPS. In all samples, Fe 2p spectrum contains a main peak at 711.0 eV which is typical for iron oxides [47] (Figure 5a for SiO_2-Fe_3O_4-PPP-Cr-Zn, and SI for other Fe_3O_4-containig composites). A region between Fe $2p_{3/2}$ and Fe $2p_{1/2}$ displays a plateau indicating the formation of Fe_3O_4. In the opposite case (formation of Fe_2O_3), the prevalence of Fe^{3+} ions would result in a satellite peak with binding energy 8 eV higher than the main peak [47]. Importantly, impregnation by Cr, Zn, and Ni species do not disturb the Fe^{3+}/Fe^{2+} ratio. If any substitution of Fe^{2+} by Zn^{2+} or Ni^{2+} would appear, which took place upon the formation of ferrites structures like $ZnFe_2O_4$, $NiFe_2O_4$, $ZnCrFeO_4$, or chromite, this would influence the Fe^{3+}/Fe^{2+} ratio and lead to the occurrence of additional signals, in particular the satellite peak, or peak shifting (See Figures) [48,49]. However, the Fe 2p spectra before and after incorporation of metal compounds are similar (See Figure 5a and Figure S9).

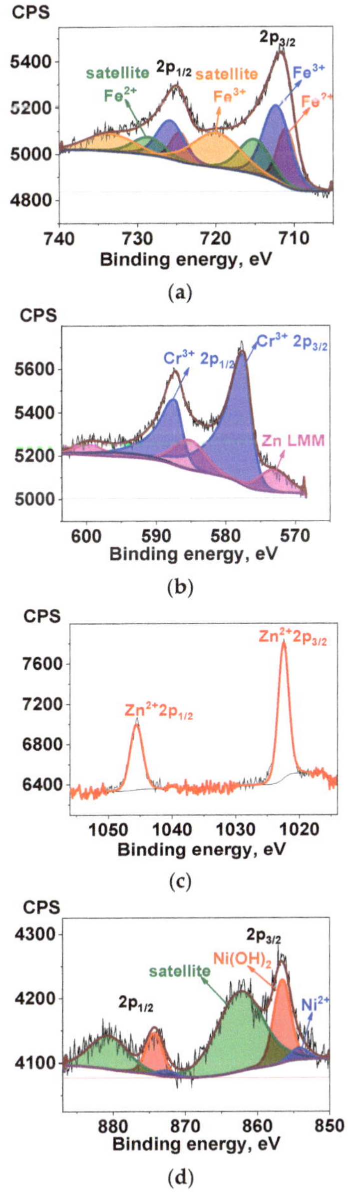

Figure 5. HR XPS spectra of SiO_2-Fe_3O_4-PPP-Cr-Zn in the Fe 2p (**a**); Cr 2p (**b**); Zn 2p (**c**) regions; and Ni 2p (**d**) of SiO_2-Fe_3O_4-PPP-Cr-Ni.

The analysis of Cr 2p XPS spectra (Figure 5b) revealed that Cr solely exists in the form of Cr^{3+}. The spin-orbit splitting energy (Cr $2p_{3/2}$—Cr $2p_{1/2}$ = 9.76 eV) coincides with that of Cr_2O_3 [50]. The positions of peaks at Zn 2p XPS spectra with binding energies of 1022.8 and 1045.9 eV for $2p_{3/2}$ and $2p_{1/2}$, respectively, correspond to Zn^{2+} typical for ZnO (Figure 5c) [51]. For Ni-containing composite, the positions of peaks at HR XPS spectrum match Ni^{2+} state in oxide-hydroxide form [49,52,53] (Figure 5d). The deconvolution parameters for all spectra are summarized in Tables S2–S7. HR XPS of Fe, Cr, and Zn for all nanocomposites are presented in SI, Figures S9–S13. The positions of the main peaks are consistent with those described in Figure 5.

Thus, one can conclude that formation of ferrites is not observed, and decomposition of Cr, Zn, and Ni precursors leads to formation of amorphous metal oxides. However, considering the peak shifting (Figure S6) and increase in the values of lattice constants revealed by XRD, we assumed the possible diffusion of Cr, Zn, and Ni species into magnetic NPs. This assumption is based on XPS data for 2p and 3p electrons. 2p electrons are known to possess considerably lower kinetic energy than that of 3p electrons. Thus, the contribution of 2p electrons in the spectrum reflects the surface concentration of the element, while 3p electrons characterize the deeper layers of the material. Data on 2p and 3p content of the metals in the nanocomposites are presented in Table 1. As one can see, the enrichment of the surface with metal is dependent on the metal type. For all samples, the composites surface was enriched with Cr^{3+} species since the chrome content for 2p electrons was higher than that of 3p electrons. However, small amount of Cr^{3+} species were able to penetrate the deeper magnetite NP level as well. For Ni-containing samples, XPS data show the surface accumulation of Ni species. Surprisingly, Zn content was higher for 3p electrons which means a higher fraction of Zn in the deeper layers rather than at the surface. This phenomenon could be explained by a better compatibility of Zn^{2+} with Fe_3O_4 in comparison with that of Cr^{3+} and Ni^{2+}. Although, the possibility of solid solution and alloy formation is governed by Hume-Rothery rules, small metal clusters do not always follow the rules [54]. In addition, the thermal decomposition of zinc acetyl acetonate starts at 190 °C and ZnO NPs formation was reported to proceed at 195 °C [55,56]. For chromium acetyl acetonate, the decomposition temperature is much higher (250 °C) [57]. The differences in decomposition temperature and consequent earlier destruction of Zn precursor are another reason for deeper migration of Zn^{2+} into a magnetite structure in comparison with Cr^{3+}.

Table 1. Atomic concentrations of metals for Cr-, Zn-, and Ni-containing magnetic nanocomposites determined by HR XPS for 2p and 3p transitions.

Sample	Relative Content by XPS, Atomic %			
	Fe 2p/3p	Cr 2p/3p	Zn 2p/3p	Ni 2p/3p
SiO_2-Fe_3O_4-PPP-Cr-Zn	0.50/0.30	0.41/0.07	0.23/0.32	-
SiO_2-Fe_3O_4-PPP-Cr-Ni	0.55/0.26	0.54/0.09	-	0.45/0
Al_2O_3-Fe_3O_4-PPP-Cr-Zn	0.52/0.31	0.41/0.08	0.25/0.31	-

To conclude, most probably the decomposition of Cr, Zn, and Ni acetyl acetonates, in the presence of magnetite-containing solid supports covered by PPP, results in the formation of thin metal oxide layers on SiO_2(Al_2O_3)-Fe_3O_4-PPP composites together with a partial diffusion of metal oxides into magnetic NP. The diffusion was the most prominent for Zn^{2+} ions and slightly occurred for Cr^{3+}. Such intermixing does not influence the Fe^{3+}/Fe^{2+} ratio and the magnetite structure was preserved in all samples. Considering the structural data obtained, the formation of ferrite structures like $ZnFe_2O_4$, $ZnCrFeO_4$, and $NiFe_2O_4$ were not established. Therefore, intermixing of ZnO and Cr_2O_3 with magnetite NP stands at the intermediate stage before the ferrite formation which has been reported elsewhere [48].

Since the adsorption of CO_2 and H_2 plays an important role for the CO_2 hydrogenation process, the adsorption capacities of the composites have been estimated by CO_2 and H_2 temperature-programmed desorption (TPD) measurements. The results are presented in Figures 6 and 7. To elucidate the polymer ability to enhance the CO_2 adsorption, measurements were carried out for parent SiO_2-Fe_3O_4 and after deposition of PPP-SiO_2-Fe_3O_4-PPP (Figure 6a). Placement of a thin polymer layer on a solid support significantly increases the CO_2 adsorption. While parent SiO_2-Fe_3O_4 is characterized by two types of adsorption centers (moderate center at 266 °C and strong center at 478 °C) with the total amount of adsorbed CO_2 equal to 0.015 mmol/g, the addition of PPP results in noticeable changes in adsorption centers. The appearance of new desorption peaks at 244 °C, 294 °C, 366 °C, and 562 °C is accompanied by increase in CO_2 amount up to 0.035 mmol/g. The effect is undoubtedly attributed to the presence of basic pyridine centers in PPP structure. Since the CO_2 adsorption is the first step of the catalytic reaction, introduction of PPP into the catalyst structure should enhance the catalysis rate due to advanced CO_2 adsorption.

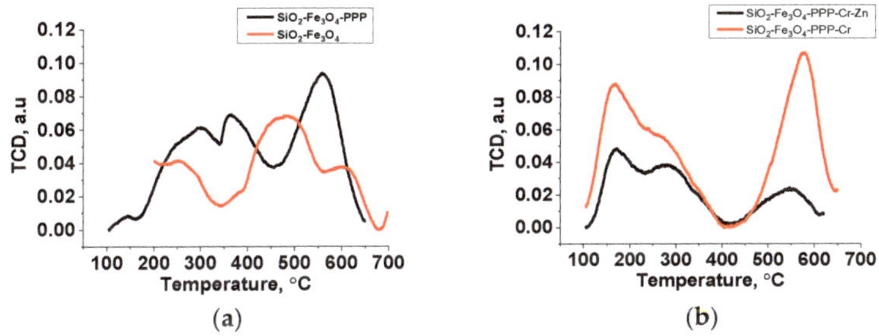

Figure 6. CO_2-TPD results of the examined catalysts: (**a**) SiO_2-Fe_3O_4 (red line) and SiO_2-Fe_3O_4-PPP (black line); and (**b**): SiO_2-Fe_3O_4-PPP-Cr (red line) and SiO_2-Fe_3O_4-PPP-Cr-Zn (black line).

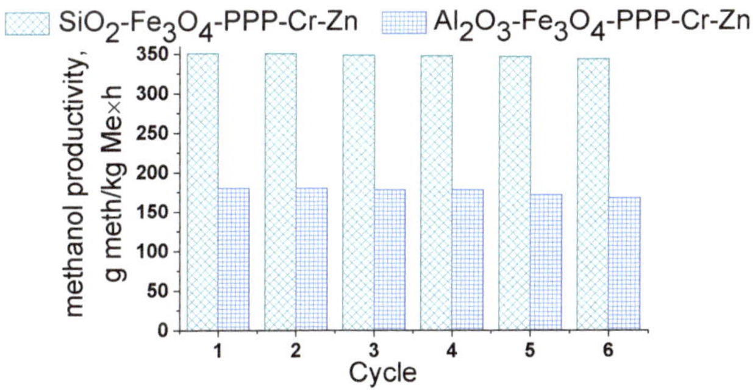

Figure 7. Recycling experiments for SiO_2-Fe_3O_4-PPP-Cr-Zn (green column) and Al_2O_3-Fe_3O_4-PPP-Cr-Zn (blue column) in CO_2 hydrogenation. Reaction conditions are: 250 °C, 5 MPa, 50 mg of the catalyst, CO_2:H_2 = 1:4, 15 mL of dodecane, 6 h.

The incorporation of metal species further influenced the CO_2 adsorption capacities of the composites (Figure 6b). Thus, monometallic SiO_2-Fe_3O_4-PPP-Cr showed the adsorption capacity of 0.073 mmol/g exceeding that of SiO_2-Fe_3O_4-PPP. It is worth noting that strong adsorption centers (at 562 °C) arising due to PPP structure are retained in composites with catalytically active metals. Formation of bimetallic composites leads to weakening of total basicity to 0.032 mmol/g. Nevertheless, the composites efficiently adsorb CO_2 which is a prerequisite for effective catalysis.

The H_2 adsorption capacities of the synthesized nanocomposites were also estimated (Figure S14a,b). Similarly, deposition of PPP leads to an increase in amount of adsorbed H_2 (0.023 mmol/g for SiO_2-Fe_3O_4 vs 0.03 mmol/g for SiO_2-Fe_3O_4-PPP). Incorporation of metal species and formation of bimetallic nanocomposite increases the total amount of adsorbed H_2 to 0.043 mmol/g.

2.2. Catalytic Properties of Nanocomposites in CO_2 Hydrogenation

Cr-, Cr-Zn-, and Cr-Ni-containing nanocomposites have been tested in CO_2 hydrogenation to methanol. The process is challenging due to the high thermodynamic stability of CO_2 molecule. The formation of methanol from CO_2 is energetically undesired, implying the high temperature and pressure along with structural promotion of the catalyst should be applied to obtain the sufficient methanol productivity. Table 2 displays the results of the catalyst performance. The widely accepted approach to assess the catalyst activity in this reaction is to calculate the methanol productivity rate per the catalyst amount and per the metal content. The activity of the catalysts was compared with commercial catalyst MegaMax 800. To elucidate the optimal ratio of the metals, nanocomposites with different metal loadings have also been synthesized. Here we focused on the best catalytic results obtained, while the activity of the composites differed in metal ratio is given in SI, Table S8.

Table 2. Catalytic properties of the nanocomposites in methanol synthesis from CO_2 + H_2.

$$CO_2 + 4H_2 \longrightarrow CH_3OH + 2H_2O$$

Sample Notation	Content by Elemental Analysis, wt%				Conversion, %	Selectivity, %	Methanol Productivity, g Methanol/kg Me ×·h
	Fe	Cr	Zn	Ni			
SiO_2-Fe_3O_4-PPP	9.1	-	-	-	-	-	
SiO_2-Fe_3O_4-PPP-Cr	6.8	2.1	-	-	4.2	98.3	115
SiO_2-Fe_3O_4-PPP-Cr-Ni	4.3	2.2	-	2.3	6.0	98.5	172
SiO_2-Fe_3O_4-PPP-Cr-Zn	4.1	2.0	2.1	-	9.4	99.1	350
Al_2O_3-Fe_3O_4-PPP-Cr-Zn	4.3	2.2	2.1	-	6.2	98.6	195
SiO_2-PPP-Cr-Zn	-	2.1	2.0	-	4.1	99.0	120
MegaMax 800					4.8	87.5	78

Reaction conditions: 250 °C, 5 MPa, CO_2:H_2 = 1:4, 15 mL of dodecane, 50 mg of the catalyst, 6 h.

The magnetic silica with deposited PPP expectedly shows no catalytic activity. The monometallic composite SiO_2-Fe_3O_4-PPP-Cr provides the methanol productivity comparable with that of a commercial catalyst. The result confirms the ability of Cr_2O_3 to catalyze the methanol synthesis from CO_2 and H_2. The incorporation of additional metals leads to an increase in the catalyst activity with the highest methanol productivity observed for the Cr-Zn-containing composite. Notably, the SiO_2-based, Cr-Zn-containing composite was in two orders of magnitude more active than Al_2O_3-based system prepared with the same metal loadings. The catalytic activity of the composite without magnetic NPs-SiO_2-PPP-Cr-Zn-was considerably lower than that of the Fe_3O_4-containing sample, revealing the promoting effect provided by Fe_3O_4 NPs. It also should be noted that a decrease in metal loading leads to a decrease in catalytic activity (Table S8).

To analyze the enhanced catalytic activity of SiO_2-based systems, the nitrogen adsorption-desorption measurements were carried out. The adsorption isotherms are presented in Figure S15. The analysis revealed the distinct textural properties of SiO_2 vs Al_2O_3 composites with the smaller BET surface area and pore sizes of the latter. This could have a detrimental effect on the catalyst activity due to worse accessibility of active catalytic species and diffusion limitations. Several studies also report a positive influence of SiO_2 on catalytic activity in CO_2 hydrogenation reaction. The addition of SiO_2 as itself to convenient Cu-ZnO-Al_2O_3 catalyst increased the activity and enhanced CuO dispersion [58]. The similar promoting effect has been observed in another study [59]. SiO_2-modified Cu-ZnO-ZrO_2 catalyst exhibits a better catalytic activity and higher long-term stability in comparison with a non-modified system due to a better intermixing of metal oxide components [59]. SiO_2 has also been shown to accelerate the selective hydrogenation of CO_2 to methanol for Co-based catalysts [60]. The effect was attributed to efficient stabilization of methoxy *CH_3O species being intermediates in the catalytic process by interface formed between SiO_2 and catalytic metal—Me-O-SiO_n. Our results support the data on the promotional effect of silica on methanol synthesis from CO_2, and highlight the importance of careful adjustment and optimization of catalyst composition by the appropriate choice of support.

Our results demonstrated that the introduction of magnetic NPs in catalyst structure along with convenience and simplicity of magnetic separation provided the promoting effect on methanol productivity. We assume that Fe_3O_4 NPs serve as a reservoir for a better intermixing and dispersion of ZnO and Cr_2O_3 components. Such assumption is supported by XPS data. This led to a closer contact between catalytic species being a prerequisite for effective catalytic performance. The positive effect of spinel structures of Fe_3O_4 on methanol synthesis has also been reported previously [42,61]. The addition of Fe_3O_4 to Cu-Zn-based catalyst results in improved methanol selectivity and enhanced catalyst activity. Magnetic NPs have also been shown to induce a strong metal support interaction (SMSI) effect due to formation of oxygen vacancies that stabilize the reaction intermediates and activate CO_2 molecule [42,62,63].

To conclude, each structural element in the designed nanocomposites influences the catalytic activity and the synergistic effect between catalyst constituents such as: support, polymer layer, magnetic NPs, and metal species ensures high catalytic activity and stability. The type of support determines distribution of metal species, accessibility of active centers, and occurrence of strong metal-support interactions which were shown to positively influence the catalysis. Here we demonstrated the advantages of SiO_2 support over Al_2O_3. Magnetic NPs boost the catalyst activity due to metal intermixing and migration of catalytic species onto Fe_3O_4 surface, as was observed by XPS data on 2p/3p electron content and XRD analysis. The precious miscibility of metal species allows for better catalytic performance, as was evidenced by the higher activity of Cr-Zn-containing catalyst over Cr-Ni-one. Bimetallic Cr-containing systems were shown to outperform monometallic ones. The results are consistent with the data on the activity of other heterogeneous catalysts for CO_2 hydrogenation, and bimetallic systems typically demonstrate a better catalytic performance [4,10,13,14]. The formation of bimetallic interfaces notably lowers the activation barrier of CO_2 to methanol hydrogenation compared to monometallic systems [64]. For industrial Cu-ZnO-Al_2O_3 catalyst, Cu-ZnO interface is believed to be the active site [65]. Moreover, the formation of bimetallic interfaces is usually followed by the formation of oxygen vacancies which are crucial for CO_2 activation and intermediate stabilization [14,63]. In particular, chromium doping with metals has been shown to induce the oxygen vacancy formation [41,66].

Comparison of the catalytic results with the literature data undoubtedly demonstrated the advantages of the proposed approach to creation of effective catalysts for the synthesis of methanol from CO_2 [67–75]. The results of recycling experiments revealed the exceptional stability of the catalysts after six repetitive uses. The methanol productivity rate was preserved at nearly 98% for SiO_2-based catalysts, while 5% drop of activity was observed for Al_2O_3-containing systems starting from the fifth use (Figure 7). We believe that superior catalyst stability is due to the protective role of the hydrophobic polymer layer. In contrast to the PPP-covered SiO_2/Al_2O_3-catalysts, the conventional catalytic systems show 11% drop of activity after 36 h of catalytic experiment [30]. The deactivation was induced by agglomeration of ZnO species and sintering of Cu particles for CuO-ZnO-Al_2O_3 system [30]. Sintering and NP aggregation have also been proved to be the main reason of deactivation for other catalytic systems [29].

To elucidate the polymer layer role, the catalysts after catalysis have been separated from the reaction mixture with external magnet and analyzed by STEM EDS. SiO_2-Fe_3O_4-PPP-Cr-Zn and Al_2O_3-Fe_3O_4-PPP-Cr-Zn have been chosen for analysis. The results are presented in Figures S16 and S17. For both samples, Cr- and Zn-species remained evenly distributed over all samples with no visible bulk NPs which could be due to the aggregation of metal species during catalysis. The results confirm the outstanding stability of the catalytic systems. Therefore, the deposition of PPP not only facilitates the catalytic reaction through the enhanced adsorption of CO_2, but also ensures the structural separation of NPs against sintering.

3. Materials and Methods

3.1. Materials

Iron (III) nitrate nonahydrate (ABCR, 98%), mesoporous silica gels (Sigma-Aldrich, Darmstadt, Germany, 6 nm, 200–425 mesh), aluminum oxide (Sigma-Aldrich, 5.8 nm, 150 mesh), zinc acetylacetonate hydrate (Sigma-Aldrich), chromium (III) acetylacetonate

(Sigma-Aldrich, 97%), nickel (II) acetylacetonate (ABCR, 98%), ethylene glycol (Sigma-Aldrich, 99%), and diphenyl ether (Sigma-Aldrich, 99%) were used as received. Acetone (99.5%) and ethanol (96%) were purchased from "Component-reactive" and used as received. Magnetic NPs in the pores of mesoporous silica gel and aluminum oxide were synthesized according to procedures published in [42,43], respectively. Deposition of polymer PPP layer on the surface of magnetic silica gel was performed as described previously [36]. Monomers for the synthesis of PPP in the presence of solid support (see Figure S1) were synthesized according to [76].

3.2. Synthesis of Al_2O_3-Fe_3O_4-PPP

Al_2O_3-Fe_3O_4 (0.2027 g) was heated in a Schlenk reaction flask under flowing argon atmosphere at 120 °C for 2 h to remove moisture. Monomer A6 (50 mg, 0.037 mmol) was dissolved in 3 mL of dichloromethane. This solution was added to Al_2O_3-Fe_3O_4 and sonicated for 20 min to allow the monomer to adsorb on the Al_2O_3-Fe_3O_4 surface. The flask was placed in a rotary evaporator and the solvent was evacuated. B2 (0.087 g, 0.111 mmol) was dissolved in 3 mL of diphenyl ether and added to the reaction flask containing A6 deposited on Al_2O_3-Fe_3O_4. The reaction flask was filled with argon and heated at 160 °C for 10 h upon stirring. The final material (Al_2O_3-Fe_3O_4-PPP) was collected from the suspension using a rare earth magnet, washed with dichloromethane (8 × 10 mL), and dried at room temperature overnight in vacuo. The elemental analysis data: N 1.23%, C 25.45%, H 1.89%.

3.3. Synthesis of Catalytically Active Nanocomposites

In a typical experiment, 0.23 g of zinc acetylacetonate were dissolved in 10 mL of acetone, followed by a dropwise addition to 2.57 g of Fe_3O_4-SiO_2-PPP (or Al_2O_3-Fe_3O_4-PPP). After acetone evaporation, the solution containing 0.70 g of chromium (III) acetylacetonate dissolved in 10 mL of acetone was added to Fe_3O_4-SiO_2-PPP (or Al_2O_3-Fe_3O_4-PPP) coated with zinc acetylacetonate. The mixture was allowed to stir for 3 h in air for acetone evaporation. The sample was then dried in a vacuum oven at room temperature for 12 h. Then the sample was heated in a tube furnace under argon with 7% of H_2 to 350 °C with a heating rate of 2 °C/min. The temperature was held at 350 °C for 3 h and then the sample was cooled to room temperature. For the Cr-Ni-containing sample (SiO_2-Fe_3O_4-PPP- Cr-Ni), 0.22 g of nickel acetyl acetonate was added instead of zinc precursor. For SiO_2-Fe_3O_4-PPP-Cr nanocomposite, only 0.70 g of chromium (III) acetylacetonate in 10 mL of acetone was used. To assess the metal content in the resulted nanocomposites, X-ray fluorescence (XRF) measurements were applied.

3.4. Characterization

Transmission electron microscopy (TEM), scanning transmission electron microscopy (STEM), and energy-dispersive X-ray (EDX) microanalysis was carried out in an Osiris TEM/STEM (Thermo Fisher Scientific, Waltham, MA, USA) equipped with a high angle annular dark field detector (HAADF) (Fischione, Export, PA, USA) and an X-ray energy dispersive spectrometer Super X (ChemiSTEM, Bruker, Bradford County, FL, USA) at an accelerating voltage of 200 kV. Specimens for TEM, STEM, and EDXS studies were prepared by placement of the Lacey carbon film on Cu grid into the vial with the suspension of the nanocomposite in CH_2Cl_2.

Powder X-ray diffraction patterns were recorded using Proto AXRD Θ-2Θ diffractometer (Detroit, MI, USA) with copper anode (Kα = 1.541874 Å, Ni-Kß filter) and 1D-detector Dectris Mythen 1K in the angular range 2θ = 5–100°. A scanning step was set to be 0.02° and the speed was 0.5°/min. Identification was performed with the PDXL software (Rigaku Corporation, Tokyo, Japan) using the ICDD PDF-2 database (2017).

X-ray photoelectron spectroscopy (XPS) data were obtained using Axis Ultra DLD (Kratos) spectrometer (Kyoto, Japan) with a monochromatic Al Kα radiation. All the data were acquired at X-ray power of 150 W. Survey spectra were recorded at an energy step of

1 eV with an analyzer pass energy 160 eV, and high-resolution spectra were recorded at an energy step of 0.1 eV with an analyzer pass energy 40 eV. Samples were out-gassed for 180 min before analysis. The data analysis was performed by CasaXPS.

Carbon dioxide temperature programmed desorption experiments were made using AutoChem HP chemisorption analyzer (Norcross, GA, USA). For carbon dioxide desorption experiments synthesized samples were placed in quartz cuvette and placed in analyzer module. The sample was heated in helium atmosphere up to 700 °C, then cooled down to 105 °C, and flashed with carbon dioxide for one hour followed by flashing with pure helium for one hour. Afterwards, the sample was heated to 700 °C with a temperature gradient of 10 °C/min and carbon dioxide desorption curve was recorded. Quantity of basic sites were calculated according to quantity of chemosorbed carbon dioxide using preliminary made calibration curve.

Hydrogen temperature programmed desorption experiments were made using AutoChem HP chemosorption analyzer. For hydrogen desorption experiments synthesized samples were placed in quartz cuvette and placed in analyzer module. The sample was heated in argon atmosphere up to 700 °C, then cooled down to ambient temperature, and flashed with mixture of 10 v.% of hydrogen in argon for one hour followed by flashing with pure argon for one hour. Afterwards, the sample was heated to 700 °C with a temperature gradient of 10 °C/min and hydrogen desorption curves were recorded. Quantity of hydrogen-capable adsorption sites were calculated according to quantity of chemosorbed hydrogen using preliminary made calibration curve.

Zn, Cr, and Ni content of the composites was obtained from X-ray fluorescence (XRF) measurements using a Zeiss Jena VRA-30 spectrometer (Oberkochen, Germany) equipped with a Mo anode, a LiF200 crystal analyzer, and a SD detector.

3.5. Catalytic Study

In a typical experiment, 50 mg of the catalyst and 15 mL of dodecane were loaded in the stainles-steel reactor (7) (internal volume of 50 mL) equipped with a propeller mixer (stirring rate 250 rpm). Then the hydrogen pressure was set to 5.0 MPa, the reactor was purged with hydrogen for 3 times, and the mixture was heated up to 250 °C. When the above temperature was achieved, the catalyst was reduced for 1 h, then the hydrogen was substituted for the gas mixture ($H_2/CO_2 = 4/1$) and stirring rate was increased up to 750 rpm. Reaction was provided for six hours followed by cooling to ambient temperature. Gas phase was directed into the chromatographic system of on-line analysis along the heated line through the return pressure valve and removed from the system through a flowmeter. Liquid phase was analyzed using GS-MS Shimadzu 2010 (Kyotocity, Japan) gas chromatomass spectrometer using preliminary made calibration curves. Methanol accumulation rate was calculated taking into account mass of formed methanol in reaction media, catalysts mass and reaction time.

For repeated catalytic experiments, the catalyst after reaction was collected with an external magnet in a vial, washed thoroughly with ethanol, and dried in vacuo until the constant weight. The reaction conditions for recycling experiments were similar to those described above.

4. Conclusions

In this work, novel SiO_2/Al_2O_3-Fe_3O_4-PPP-Cr-Me (Me = Zn, Ni) catalysts possessing high activity and exceptional stability in CO_2 hydrogenation reaction to methanol have been synthesized. The developed approach to synthesis of the catalysts offers several possibilities for structural adoption that boost the catalytic activity due to the synergetic effect Synthesized SiO_2/Al_2O_3-Fe_3O_4-PPP composites providing a carrier for deposition and dispersion of catalytically active metal species as well as a structural barrier against sintering. Presence of a hydrophobic heteroaromatic polymer layer on the silica and alumina supports prevents the aggregation of catalytically active species, elevates the stability of the catalyst, and facilitates CO_2 hydrogenation through enhanced adsorption of

CO_2 and H_2. Magnetic NPs facilitate the reaction rate through the structural promotion, probably due to formation of oxygen vacancies, and allow for fast and simple magnetic separation of the catalyst for repeated uses as well. Creation of bimetallic interface by combination of two metals (Cr-Zn and Cr-Ni) enhances the catalytic activity with the highest methanol productivity achieved for the Cr-Zn-containing composite. Considering the robust synthetic procedures, high activity attained for non-noble metals, easy catalyst recovery, and excellent stability over six consecutive catalytic cycles, the proposed approach can be considered as a promising strategy for improving the performance of the catalysts in production of methanol from $CO_2 + H_2$.

Supplementary Materials: The following supporting information can be downloaded at: https://www.mdpi.com/article/10.3390/catal13010001/s1, Figure S1: Schematic presentation of deposition of a thin layer of pyridylphenylene polymer on a solid support; Figure S2: STEM EDS maps of Al_2O_3-Fe_3O_4-PPP-Cr-Zn; Figure S3: STEM EDS maps of SiO_2-PPP-Cr-Zn. Figure S4: STEM EDS maps of SiO_2-Fe_3O_4-PPP-Cr. Figure S5: EDS spectra of SiO_2-PPP-Cr-Zn (a) and SiO_2-Fe_3O_4-PPP-Cr (b) nanocomposites; Figure S6: XRD patterns of the samples in the area of the (311) spinel peak; Figure S7: XRD patterns of SiO_2-PPP-Cr-Zn (a), Al_2O_3-Fe_3O_4-PPP (b), Al_2O_3-Fe_3O_4-PPP-Cr-Zn (c); Figure S8: XPS survey spectra of the SiO_2-Fe_3O_4-PPP-Cr-Zn (a), SiO_2-Fe_3O_4-PPP-Cr (b), SiO_2-PPP-Cr-Zn (c), SiO_2-Fe_3O_4-PPP-Cr-Ni (d), Al_2O_3-Fe_3O_4-PPP-Cr-Zn (e) nanocomposites; Figure S9: HR XPS of Fe 2p of SiO_2-Fe_3O_4-PPP before incorporation of metal oxides in the nanocomposite structure; Figure S10: HR XPS of Fe 2p (a) and Cr 2p (b) regions of SiO_2-Fe_3O_4-PPP-Cr; Figure S11: HR XPS of Cr 2p (a) and Zn 2p (b) regions of SiO_2-PPP-Cr-Zn; Figure S12: HR XPS of Fe 2p (a), Cr 2p (b) and Ni 2p (c) regions of SiO_2-Fe_3O_4-PPP-Cr-Ni; Figure S13: HR XPS of Fe 2p (a), Cr 2p (b) and Zn 2p (c) regions of Al_2O_3-Fe_3O_4-PPP-Cr-Zn; Figure S14. H_2-TPD results of the examined catalysts: (a) SiO_2-Fe_3O_4 (red line) and SiO_2-Fe_3O_4-PPP (black line) and (b): SiO_2-Fe_3O_4-PPP-Cr (red line) and SiO_2-Fe_3O_4-PPP-Cr-Zn (black line). Figure S15: N_2 adsorption-desorption isotherms (a,c) and pore sizes distributions (b,d) of Al_2O_3-Fe_3O_4-PPP (a,c) and SiO_2-Fe_3O_4-PPP (c,d). $S_{BET} = 164\ m^2/g$ (Al_2O_3-Fe_3O_4-PPP) and $S_{BET} = 246\ m^2/g$ (SiO_2-Fe_3O_4-PPP); Figure S16: STEM dark field image (a) EDS maps of Si (b), Cr (c), C (d), Fe (e) and Zn (f) of SiO_2-Fe_3O_4-PPP-Cr-Zn nanocomposite after catalysis; Figure S17: STEM dark field image (a) EDS maps of Al (b), Zn (c), C (d), Fe (e) and Cr (f) of Al_2O_3-Fe_3O_4-PPP-Cr-Zn nanocomposite after catalysis. Table S1: Atomic percentage of elements in samples; Table S2: Fitting parameters for HR XPS of Fe 2p, Cr 2p and Zn 2p of SiO_2-Fe_3O_4-PPP-Cr-Zn; Table S3: Fitting parameters for HR XPS of Fe 2p of SiO_2-Fe_3O_4-PPP; Table S4: Fitting parameters for HR XPS of Fe 2p and Cr 2p of SiO_2-Fe_3O_4-PPP-Cr; Table S5: Fitting parameters for HR XPS Cr 2p and Zn 2p of SiO_2-PPP-Cr-Zn; Table S6: Fitting parameters for HR XPS of Fe 2p, Cr 2p and Ni 2p of SiO_2-Fe_3O_4-PPP-Cr-Ni; Table S7: Fitting parameters for HR XPS of Fe 2p, Cr 2p and Zn 2p of Al_2O_3-Fe_3O_4-PPP-Cr-Zn. Table S8. Catalytic properties of the nanocomposites in methanol synthesis from $CO_2 + H_2$.

Author Contributions: S.A.S., methodology, writing—original draft preparation, N.V.K., methodology, visualization, M.E.G., investigation, A.V.B., resources, A.K.R., visualization, investigation, V.Y.D., investigation, M.G.S., conceptualization, Z.B.S., conceptualization, writing—review and editing, supervision, data curation, project administration. All authors have read and agreed to the published version of the manuscript.

Funding: This research was funded by Russian Science Foundation, grant number 22-43-02025.

Data Availability Statement: Not applicable.

Acknowledgments: The assistance of A.L. Vasiliev, Federal Scientific Research Centre, Shubnikov Institute of Crystallography, in recording of STEM EDS images is gratefully acknowledged. The contribution of the Centre for Molecular Composition Studies of the INEOS RAS (elemental analysis) with financial support from Ministry of Science and Higher Education of the Russian Federation is gratefully acknowledged.

Conflicts of Interest: The authors declare no conflict of interest.

References

1. Dudley, B. *BP Statistical Review of World Energy*; BP Statistical Review: London, UK, 2019.

2. Friedlingstein, P; Jones, M.W; O'Sullivan, M; Andrew, R.M; Bakker, D.C.E; Hauck, J; Le Quéré, C; Peters, G.P; Peters, W; Pongratz, J; et al. Global Carbon Budget 2021. *Earth Syst. Sci. Data* **2022**, *14*, 1917–2005. [CrossRef]
3. Aresta, M; Dibenedetto, A; Angelini, A. Catalysis for the Valorization of Exhaust Carbon: From CO_2 to Chemicals, Materials, and Fuels. Technological Use of CO_2. *Chem. Rev.* **2014**, *114*, 1709–1742. [CrossRef] [PubMed]
4. Ye, R.-P; Ding, J; Gong, W; Argyle, M.D; Zhong, Q; Wang, Y; Russell, C.K; Xu, Z; Russell, A.G; Li, Q; et al. CO_2 hydrogenation to high-value products via heterogeneous catalysis. *Nat. Commun.* **2019**, *10*, 5698. [CrossRef]
5. Sempuga, B.C; Yao, Y. CO_2 hydrogenation from a process synthesis perspective: Setting up process targets. *J. CO_2 Util.* **2017**, *20*, 34–42. [CrossRef]
6. Saeidi, S; Amin, N.A.S; Rahimpour, M.R. Hydrogenation of CO_2 to value-added products—A review and potential future developments. *J. CO_2 Util.* **2014**, *5*, 66–81. [CrossRef]
7. Ni, Y; Shi, L; Liu, H; Zhang, W; Liu, Y; Zhu, W; Liu, Z. A green route for methanol carbonylation. *Catal. Sci. Technol.* **2017**, *7*, 4818–4822. [CrossRef]
8. Wingad, R.L; Bergström, E.J.E; Everett, M; Pellow, K.J; Wass, D.F. Catalytic conversion of methanol/ethanol to isobutanol–a highly selective route to an advanced biofuel. *Chem. Commun.* **2016**, *52*, 5202–5204. [CrossRef]
9. Li, T; Shoinkhorova, T; Gascon, J; Ruiz-Martínez, J. Aromatics Production via Methanol-Mediated Transformation Routes. *ACS Catal.* **2021**, *11*, 7780–7819. [CrossRef]
10. Zhang, X; Zhang, G; Song, C; Guo, X. Catalytic Conversion of Carbon Dioxide to Methanol: Current Status and Future Perspective. *Front. Energy Res.* **2021**, *8*, 621119. [CrossRef]
11. Behrens, M; Studt, F; Kasatkin, I; Kühl, S; Hävecker, M; Abild-Pedersen, F; Zander, S; Girgsdies, F; Kurr, P; Kniep, B.-L; et al. The Active Site of Methanol Synthesis over Cu/ZnO/Al_2O_3 Industrial Catalysts. *Science* **2012**, *336*, 893–897. [CrossRef]
12. Olah, G.A. Beyond Oil and Gas: The Methanol Economy. *Angew. Chem. Int. Ed.* **2005**, *44*, 2636–2639. [CrossRef]
13. Álvarez, A; Bansode, A; Urakawa, A; Bavykina, A.V; Wezendonk, T.A; Makkee, M; Gascon, J; Kapteijn, F. Challenges in the Greener Production of Formates/Formic Acid, Methanol, and DME by Heterogeneously Catalyzed CO_2 Hydrogenation Processes. *Chem. Rev.* **2017**, *117*, 9804–9838. [CrossRef] [PubMed]
14. Ren, M; Zhang, Y; Wang, X; Qiu, H. Catalytic Hydrogenation of CO_2 to Methanol: A Review. *Catalysts* **2022**, *12*, 403. [CrossRef]
15. Shi, Z; Tan, Q; Tian, C; Pan, Y; Sun, X; Zhang, J; Wu, D. CO_2 hydrogenation to methanol over Cu-In intermetallic catalysts: Effect of reduction temperature. *J. Catal.* **2019**, *379*, 78–89. [CrossRef]
16. Bahruji, H; Bowker, M; Hutchings, G; Dimitratos, N; Wells, P; Gibson, E; Jones, W; Brookes, C; Morgan, D; Lalev, G. Pd/ZnO catalysts for direct CO_2 hydrogenation to methanol. *J. Catal.* **2016**, *343*, 133–146. [CrossRef]
17. Díez, J; Valverde, J; sánchez, P; Dorado, F. CO_2 Hydrogenation to Methanol at Atmospheric Pressure: Influence of the Preparation Method of Pd/ZnO Catalysts. *Catal. Lett.* **2016**, *146*, 373–382. [CrossRef]
18. Xu, J; Su, X; Liu, X; Pan, X; Pei, G; Huang, Y; Wang, X; Zhang, T; Geng, H. Methanol synthesis from CO_2 and H_2 over Pd/ZnO/Al2O3: Catalyst structure dependence of methanol selectivity. *Appl. Catal. A Gen.* **2016**, *514*, 51–59. [CrossRef]
19. Wang, J; Sun, K; Jia, X; Liu, C.-j. CO_2 hydrogenation to methanol over Rh/In2O3 catalyst. *Catal. Today* **2021**, *365*, 341–347. [CrossRef]
20. Jiang, X; Koizumi, N; Guo, X; Song, C. Bimetallic Pd–Cu catalysts for selective CO_2 hydrogenation to methanol. *Appl. Catal. B Environ.* **2015**, *170–171*, 173–185. [CrossRef]
21. Gallo, A; Snider, J.L; Sokaras, D; Nordlund, D; Kroll, T; Ogasawara, H; Kovarik, L; Duyar, M.S; Jaramillo, T.F. Ni_5Ga_3 catalysts for CO_2 reduction to methanol: Exploring the role of Ga surface oxidation/reduction on catalytic activity. *Appl. Catal. B Environ.* **2020**, *267*, 118369. [CrossRef]
22. Studt, F; Sharafutdinov, I; Abild-Pedersen, F; Elkjær, C.F; Hummelshøj, J.S; Dahl, S; Chorkendorff, I; Nørskov, J.K. Discovery of a Ni-Ga catalyst for carbon dioxide reduction to methanol. *Nat. Chem.* **2014**, *6*, 320–324. [CrossRef] [PubMed]
23. Kattel, S; Liu, P; Chen, J.G. Tuning Selectivity of CO_2 Hydrogenation Reactions at the Metal/Oxide Interface. *J. Am. Chem. Soc.* **2017**, *139*, 9739–9754. [CrossRef] [PubMed]
24. Elnabawy, A.O; Schimmenti, R; Cao, A; Nørskov, J.K. Why ZnO is the Support for Cu in Methanol Synthesis? A Systematic Study of the Strong Metal Support Interactions. *ACS Sustain. Chem. Eng.* **2022**, *10*, 1722–1730. [CrossRef]
25. Kuld, S; Thorhauge, M; Falsig, H; Elkjær, C.F; Helveg, S; Chorkendorff, I; Sehested, J. Quantifying the promotion of Cu catalysts by ZnO for methanol synthesis. *Science* **2016**, *352*, 969–974. [CrossRef]
26. Stangeland, K; Li, H; Yu, Z. CO_2 hydrogenation to methanol: The structure–activity relationships of different catalyst systems. *Energy Ecol. Environ.* **2020**, *5*, 272–285. [CrossRef]
27. Behrens, M. Promoting the Synthesis of Methanol: Understanding the Requirements for an Industrial Catalyst for the Conversion of CO_2. *Angew. Chem. Int. Ed.* **2016**, *55*, 14906–14908. [CrossRef]
28. Gao, P; Zhong, L; Zhang, L; Wang, H; Zhao, N; Wei, W; Sun, Y. Yttrium oxide modified Cu/ZnO/Al2O3 catalysts via hydrotalcite-like precursors for CO_2 hydrogenation to methanol. *Catal. Sci. Technol.* **2015**, *5*, 4365–4377. [CrossRef]
29. Frei, M.S; Mondelli, C; García-Muelas, R; Kley, K.S; Puértolas, B; López, N; Safonova, O.V; Stewart, J.A; Curulla Ferré, D; Pérez-Ramírez, J. Atomic-scale engineering of indium oxide promotion by palladium for methanol production via CO_2 hydrogenation. *Nat. Commun.* **2019**, *10*, 3377. [CrossRef]
30. Liang, B; Ma, J; Su, X; Yang, C; Duan, H; Zhou, H; Deng, S; Li, L; Huang, Y. Investigation on Deactivation of Cu/ZnO/Al2O3 Catalyst for CO_2 Hydrogenation to Methanol. *Ind. Eng. Chem. Res.* **2019**, *58*, 9030–9037. [CrossRef]

31. Baird, N; Losovyj, Y; Yuzik-Klimova, E.Y; Kuchkina, N.V; Shifrina, Z.B; Pink, M; Stein, B.D; Morgan, D.G; Wang, T; Rubin, M.A; et al. Zinc-Containing Magnetic Oxides Stabilized by a Polymer: One Phase or Two? *ACS Appl. Mater. Interfaces* **2016**, *8*, 891–899. [CrossRef]
32. Baird, N; Dittmar, J.W; Losovyj, Y.B; Morgan, D.G; Stein, B.D; Pink, M; Kuchkina, N.V; Serkova, E.S; Lependina, O.L; Grigoriev, M.E; et al. Enhancing the Catalytic Activity of Zn-Containing Magnetic Oxides in a Methanol Synthesis: Identifying the Key Factors. *ACS Appl. Mater. Interfaces* **2017**, *9*, 2285–2294. [CrossRef]
33. Baird, N; Dittmar, J.W; Losovyj, Y.B; Pink, M; Morgan, D.G; Stein, B.D; Torozova, A.S; Krasnova, I.Y; Grigoriev, M.E; Sidorov, A.I; et al. Cr–Containing Magnetic Oxides in a Methanol Synthesis: Does Cr Ion Distribution Matter? *Chem. Sel.* **2017**, *2*, 6269–6276. [CrossRef]
34. Sorokina, S.A; Mikhailov, S.P; Kuchkina, N.V; Bykov, A.V; Vasiliev, A.L; Ezernitskaya, M.G; Golovin, A.L; Nikoshvili, L.Z; Sulman, M.G; Shifrina, Z.B. Ru@hyperbranched Polymer for Hydrogenation of Levulinic Acid to Gamma-Valerolactone: The Role of the Catalyst Support. *Int. J. Mol. Sci.* **2022**, *23*, 799. [CrossRef] [PubMed]
35. Alibegovic, K; Morgan, D.G; Losovyj, Y; Pink, M; Stein, B.D; Kuchkina, N.V; Serkova, E.S; Salnikova, K.E; Shifrina, Z.B; Matveeva, V.G; et al. Efficient Furfuryl Alcohol Synthesis from Furfural over Magnetically Recoverable Catalysts: Does the Catalyst Stabilizing Medium Matter? *Chem. Sel.* **2017**, *2*, 5485–5491. [CrossRef]
36. Kuchkina, N.V; Haskell, A.K; Sorokina, S.A; Torozova, A.S; Nikoshvili, L.Z; Sulman, E.M; Stein, B.D; Morgan, D.G; Bronstein, L.M; Shifrina, Z.B. Pd Catalyst Based on Hyperbranched Polypyridylphenylene Formed In Situ on Magnetic Silica Allows for Excellent Performance in Suzuki–Miyaura Reaction. *ACS Appl. Mater. Interfaces* **2020**, *12*, 22170–22178. [CrossRef]
37. Wang, Y; Tan, L; Tan, M; Zhang, P; Fang, Y; Yoneyama, Y; Yang, G; Tsubaki, N. Rationally Designing Bifunctional Catalysts as an Efficient Strategy To Boost CO_2 Hydrogenation Producing Value-Added Aromatics. *ACS Catal.* **2019**, *9*, 895–901. [CrossRef]
38. Bradford, M; Konduru, M; Fuentes, D. Characterization and application of Cr2O3-ZnO catalysts for methanol synthesis. *Fuel Process. Technol.* **2003**, *83*, 11–25. [CrossRef]
39. Wen-De, H; Jun, K; Yang-Dong, W; Chuan-Ming, W; Wei-Min, Y. Insights into Syngas to Methanol Conversion on Cr₂O₃ Oxide from First-Principles-based Microkinetic Simulations. *Chin. J. Chem. Phys.* **2022**, *35*, 655–663. [CrossRef]
40. Wang, Y; Gao, W; Wang, K; Gao, X; Zhang, B; Zhao, H; Ma, Q; Zhang, P; Yang, G; Wu, M; et al. Boosting the synthesis of value-added aromatics directly from syngas via a Cr2O3 and Ga doped zeolite capsule catalyst. *Chem. Sci.* **2021**, *12*, 7786–7792. [CrossRef]
41. Carey, J.J; Legesse, M; Nolan, M. Low Valence Cation Doping of Bulk Cr2O3: Charge Compensation and Oxygen Vacancy Formation. *J. Phys. Chem. C* **2016**, *120*, 19160–19174. [CrossRef]
42. Oracko, T; Jaquish, R; Losovyj, Y.B; Morgan, D.G; Pink, M; Stein, B.D; Doluda, V.Y; Tkachenko, O.P; Shifrina, Z.B; Grigoriev, M.E; et al. Metal-Ion Distribution and Oxygen Vacancies That Determine the Activity of Magnetically Recoverable Catalysts in Methanol Synthesis. *ACS Appl. Mater. Interfaces* **2017**, *9*, 34005–34014. [CrossRef] [PubMed]
43. Jaquish, R; Reilly, A.K; Lawson, B.P; Golikova, E; Sulman, A.M; Stein, B.D; Lakina, N.V; Tkachenko, O.P; Sulman, E.M; Matveeva, V.G; et al. Immobilized glucose oxidase on magnetic silica and alumina: Beyond magnetic separation. *Int. J. Biol. Macromol.* **2018**, *120*, 896–905. [CrossRef] [PubMed]
44. Saedy, S; Newton, M.A; Zabilskiy, M; Lee, J.H; Krumeich, F; Ranocchiari, M; van Bokhoven, J.A. Copper–zinc oxide interface as a methanol-selective structure in Cu–ZnO catalyst during catalytic hydrogenation of carbon dioxide to methanol. *Catal. Sci. Technol.* **2022**, *12*, 2703–2716. [CrossRef]
45. Nguyen, X.S; Zhang, G; Yang, X. Mesocrystalline Zn-Doped Fe3O4 Hollow Submicrospheres: Formation Mechanism and Enhanced Photo-Fenton Catalytic Performance. *ACS Appl. Mater. Interfaces* **2017**, *9*, 8900–8909. [CrossRef] [PubMed]
46. Raju, G; Murali, N; Prasad, M.S.N.A; Suresh, B; Apparao Babu, D; Gnana Kiran, M; Ramakrishna, A; Tulu Wegayehu, M; Kishore Babu, B. Effect of chromium substitution on the structural and magnetic properties of cobalt ferrite. *Mater. Sci. Energy Technol.* **2019**, *2*, 78–82. [CrossRef]
47. Yamashita, T; Hayes, P. Analysis of XPS spectra of Fe^{2+} and Fe^{3+} ions in oxide materials. *Appl. Surf. Sci.* **2008**, *254*, 2441–2449. [CrossRef]
48. Bera, S; Prince, A.A.M; Velmurugan, S; Raghavan, P.S; Gopalan, R; Panneerselvam, G; Narasimhan, S.V. Formation of zinc ferrite by solid-state reaction and its characterization by XRD and XPS. *J. Mater. Sci.* **2001**, *36*, 5379–5384. [CrossRef]
49. Biesinger, M.C; Payne, B.P; Grosvenor, A.P; Lau, L.W.M; Gerson, A.R; Smart, R.S.C. Resolving surface chemical states in XPS analysis of first row transition metals, oxides and hydroxides: Cr, Mn, Fe, Co and Ni. *Appl. Surf. Sci.* **2011**, *257*, 2717–2730. [CrossRef]
50. Biesinger, M.C; Brown, C; Mycroft, J.R; Davidson, R.D; McIntyre, N.S. X-ray photoelectron spectroscopy studies of chromium compounds. *Surf. Interface Anal.* **2004**, *36*, 1550–1563. [CrossRef]
51. Pauly, N; Yubero, F; Espinós, J.P; Tougaard, S. XPS primary excitation spectra of Zn 2p, Fe 2p, and Ce 3d from ZnO, α-Fe2O3, and CeO_2. *Surf. Interface Anal.* **2019**, *51*, 353–360. [CrossRef]
52. Biesinger, M.C; Payne, B.P; Lau, L.W.M; Gerson, A; Smart, R.S.C. X-ray photoelectron spectroscopic chemical state quantification of mixed nickel metal, oxide and hydroxide systems. *Surf. Interface Anal.* **2009**, *41*, 324–332. [CrossRef]
53. Peck, M.A; Langell, M.A. Comparison of Nanoscaled and Bulk NiO Structural and Environmental Characteristics by XRD, XAFS, and XPS. *Chem. Mater.* **2012**, *24*, 4483–4490. [CrossRef]

54. Choudhury, S; Aguiar, J.A; Fluss, M.J; Hsiung, L.L; Misra, A; Uberuaga, B.P. Non-uniform Solute Segregation at Semi-Coherent Metal/Oxide Interfaces. *Sci. Rep.* **2015**, *5*, 13086. [CrossRef] [PubMed]
55. Rudolph, G; Henry, M.C. The Thermal Decomposition of Zinc Acetylacetonate Hydrate. *Inorg. Chem.* **1964**, *3*, 1317–1318. [CrossRef]
56. Musić, S; Šarić, A; Popović, S. Formation of nanosize ZnO particles by thermal decomposition of zinc acetylacetonate monohydrate. *Ceram. Int.* **2010**, *36*, 1117–1123. [CrossRef]
57. Siddiqi, M.A; Siddiqui, R.A; Atakan, B. Thermal stability, sublimation pressures and diffusion coefficients of some metal acetylacetonates. *Surf. Coat. Technol.* **2007**, *201*, 9055–9059. [CrossRef]
58. Zhang, L; Zhang, Y; Chen, S. Effect of promoter SiO_2, TiO_2 or SiO_2-TiO_2 on the performance of CuO-ZnO-Al_2O_3 catalyst for methanol synthesis from CO_2 hydrogenation. *Appl. Catal. A Gen.* **2012**, *415–416*, 118–123. [CrossRef]
59. Phongamwong, T; Chantaprasertporn, U; Witoon, T; Numpilai, T; Poo-arporn, Y; Limphirat, W; Donphai, W; Dittanet, P; Chareonpanich, M; Limtrakul, J. CO_2 hydrogenation to methanol over CuO–ZnO–ZrO_2–SiO_2 catalysts: Effects of SiO_2 contents. *Chem. Eng. J.* **2017**, *316*, 692–703. [CrossRef]
60. Wang, L; Guan, E; Wang, Y; Wang, L; Gong, Z; Cui, Y; Meng, X; Gates, B.C; Xiao, F.-S. Silica accelerates the selective hydrogenation of CO_2 to methanol on cobalt catalysts. *Nat. Commun.* **2020**, *11*, 1033. [CrossRef]
61. Liu, T; Xu, D; Wu, D; Liu, G; Hong, X. Spinel ZnFe2O4 Regulates Copper Sites for CO_2 Hydrogenation to Methanol. *ACS Sustain. Chem. Eng.* **2021**, *9*, 4033–4041. [CrossRef]
62. Singh, R; Tripathi, K; Pant, K.K. Investigating the role of oxygen vacancies and basic site density in tuning methanol selectivity over Cu/CeO_2 catalyst during CO_2 hydrogenation. *Fuel* **2021**, *303*, 121289. [CrossRef]
63. Rui, N; Wang, Z; Sun, K; Ye, J; Ge, Q; Liu, C.-j. CO_2 hydrogenation to methanol over Pd/In2O3: Effects of Pd and oxygen vacancy. *Appl. Catal. B Environ.* **2017**, *218*, 488–497. [CrossRef]
64. Wu, P.P; Yang, B. Significance of Surface Formate Coverage on the Reaction Kinetics of Methanol Synthesis from CO_2 Hydrogenation over Cu. *ACS Catal.* **2017**, *7*, 7187–7195. [CrossRef]
65. Kattel, S; Ramirez, P.J; Chen, J.G; Rodriguez, J.A; Liu, P. Active sites for CO_2 hydrogenation to methanol on Cu/ZnO catalysts. *Science* **2017**, *355*, 1296–1299. [CrossRef] [PubMed]
66. Carey, J.J; Nolan, M. Enhancing the oxygen vacancy formation and migration in bulk chromium(iii) oxide by alkali metal doping: A change from isotropic to anisotropic oxygen diffusion. *J. Mater. Chem. A* **2017**, *5*, 15613–15630. [CrossRef]
67. Chen, S; Zhang, J; Song, F; Zhang, Q; Yang, G; Zhang, M; Wang, X; Xie, H; Tan, Y. Induced high selectivity methanol formation during CO_2 hydrogenation over a CuBr2-modified CuZnZr catalyst. *J. Catal.* **2020**, *389*, 47–59. [CrossRef]
68. Xu, D; Hong, X; Liu, G. Highly dispersed metal doping to ZnZr oxide catalyst for CO_2 hydrogenation to methanol: Insight into hydrogen spillover. *J. Catal.* **2021**, *393*, 207–214. [CrossRef]
69. Mureddu, M; Lai, S; Atzori, L; Rombi, E; Ferrara, F; Pettinau, A; Cutrufello, M.G. Ex-LDH-Based Catalysts for CO_2 Conversion to Methanol and Dimethyl Ether. *Catalysts* **2021**, *11*, 615. [CrossRef]
70. Martin, O; Martín, A.J; Mondelli, C; Mitchell, S; Segawa, T.F; Hauert, R; Drouilly, C; Curulla-Ferré, D; Pérez-Ramírez, J. Indium Oxide as a Superior Catalyst for Methanol Synthesis by CO_2 Hydrogenation. *Angew. Chem. Int. Ed.* **2016**, *55*, 6261–6265. [CrossRef]
71. Yao, L; Pan, Y; Wu, D; Li, J; Xie, R; Peng, Z. Approaching full-range selectivity control in CO_2 hydrogenation to methanol and carbon monoxide with catalyst composition regulation. *Inorg. Chem. Front.* **2021**, *8*, 2433–2441. [CrossRef]
72. Fang, T; Liu, B; Lian, Y; Zhang, Z. Selective Methanol Synthesis from CO_2 Hydrogenation over an In_2O_3/Co/C-N Catalyst. *Ind. Eng. Chem. Res.* **2020**, *59*, 19162–19167. [CrossRef]
73. Ahmad, K; Upadhyayula, S. Kinetics of CO_2 hydrogenation to methanol over silica supported intermetallic Ga3Ni5 catalyst in a continuous differential fixed bed reactor. *Int. J. Hydrogen Energy* **2020**, *45*, 1140–1150. [CrossRef]
74. Temvuttirojn, C; Poo-arporn, Y; Chanlek, N; Cheng, C.K; Chong, C.C; Limtrakul, J; Witoon, T. Role of Calcination Temperatures of ZrO2 Support on Methanol Synthesis from CO_2 Hydrogenation at High Reaction Temperatures over ZnOx/ZrO2 Catalysts. *Ind. Eng. Chem. Res.* **2020**, *59*, 5525–5535. [CrossRef]
75. Kuwahara, Y; Mihogi, T; Hamahara, K; Kusu, K; Kobayashi, H; Yamashita, H. A quasi-stable molybdenum sub-oxide with abundant oxygen vacancies that promotes CO_2 hydrogenation to methanol. *Chem. Sci.* **2021**, *12*, 9902–9915. [CrossRef] [PubMed]
76. Kuchkina, N.V; Zinatullina, M.S; Serkova, E.S; Vlasov, P.S; Peregudov, A.S; Shifrina, Z.B. Hyperbranched pyridylphenylene polymers based on the first-generation dendrimer as a multifunctional monomer. *RSC Adv.* **2015**, *5*, 99510–99516. [CrossRef]

catalysts

MDPI

Article

Investigation of CeO_2, MoO_3, and $Ce_2(MoO_4)_3$, Synthesized by the Pechini Method, as Catalysts for Fructose Conversion

Dhara Beatriz de Amorim Pryston, Thatiane Veríssimo dos Santos Martins, Jailton Alves de Vasconcelos Júnior, Débora Olimpio da Silva Avelino, Mario Roberto Meneghetti and Simoni Margareti Plentz Meneghetti *

Group of Catalysis and Chemical Reactivity, Institute of Chemistry and Biotechnology, Federal University of Alagoas, Av. Lourival Melo Mota, Tabuleiro do Martins, Maceió 57072-970, AL, Brazil
* Correspondence: simoni.plentz@gmail.com; Tel.: +55-82-3214-1373

Abstract: Cerium oxide (Ce100), molybdenum oxide (Mo100), and a material containing Ce and Mo (CeMo) were synthesized by the Pechini method, using glycerol as a polyol. These materials were applied for fructose conversion in an aqueous medium. The characterization results show the formation of cerium molybdate ($Ce_2(MoO_4)_3$) for CeMo. Ce100 presented good thermal stability, and Mo100 sublimation of MoO_3 and polymolybdates was verified. CeMo exhibited a mass loss of 19%, associated with the sublimation of MoO_3 and polymolybdate species. Additionally, the existence of Bronsted and Lewis acid sites was confirmed, and the addition of Mo to Ce was an efficient strategy to increase the acidity. Regarding the catalytic activity (150 °C and 0.5 to 6 h), Ce100 exhibited low conversions and high selectivity to 5-hydroxymethylfurfural (5-HMF). For Mo100, high conversions, with a significant formation of insoluble materials, were detected. For CeMo, beyond the high activity, a lower formation of insoluble materials was noted. In this case, selectivity toward products from the retro–aldolic route and 5-HMF were obtained. These results indicate that the main factor influencing fructose conversion is an adequate combination of the acid sites. Recycling experiments were carried out, and stability was observed for four cycles, confirming the robustness of this system.

Keywords: cerium oxide; molybdenum oxide; cerium molybdate; biomass; biorefinery

1. Introduction

The environmental and social problems caused by the dependence of the global energy matrix and the chemical synthesis on fossil fuels (oil, coal, and natural gas) motivated the search for sustainable energy sources, with an emphasis on biomass [1–3]. Regarding biomass-derived inputs, special attention is directed to carbohydrates, such as fructose, since the conversion of these molecules provides the formation of a spectrum of valuable chemicals [4], including 5-hydroxymethylfurfural [5] and lactic acid [6].

Catalysis is strategic in the transformation of biomass, mainly because it influences the selectivity to the desired product and thus minimizes the generation of byproducts. The development of heterogeneous catalysts is a key resource for sustainable advancement due to its associated advantages, such as easy recovery and reuse. From this perspective, metal oxides have been widely explored due to their acid/base and redox characteristics, as well as the presence of structural defects. These parameters are very important for academic and industrial applications; moreover, these materials are affordable [3,7–12].

The unique properties of cerium- and molybdenum-based materials justify their various catalytic applications. Cerium oxide exhibits outstanding physical and chemical stability, ease of reduction or oxidation ($Ce^{3+} \leftrightarrow Ce^{4+}$), and high electron mobility in the crystal lattice [8,13,14]. Molybdenum oxide is known to be an efficient modulator of acidity [15]. The presence of Lewis and Bronsted acid sites directly influences the conversion of carbohydrates [11].

The junction of metal oxides can lead to attractive changes in physical, chemical, and catalytic properties [8]. This fact allows for the development in this study of a promising catalytic system using cerium and molybdenum for carbohydrate conversion. Although the application of cerium molybdate—exhibiting contributions in photocatalysis [16], gas detection [12], antiviral activity [17], and catalysis [18]—in diverse areas is extensive, to date, no information has been found on its evaluation in the conversion of hexoses. Several synthesis routes, including the precipitation method [19], wet impregnation [15,20], and hydrothermal methods [21], can be used for the preparation of metal oxides. However, the Pechini method has been noted as one of the most promising methodologies for obtaining heterogeneous catalysts because it requires low cost and the use of mild temperatures, as well as being considered a simple process [22]. The basic procedure of the synthesis consists of the formation of a polymeric resin in which the cations are homogeneously dispersed through a reaction between a polycarboxylic acid (such as citric acid), metal ions, and a polyol [23–26]. Ethylene glycol is commonly used as a polymerizing agent in the Pechini method [22,27]. However, this compound presents toxicity and is generated largely by fossil fuel sources [28,29]. On the other hand, glycerol is a nontoxic product, and it can be obtained renewably; a possible replacement of polyol may assign a more sustainable footprint to this route [24,30,31].

Therefore, this study proposed to synthesize cerium oxide (CeO_2), molybdenum oxide (MoO_3), and a mixed oxide based on Mo and Ce through the modified Pechini method, using glycerol as a polymerizing agent. For the first time, such materials, after characterization, were evaluated in the conversion of fructose to chemical inputs of industrial interest.

2. Results and Discussion

The catalysts Ce100, Mo100, and Ce Mo were used in the conversion of fructose in an aqueous medium, and the results are shown in Figure 1. Initially, the experiments showed that Ce100 leads to low conversions, comparable to those obtained in the reaction carried out without the addition of a catalyst. However, in the presence of Mo100 and CeMo, higher conversions were observed, especially for Mo100. For example, at 3 h of reaction, the observed values of conversion were 97.4 and 91.8% for Mo100 and CeMo, respectively. After 4 h of reaction, the two systems led to very close conversions, and the system tended to reach equilibrium.

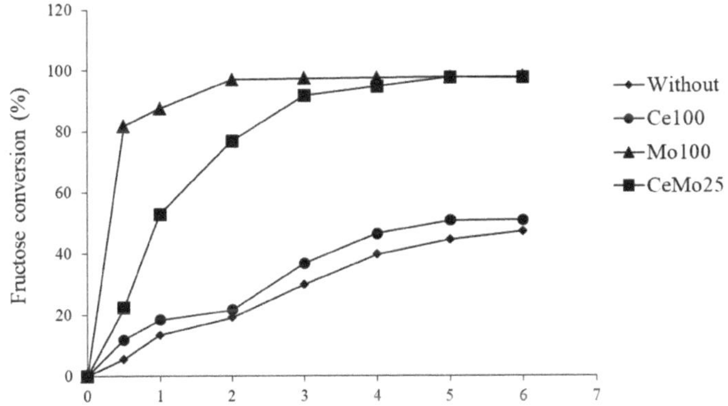

Figure 1. Fructose conversion at 150 °C without a catalyst or with 1.5×10^{-3} g of Ce100, Mo100, or CeMo.

Such conversions may be related to the properties of these catalytic systems, and Table 1 presents some results regarding the characterization and evaluation of the acid sites.

Table 1. Textural and structural properties of the solids investigated in this study.

Sample	S_{BET} [a] ($m^2 \cdot g^{-1}$)	Crystallite Size (nm) [b]	Amount Bronsted [c] ($mmol \cdot g^{-1}$)	Amount Lewis [c] ($mmol \cdot g^{-1}$)	A_B/A_L [d]
Ce100	70.6	10.1	157.5	120.0	1.3
Mo100	>5.0	133.6	700.0	531.8	1.3
CeMo	>5.0	40.5	490.7	291.1	1.7

[a] S_{BET}, specific surface area, calculated by the BET equation. [b] Crystallite size (nm) calculated by the Debye–Scherrer equation. [c] Number of Lewis and Bronsted acid sites ($mmol \cdot g^{-1}$), obtained from FTIR spectra using pyridine as the probe molecule. [d] Ratio of the number of Lewis and Bronsted acid sites.

The low conversion obtained in the presence of Ce100 can be explained by the fact that, even with a surface area much larger than that observed for the other systems investigated here, there is a smaller number of Bronsted and Lewis acid sites in the structure of this material. In the case of the other two systems, higher numbers of acid sites were observed, which justifies the higher conversions, despite the low surface areas. Furthermore, it is important to note that in the CeMo catalyst, the A_B/A_L ratio is superior to the other pure oxides, indicating a higher incidence of Bronsted acid sites in this material.

Previous study data indicate that the number of Lewis and Bronsted acid sites available in the catalyst is related to the success of this type of conversion, and the synergism resulting from the combination between these sites is responsible for obtaining promising results in terms of converting carbohydrates into molecules of interest [8,32,33]. For example, in the synthesis of the mixed oxides of molybdenum and tin, increasing the content of molybdenum in relation to tin increases the acidity, leading to high fructose conversions, with selectivity to the products obtained by retro-aldol and dehydration reactions [11].

Regarding the soluble products identified (see Figure 2) in the presence of Mo100, there is no defined selectivity, although in the first few hours of reaction, this system is the most active of all. In contrast, in the presence of Ce100, a great selectivity to HMF and to products and intermediates of the retro–aldol pathway is observed (glyceraldehyde, dihydroxyacetone, pyruvaldehyde, and lactic acid). However, the latter showed maximum conversions of 50% between 5 and 6 h of reaction. For CeMo, a change in the selectivity profile is observed, with a majority targeting the retro–aldol pathways (pyruvaldehyde, glyceraldehyde, lactic acid, and acetic acid), isomerization (glucose), and dehydration (5-HMF).

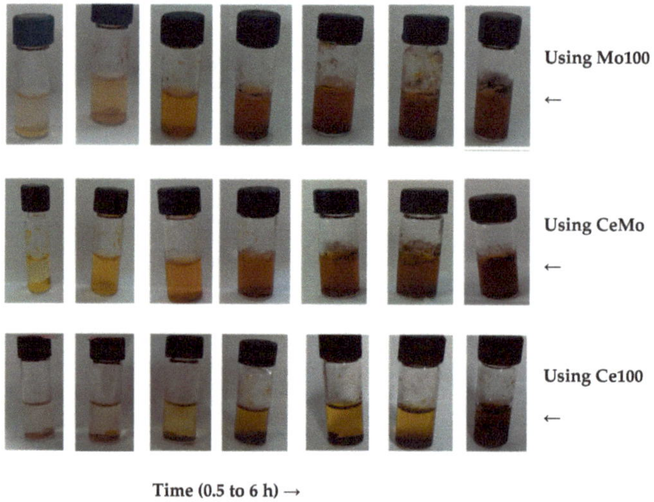

Figure 2. Visual aspects of the samples during the reaction, using catalysts (1.5×10^{-3} g) at 150 °C (0.5 to 6 h).

The formation of 5-HMF follows the route of dehydration of the fructose molecule, which occurs through the consecutive loss of three water molecules. However, due to the high instability of 5-HMF in aqueous media, organic acids are formed, as well as humins [34,35]. The Ce100 catalyst showed high selectivity (70.2% at 6 h) for 5-HMF. On the other hand, the Mo100 and CeMo systems exhibited lower selectivity for this product under the same conditions, 14.5% and 27.5%, respectively. This fact can be explained by the formation of organic acids (levulinic and formic) through the hydration of 5-HMF [34]. It should be noted that for Ce100, no levulinic or formic acid was identified, and this fact can be associated with the nature of the acid sites associated with this catalyst [36]. The literature reports that the decomposition of 5-HMF into organic acids and other side reactions are favored when there are higher numbers of acidic sites in the catalyst [11,37,38]. As shown in Table 2, CeMo and Mo100 contain more of these sites when compared to Ce100. However, for Ce100 and CeMo, the increase in fructose conversion implies an increase in 5-HMF selectivity, a characteristic that is not observed for Mo100.

Table 2. Ranges of thermal decomposition for Ce100, Mo100, and CeMo.

Range T (°C)	Ce100	Mo100	CeMo
100–150	2.3	nd	nd
700–850	2.7	78.2	nd
850–1000	nd	17.5	nd
>1000	2.4	nd	19.0

nd = not detected.

The reduced selectivity to 5-HMF, mainly with the use of Mo100 (14.5%) and CeMo (27.5%) compared to Ce100 (70.2%), may also be associated with the formation of humins, promoted by the condensation between 5-HMF and other reaction intermediates and which is closely related to the conditions used and the acidity of the catalyst [36,39]. The presence of this byproduct can be quantitatively identified through the color change of the reaction medium, that is, the evolution to a dark brown color [36]. As shown in Figure 3, there was a greater formation of humins for Mo100 due to the high acidity of the material (Table 2). For CeMo, a lower and adequate acidity inhibited the high generation of these byproducts. The same justification can be used for Ce100; however, the lower number of acid sites also directly influenced the catalyst activity (see Figure 1). Such observations suggest that a possible structural change, in relation to the pure oxides, obtained in the case of CeMo leads to a better conversion of fructose, maintaining the advantage in terms of the formation of a smaller amount of insoluble products when compared to Mo100.

Lactic acid is a product widely used in the food, chemical, pharmaceutical, and cosmetics industries [40]. When using CeMo and Ce100, a high selectivity was observed for pyruvaldehyde (lactic acid precursor) in the initial reaction times up to 2 h. With longer reaction times, the selectivity to pyruvaldehyde decreases as lactic acid is formed. Interestingly, for Ce100, the presence of glyceraldehyde is notable at lower levels than those of the other precursors, suggesting that the conversion reaction of the product to pyruvaldehyde is very rapid [41]. Previous studies report that the Bronsted and Lewis acid sites are fundamental in promoting the formation of lactic acid, with the first acting on the dehydration of triose to pyruvaldehyde and the second promoting the conversion of pyruvaldehyde to lactic acid through the 1,2 displacement of hydride [42,43].

As shown in Figure 2, Mo100 showed greater selectivity for lactic acid and its intermediates due to the greater number of Lewis acid sites in its structure, which enabled the conversion of pyruvaldehyde. Similarly, this explanation is also useful in relation to the greater selectivity of CeMo to lactic acid when compared to Ce100. This suggests that the lower Lewis acidity and considerable number of Bronsted acid sites present in CeMo and Ce100 led to higher selectivities to pyruvaldehyde in the initial reaction times.

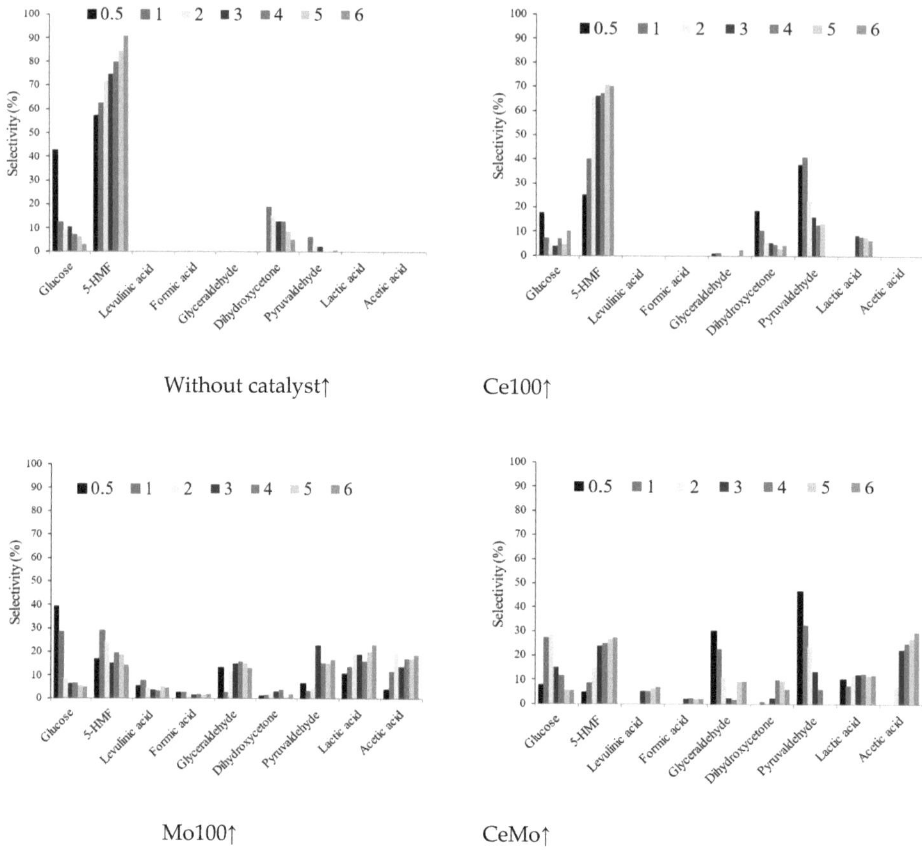

Figure 3. Selectivity for soluble products identified in the fructose conversion at 150 °C, without a catalyst and with 1.5×10^{-3} g of Ce100, Mo100, and CeMo.

Acetic acid is one of the products in high demand in the chemical industry, and it is used as a precursor for various chemical compounds [44,45]. Its most common formation route from the biomass is the retro–aldolic pathway [11,24]. When Ce100 was used as a catalyst, the presence of this compound was not observed, since the decomposition route for the formation of acetic acid is directed by the Lewis acid sites, which were also not identified when Ce100 was used [46]. Furthermore, a greater selectivity of CeMo (29.9%) to acetic acid was observed in relation to Mo100 (18.8%) at 6 h. This result suggests a greater suitability of sites in CeMo, while the greater acidity of Mo100 led to other parallel reactions.

The characterization of these materials, obtained through the Pechini method by using glycerol instead of ethylene glycol, by XRD revealed that cerium molybdate ($Ce_2(MoO_4)_3$) was formed in the case of the CeMo material (Figure 4), and no signals referring to the individual oxides were identified. The formation of ($Ce_2(MoO_4)_3$) was confirmed by the crystalline planes assigned to the reflection lines (002) (100) (101) (112) (004) (200) (211) (204) (220) (116) (312) (224) (card # 30–0303) [17]. As shown in Figure 4, Ce100 exhibits a fluorite-like (CaF_2) face-centered cubic structure, according to the JCPDS crystallographic sheet No. 34-0394 [8,13,21]. Mo100 showed all peaks perfectly indexed, suggesting the formation of the orthorhombic phase (JCPDS No. 05-0508).

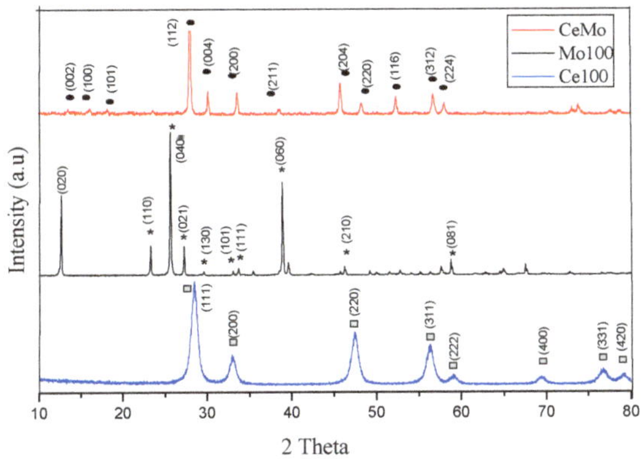

Figure 4. XRD patterns of Ce100 (● = fluorite-like face-centered cubic structure (JCPDS 34-0394)), Mo100 (* = orthorhombic phase (JCPDS No. 05-0508)), and CeMo (□ = $Ce_2(MoO_4)_3$ (JCPDS 30–0303).

Based on the information extracted from the diffractograms, the crystallite size (see Table 1) was calculated using the Debye–Scherrer equation, and for this, the most intense diffraction peaks of CeMo, Ce100, and Mo100 were used, e.g., (112), (040), and (111), respectively. Therefore, the diffractograms show that the addition of molybdenum leads to an increase in crystallite size, suggesting crystal growth by particle aggregation because of mixed oxide formation [15]. This fact also corroborates the result obtained by the BET method, from which a pronounced reduction in the specific surface area was observed (see Table 1).

The structural changes that occurred by the formation of $Ce_2(MoO_4)_3$, in relation to the individual oxides, were confirmed by the evaluation of thermal behavior. In the thermogram obtained for CeMo (Table 2 and Figure S1, Supplementary Materials), only one thermal decomposition event was observed above 1000 °C (with a mass loss of 19%), which clearly continues to occur, but which is not quantified due to the limitation of the equipment, which operates at a maximum temperature of 1500 °C. This suggests that the interaction between the molybdenum and cerium ions for the formation of cerium molybdate leads to an increase in thermal stability, and the mass loss above 850 °C is associated with the decomposition of the material into cerium and molybdenum oxide and the subsequent sublimation of MoO_3 [47–49]. Reports indicate that in this type of material, when subjected to temperatures above 700 °C, defects are formed that favor greater mobility of the molybdenum and cerium ions, consequently leading to changes in the mutual interaction between them, leading to increased thermal stability [48].

Three thermal decomposition events were observed in Ce100 (Figure S1, Supplementary Materials). The first event (mass loss of 2.3%) is attributed to the desorption of water molecules (Khalil et al., 2005). The second event starts at 730 °C (mass loss of 2.7%) and can be related to the oxygen output from the oxide structure [50,51]. The total mass loss recorded after heating from 25 °C to 1500 °C was 7.4%, suggesting good thermal stability. In the case of Mo100 (see Figure S1, Supplementary Materials), two thermal decomposition events were observed. The first, which occurs in the range of 700–850 °C (mass loss of 78.2%), corresponds to the sublimation of MoO_3. The second occurs between 850–1000 °C (mass loss of 17.5%) and can be associated with the output of polymolybdates [11].

Therefore, it can be inferred that the synthesis of the material from the molybdenum and cerium precursors, under the conditions used here, is an efficient strategy for the modulation of acid sites, resulting in a suitable catalytic system for the conversion of fructose into an aqueous medium. The formation of insoluble byproducts is minimized when compared to Mo100, leading to promising conversions, mainly after 3 h of reaction under the conditions investigated here. In addition, a superior thermal stability of CeMo was observed in relation to Mo100.

Catalyst recyclability is of great importance in biorefinery processes [11]. In this study, the catalyst was calcined to eliminate organic fractions and perform reuse tests (1 h, 1.5×10^{-3} g of catalyst, 150 °C). The results of the recycling experiments are shown in Figure 5, and only a small variation in conversion and selectivity values was observed for CeMo for each of the 4 cycles tested (approximately 72.0% of conversion, on average). The liquid samples collected after the first and the last (4th) cycle were subjected to X-ray dispersive energy spectrometry (EDX) analysis, and no significant amount of molybdenum or cerium was detected, suggesting that leaching did not occur.

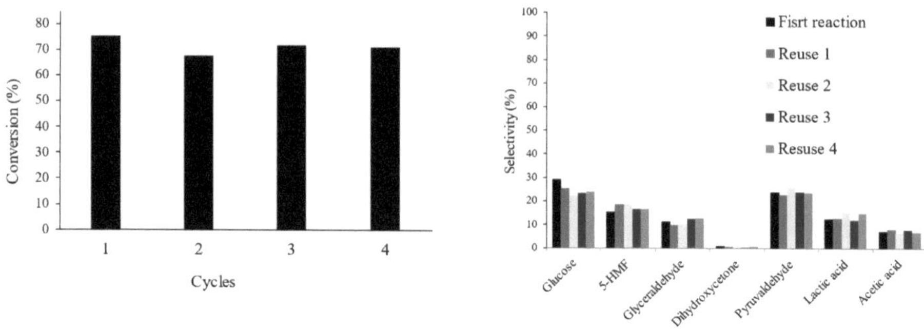

Figure 5. Results of reuse for the CeMo system (150 °C, 1 h).

3. Materials and Methods

3.1. Materials

Ammonium heptamolybdate tetrahydrate (99.9%, Sigma-Aldrich, St. Louis, MI, USA), cerium(III) nitrate hexahydrate (99.9%, Sigma-Aldrich), citric acid (>99.0%, Sigma-Aldrich), ammonium hydroxide (Sigma-Aldrich), nitric acid (>65.0%, Sigma-Aldrich), glycerol (>99.0%, Sigma-Aldrich), and fructose (>99.0%, Sigma-Aldrich) were obtained commercially and used as received.

3.2. Synthesis and Characterization

Initially, citric acid and the metal precursor were weighed in a molar ratio of 3:1. Then, both were dissolved separately in deionized water at 30 °C under constant stirring and mixed for 30 min. The next step was the addition of glycerol to the prepared solution in a molar ratio of 60:40, based on the amount of citric acid, maintaining vigorous stirring at 70 °C for 3 h to ensure the homogeneity of the system. Subsequently, the temperature was increased to 130 °C to promote the polyesterification reaction and the subsequent formation of the polymeric resin. The metal precursors used were cerium(III) nitrate hexahydrate and ammonium heptamolybdate [24,31]. The mixed catalyst was prepared in a mass ratio of 75% cerium to 25% molybdenum, and cerium(III) nitrate hexahydrate and ammonium heptamolybdate were weighed and dissolved separately in deionized water at 30 °C under constant stirring. Soon after, both were mixed and added to a beaker containing a solution of citric acid, maintained under vigorous stirring for 30 min. All the following procedures were identical to those described above for the pure oxides. The materials were subjected to heat treatment at 280 °C for 2 h, using a heating rate of 10 °C/min to break the polymer chains and form a semicarbonized black powder. Subsequently, they were macerated in a mortar and sieved (100 mesh) to be placed again in the muffle furnace at 550 °C for 4 h. Finally, all materials were stored under an argon atmosphere. Table 3 presents the description of the catalysts and the codes used.

Table 3. Code of the catalysts and the respective descriptions.

Code	Description
Ce100	Cerium oxide(IV) (CeO_2)
Mo100	Molybdenum oxide(VI) (MoO_3)
CeMo	Cerium molybdate ($Ce_2(MoO_4)_3$)

For material characterization, X-ray diffraction (XRD) measurements were performed using a Shimadzu diffractometer, model XRD-6000, and a Cu Kα radiation source (1.5418 Å) with a nickel filter to obtain wide-angle diffraction patterns in the 2θ = 3–40° range. Crystallite size (nm) was calculated using the Debye–Scherrer equation.

Nitrogen adsorption measurements were performed at 77.15 K using a gas adsorption analyzer (Micromeritics, model ASAP-2020). The textural properties were estimated from the N_2 adsorption isotherms using the Brunauer–Emmett–Teller (BET) equation and the Barrett–Joyner–Halenda (BJH) method. Before the analysis, all the samples were pretreated under vacuum at 423.15 K for 24 h. Thermogravimetric (TG/DTG) analyses were performed using a Shimadzu analyzer, model DTG-60H, with an N_2 flow rate of 40 mL·min^{-1} in the temperature range of 30–800 °C at a heating rate of 10 °C·min^{-1}. Fourier transform infrared (FTIR) spectra were obtained with a Shimadzu IR Prestige 21 infrared spectrophotometer using potassium bromide tablets (KBr). A total of 80 scans were performed in the transmittance mode in the spectral range of 4000–400 cm^{-1} at a resolution of 4.0 cm^{-1}.

3.3. Conversion of Fructose

All experiments were performed in 4 mL vials under magnetic stirring and heating (150 °C), at several reaction times. The fructose solution was 0.016 g of fructose in 2 mL of deionized water, and for some experiments, catalysts (1.5×10^{-3} g) were used. Conversion, yield, and selectivity were calculated from the results of quantification by HPLC. The reuse tests of CeMo (G) were carried out at 150 °C for 2 h under the same reaction conditions. Subsequently, the catalyst was removed from the reaction medium by centrifugation and calcined at 550 °C for 4 h for subsequent reuse. The liquid samples collected after the cycles were subjected to X-ray dispersive energy spectrometry (EDX) analysis to determine whether leaching had occurred.

4. Conclusions

In the absence of a catalyst or using Ce100, low fructose conversions were observed, with the formation mainly of 5-HMF. In contrast, in the presence of Mo100 and CeMo, higher conversions are achieved, and the formation of different products was observed. However, using Mo100, which exhibits a higher number of Lewis and Bronsted acid sites than does CeMo, higher conversions were observed before the achievement of thermodynamic equilibrium, but this is accompanied by the formation of a greater amount of soluble and insoluble material, directly reflected in the color evolution of the reactions over time. In relation to the soluble products formed from fructose using CeMo, the direction to the retro–aldolic pathway was observed, evidencing the modulation of these systems. It is important to highlight that the low conversion obtained in the presence of Ce100 can be explained by the smaller number of Brostend and Lewis acid sites in the structure of this material, even with a surface area much higher than that observed for the other systems. In the case of Mo100 and CeMo, higher levels of acid sites were observed, which justifies the higher conversions, despite the low surface areas. Reuse and leaching tests indicated that the CeMo system proved to be stable and robust under the conditions tested.

Supplementary Materials: The following supporting information can be downloaded at: https://www.mdpi.com/article/10.3390/catal13010004/s1, Figure S1: Thermogravimetric analysis (TG/DTG): (**A**) Ce100, (**B**) Mo100, and (**C**) CeMo.

Author Contributions: Conceptualization, D.B.d.A.P., M.R.M. and S.M.P.M.; methodology, D.B.d.A.P., T.V.d.S.M., J.A.d.V.J. and D.O.d.S.A.; investigation, D.B.d.A.P., T.V.d.S.M., J.A.d.V.J., M.R.M. and S.M.P.M.; resources, M.R.M. and S.M.P.M.; writing—original draft preparation, D.B.d.A.P., T.V.d.S.M. and S.M.P.M.; writing—review and editing, D.B.d.A.P., T.V.d.S.M., M.R.M. and S.M.P.M.; visualization, D.B.d.A.P., T.V.d.S.M., J.A.d.V.J., M.R.M. and S.M.P.M.; supervision, M.R.M. and S.M.P.M.; project administration, M.R.M. and S.M.P.M.; funding acquisition, M.R.M. and S.M.P.M. All authors have read and agreed to the published version of the manuscript.

Funding: This research was supported by the National Council for Scientific and Technological Development (CNPq), the Brazilian Federal Agency for the Improvement of Higher Education (CAPES), the Brazilian Innovation Agency (FINEP), and the Alagoas Research Foundation (FAPEAL). G.C.A. and I.M.A.S. express their appreciation for fellowships granted by CAPES and CNPq. M.R.M. and S.M.P.M. thank CNPq for research fellowships.

Data Availability Statement: Not applicable.

Acknowledgments: The authors also thank the LSCAT/CTEC/UFAL, TecNano/ICF/UFAL, and GON/IF/UFAL teams for their contributions.

Conflicts of Interest: The authors declare no conflict of interest.

References

1. Huang, H.; Gao, Y.; Chen, H.; Wu, Y.; Wang, J.; Yu, C.; Li, J.; Zou, C. Biomass Briquette Fuel, Boiler Types and Pollutant Emissions of Industrial Biomass Boiler: A Review. *Particuology* **2022**, *77*, 79–90. [CrossRef]
2. Gani, A. Fossil Fuel Energy and Environmental Performance in an Extended STIRPAT Model. *J. Clean. Prod.* **2021**, *297*, 126526. [CrossRef]
3. Deng, W.; Feng, Y.; Fu, J.; Guo, H.; Guo, Y.; Han, B.; Jiang, Z.; Kong, L.; Li, C.; Liu, H.; et al. Catalytic Conversion of Lignocellulosic Biomass into Chemicals and Fuels. *Green Energy Environ.* 2022; *in press*. [CrossRef]
4. Bayu, A.; Abudula, A.; Guan, G. Reaction Pathways and Selectivity in Chemo-Catalytic Conversion of Biomass-Derived Carbohydrates to High-Value Chemicals: A Review. *Fuel Process. Technol.* **2019**, *196*, 106162. [CrossRef]
5. Mehta, J.; Metre, A.V.; Bhakhar, M.S.; Panwar, D.S.; Dharaskar, S. Biomass-Derived 5-Hydroxymethylfurfural (HMF) and 2,5-Dimethylfuran (DMF) Synthesis as Promising Alternative Fuel: A Prospective Review. *Mater. Today Proc.* **2022**, *62*, 6978–6984. [CrossRef]
6. Li, Y.; Ma, J.; Jin, D.; Jiao, G.; Yang, X.; Liu, K.; Zhou, J.; Sun, R. Copper Oxide Functionalized Chitosan Hybrid Hydrogels for Highly Efficient Photocatalytic-Reforming of Biomass-Based Monosaccharides to Lactic Acid. *Appl. Catal.* **2021**, *291*, 120123. [CrossRef]
7. Védrine, J.C. Metal Oxides in Heterogeneous Oxidation Catalysis: State of the Art and Challenges for a More Sustainable World. *ChemSusChem* **2019**, *12*, 577–588. [CrossRef]
8. Stošić, D.; Bennici, S.; Rakić, V.; Auroux, A. CeO_2–Nb_2O_5 Mixed Oxide Catalysts: Preparation, Characterization and Catalytic Activity in Fructose Dehydration Reaction. *Catal. Today* **2012**, *192*, 160–168. [CrossRef]
9. Gawande, M.B.; Pandey, R.K.; Jayaram, R.V. Role of Mixed Metal Oxides in Catalysis Science—Versatile Applications in Organic Synthesis. *Catal. Sci. Technol.* **2012**, *2*, 1113. [CrossRef]
10. Xu, S.; Pan, D.; Li, W.; Shen, P.; Wu, Y.; Song, X.; Zhu, Y.; Xu, N.; Gao, L.; Xiao, G. Direct Conversion of Biomass-Derived Carbohydrates to 5-Hydroxymethylfurfural Using na Efficient and Inexpensive Manganese Phosphate Catalyst. *Fuel Process. Technol.* **2018**, *181*, 199–206. [CrossRef]
11. Dos Santos, T.V.; Da Silva Avelino, D.O.; Meneghetti, M.R.; Meneghetti, S.M.P. Mixed Oxides Based on SnO_2 Impregnated with MoO_3: A Robust System to Apply in Fructose Conversion. *Catal. Commun.* **2018**, *114*, 120–123. [CrossRef]
12. Oliveira, F.K.F.; Santiago, A.A.G.; Catto, A.C.; da Silva, L.F.; Tranquilin, R.L.; Longo, E.; Motta, F.V.; Bomio, M.R.D. Cerium Molybdate Nanocrystals: Microstructural, Optical and Gas-Sensing Properties. *J. Alloy. Compd.* **2021**, *857*, 157562. [CrossRef]
13. Chen, B.; Li, P.; Wang, B.; Wang, Y. Flame-Annealed Porous TiO_2/CeO_2 Nanosheets for Enhenced CO Gas Sensors. *Appl. Surf. Sci.* **2022**, *593*, 153418. [CrossRef]
14. Benali, F.; Boukoussa, B.; Ismail, I.; Hachemaoui, M.; Iqbal, J.; Taha, I.; Cherifi, Z.; Mokhtar, A. One Pot Preparation of CeO_2@Alginate Composite Beads for the Catalytic Reduction of MB Dye: Effect of Cerium Percentage. *Surf. Interfaces* **2021**, *26*, 101306. [CrossRef]
15. Assis, G.C.; Silva, I.M.A.; Dos Santos, T.V.; Meneghetti, M.R.; Meneghetti, S.M.P. Photocatalytic Properties of SnO_2/MoO_3 Mixed Oxides and Their Relation to the Electronic Properties and Surface Acidity. *J. Photochem. Photobiol.* **2021**, *407*, 113035. [CrossRef]
16. Muthuvel, I.; Sathyapriya, S.; Suguna, S.; Gowthami, K.; Thirunarayanan, G.; Rajalakshmi, S.; Sundaramurthy, N.; Dinesh Karthik, A.; Rajachandrasekar, T. Solar Light Driven Cerium Molybdate Nanocatalyst for Effective Photodecomposition of Fuchsin Basic Dye. *Mater. Today Proc.* **2021**, *43*, 2274–2279. [CrossRef]

17. Ito, T.; Sunada, K.; Nagai, T.; Ishiguro, H.; Nakano, R.; Suzuki, Y.; Nakano, A.; Yano, H.; Isobe, T.; Matsushita, S.; et al. Preparation of Cerium Molybdates and Their Antiviral Activity against Bacteriophage Φ6 and SARS-CoV-2. *Mater. Lett.* **2021**, *290*, 129510. [CrossRef]
18. Liu, X.-Y.; Li, X.-P.; Zhao, R.-X. $Ce_2(MoO_4)_3$ as na Efficient Catalyst for Aerobic Oxidative Desulfurization of Fuels. *Pet. Sci.* **2022**, *19*, 861–869. [CrossRef]
19. Putri, G.E.; Rilda, Y.; Syukri, S.; Labanni, A.; Arief, S. Highly Antimicrobial Activity of Cerium Oxide Nanoparticles Synthesized Using Moringa Oleifera Leaf Extract by a Rapid Green Precipitation Method. *J. Mater. Res. Technol.* **2021**, *15*, 2355–2364. [CrossRef]
20. Hadebe, N.P.; Kerru, N.; Tukulula, M.; Jonnalagadda, S.B. A Sustainable Molybdenum Oxide Loaded on Zirconia (MoO_3/ZrO_2) Catalysed Multicomponent Reaction to Synthesise Novel Dihydropyridines. *Sustain. Chem. Pharm.* **2022**, *25*, 100578. [CrossRef]
21. Afza, N.; Shivakumar, M.S.; Alam, M.W.; Kumar, A.N.; Bhatt, A.S.; Murthy, H.C.A.; Ravikumar, C.R.; Mylarappa, M.; Selvanandan, S. Facile Hydrothermal Synthesis of Cerium Oxide/RGO Nanocomposite for Photocatalytic and Supercapacitor Applications. *Appl. Surf. Sci.* **2022**, *11*, 100307. [CrossRef]
22. Garcia, A.B.S.; Bispo-Jr, A.G.; Lima, S.A.M.; Pires, A.M. Effects of the Pechini's Modified Synthetic Route on Structural and Photophysical Properties of Eu^{3+} or Tb^{3+}-Doped $LaAlO_3$. *Mater. Res. Bull.* **2021**, *143*, 111462. [CrossRef]
23. Wang, Y.; Li, H.; Huang, D.; Wang, X.; Cai, L.; Chen, Y.; Wang, W.; Song, Y.; Han, G.; Zhen, B. A High-Performance Ethanol Gas Sensor Based on Ce-Doped SnO_2 Nanomaterials Prepared by the Pechini Method. *Mater. Sci. Semicond. Process.* **2022**, *137*, 106188. [CrossRef]
24. Dos Santos, T.V.; Pryston, D.B.A.; Assis, G.C.; Meneghetti, M.R.; Meneghetti, S.M.P. Tin, Niobium and Tin-Niobium Oxides Obtained by the Pechini Method Using Glycerol as a Polyol: Synthesis, Characterization and Use as a Catalyst in Fructose Conversion. *Catal. Today* **2021**, *379*, 62–69. [CrossRef]
25. Kang, J.; Gwon, Y.R.; Cho, S.K. Photoelectrochemical Water Oxidation on $PbCrO_4$ Thin Film Photoanode Fabricated via Pechini Method: Various Solution-Processes for $PbCrO_4$ Film Synthesis. *J. Electroanal. Chem.* **2020**, *878*, 114601. [CrossRef]
26. Mahdi, M.A.; Jasim, L.S.; Ranjeh, M.; Masjedi-Arani, M.; Salavati-Niasari, M. Improved Pechini Sol-Gel Fabrication of $Li_2B_4O_7/NiO/Ni_3(BO_3)_2$ Nanocomposites to Advanced Photocatalytic Performance. *Arab. J. Chem.* **2022**, *15*, 103768. [CrossRef]
27. Aflaki, M.; Davar, F. Synthesis, luminescence and photocatalyst properties of zirconia nanosheets by modified Pechini method. *J. Mol. Liq.* **2016**, *221*, 1071–1079. [CrossRef]
28. Bayat, B.; Tahvildari, K.; Hemmati, A.; Bazyari, A.; Ghaemi, A. Production of ethylene glycol monobutyl ether through etherification of ethylene glycol using a nanostructured heterogeneous catalyst of amyberlyst-15. *Process Saf. Environ. Prot.* **2022**, *165*, 597–609. [CrossRef]
29. Mendelson, N.L.; Fitzgerald, A.; Kumar, A.; Pineda, J.A.; MacDougall, J.; Solomon, R.J.; Gibson, P.C.; Prikis, M.; Marroquin, C.E. Impact of Ethylene Glycol Toxicity on Donor Organ Viability: A Case Report. *Transplant. Proc.* **2017**, *49*, 2411–2414. [CrossRef]
30. Pirzadi, Z.; Meshkani, F. From glycerol production to its value-added uses: A critical review. *Fuel* **2022**, *329*, 125044. [CrossRef]
31. Durán, P.; Capel, F.; Gutierrez, D.; Tartaj, J.; Moure, C. Cerium (IV) Oxide Synthesis and Sinterable Powders Prepared by the Polymeric Organic Complex Solution Method. *J. Eur. Ceram. Soc.* **2002**, *22*, 1711–1721. [CrossRef]
32. Bodachivskyi, I.; Kuzhiumparambil, U.; Williams, D.B.G. Acid-Catalyzed Conversion of Carbohydrates intoValue-Added Small Molecules in Aqueous Media and Ionic Liquids. *ChemSusChem* **2018**, *11*, 642–660. [CrossRef] [PubMed]
33. Doungsri, S.; Rattanaphanee, P.; Wongkoblap, A. Production of Lactic Acid from Cellulose Using Solid Catalyst. *MATEC Web Conf.* **2019**, *268*, 07006. [CrossRef]
34. Qi, X.; Watanabe, M.; Aida, T.M.; Smith, R.L., Jr. Catalytic Dehydration of Fructose into 5-Hydroxymethylfurfural by Ion-Exchange Resin in Mixed-Aqueous System by Microwave Heating. *Green Chem.* **2008**, *10*, 799. [CrossRef]
35. Gomes, F.N.D.C.; Mendes, F.M.T.; Souza, M.M.V.M. Synthesis of 5-Hydroxymethylfurfural from Fructose Catalyzed by Phosphotungstic acid. *Catal. Today* **2017**, *279*, 296–304. [CrossRef]
36. Taghavi, S.; Ghedini, E.; Menegazzo, F.; Mäki-Arvela, P.; Peurla, M.; Zendehdel, M.; Cruciani, G.; Di Michele, A.; Murzin, D.Y.; Signoretto, M. CuZSM-5@HMS Composite as an Efficient Micro-Mesoporous Catalyst for Conversion of Sugars into Levulinic Acid. *Catal. Today* **2022**, *390*, 146–161. [CrossRef]
37. Yang, F.; Liu, Q.; Bai, X.; Du, Y. Conversion of Biomass into 5-Hydroxymethylfurfural Using Solid Acid Catalyst. *Bioresour. Technol.* **2011**, *102*, 3424–3429. [CrossRef]
38. Wei, W.; Yang, H.; Wu, S. Efficient Conversion of Carbohydrates into Levulinic Acid over Chromium Modified Niobium Phosphate Catalyst. *Fuel* **2019**, *256*, 115940. [CrossRef]
39. Tsilomelekis, G.; Orella, M.J.; Lin, Z.; Cheng, Z.; Zheng, W.; Nikolakis, V.; Vlachos, D.G. Molecular Structure, Morphology and Growth Mechanisms and Rates of 5-Hydroxymethyl Furfural (HMF) Derived Humins. *Green Chem.* **2016**, *18*, 1983–1993. [CrossRef]
40. Cantero, D.A.; Vaquerizo, L.; Martinez, C.; Bermejo, M.D.; Cocero, M.J. Selective Transformation of Fructose and High Fructose Content Biomass into Lactic Acid in Supercritical Water. *Catal. Today* **2015**, *255*, 80–86. [CrossRef]
41. Shen, D.K.; Gu, S. The Mechanism for Thermal Decomposition of Cellulose and Its Main Products. *Bioresour. Technol.* **2009**, *100*, 6496–6504. [CrossRef] [PubMed]
42. Fang, T.; Liu, M.; Li, Z.; Xiong, L.; Zhang, D.; Meng, K.; Qu, X.; Zhang, G.; Jin, X.; Yang, C. Hydrothermal Conversion of Fructose to Lactic Acid and Derivatives: Synergies of Metal and Acid/Base Catalysts. *Chin. J. Chem. Eng.* 2022; *in press*. [CrossRef]

43. Kong, L.; Shen, Z.; Zhang, W.; Xia, M.; Gu, M.; Zhou, X.; Zhang, Y. Conversion of Sucrose into Lactic Acid over Functionalized Sn-Beta Zeolite Catalyst by 3-Aminopropyltrimethoxysilane. *ACS Omega* **2018**, *3*, 17430–17438. [CrossRef] [PubMed]
44. Lomate, S.; Katryniok, B.; Dumeignil, F.; Paul, S. High Yield Lactic Acid Selective Oxidation into Acetic Acid over a Mo-V-Nb Mixed Oxide Catalyst. *Sustain. Chem. Process.* **2015**, *3*, 5. [CrossRef]
45. Fang, Y.; Zeng, X.; Yan, P.; Jing, Z.; Jin, F. An Acidic Two-Step Hydrothermal Process To Enhance Acetic Acid Production from Carbohydrate Biomass. *Ind. Eng. Chem. Res.* **2012**, *51*, 4759–4763. [CrossRef]
46. Dos Santos, T.V.; Da Silva Avelino, D.O.; Pryston, D.B.A.; Meneghetti, M.R.; Meneghetti, S.M.P. Tin, molybdenum and tin-molybdenum oxides: Influence of Lewis and Bronsted acid sites on xylose conversion. *Catal. Today* **2021**, *394*, 125–132. [CrossRef]
47. Yousefi, T.; Khanchi, A.R.; Ahmadi, S.J.; Rofouei, M.K.; Yavari, R.; Davarkhah, R.; Myanji, B. Cerium(III) Molybdate Nanoparticles: Synthesis, Characterization and Radionuclides Adsorption Studies. *J. Hazard. Mater.* **2012**, *215*, 266–271. [CrossRef]
48. Nasser, H.; Rédey, Á.; Yuzhakova, T.; Kovács, J. Thermal Stability and Surface Structure of Mo/CeO_2 and Ce-Doped Mo/$Al2O_3$ Catalysts. *J. Therm. Anal. Calorim.* **2009**, *95*, 69–74. [CrossRef]
49. Castellan, A.; Bart, J.C.J.; Bossi, A.; Perissinoto, P.; Giordano, N. On the Formation of Cerium Molybdates under different atmospheric and thermal conditions. *Z. Anorg. Allg. Chem.* **1976**, *422*, 155–172. [CrossRef]
50. Khalil, K.M.S.; Elkabee, L.A.; Murphy, B. Preparation and Characterization of Thermally Stable Porous Ceria Aggregates Formed via a Sol–Gel Process of Ultrasonically Dispersed Cerium(IV) Isopropoxide. *Micropor. Mesopor. Mater.* **2005**, *78*, 83–89. [CrossRef]
51. Suresh, R.; Ponnuswamy, V.; Mariappan, R. Effect of Annealing Temperature on the Microstructural, Optical and Electrical Properties of CeO_2 Nanoparticles by Chemical Precipitation Method. *Appl. Surf. Sci.* **2013**, *273*, 457–464. [CrossRef]

catalysts MDPI

Article

Transfer Hydrogenation of Biomass-Like Phenolic Compounds and 2-PrOH over Ni-Based Catalysts Prepared Using Supercritical Antisolvent Coprecipitation

Alexey Philippov *, Nikolay Nesterov and Oleg Martyanov

Boreskov Institute of Catalysis, Siberian Branch of the Russian Academy of Sciences, Academician Lavrentiev Avenue 5, 630090 Novosibirsk, Russia
* Correspondence: philippov@catalysis.ru; Tel.: +7-38-3326-9313

Abstract: Transfer hydrogenation (TH) is considered as one of the most promising ways to convert biomass into valuable products. This study aims to demonstrate the performance of high-loaded Ni-based catalysts in the TH of phenolic compounds such as guaiacol and dimethoxybenzenes. The experiments were carried out under supercritical conditions at 250 °C using 2-PrOH as the only hydrogen donor. Ni-SiO_2 and NiCu-SiO_2 were synthesized using the eco-friendly original method based on supercritical antisolvent coprecipitation. It has been found that guaiacol is rapidly converted into 2-methoxycyclohexanol and cyclohexanol, while the presence of Cu impedes the formation of the latter product. Transformations of dimethoxybenzene position isomers are slower and result in different products. Thus, 1,3-dimethoxybenzene loses oxygen atoms transform into methoxycyclohexane and cyclohexanol, whereas the saturation of the aromatic ring is more typical for other isomers. The Cu addition increases specific catalytic activity in the TH of 1,2-and 1,3-dimethoxybenzene compared to the Cu-free catalyst.

Keywords: transfer hydrogenation; antisolvent supercritical coprecipitation; phenolic compounds; 2-propanol; hydrodeoxygeantion; hydrodearomatization

1. Introduction

The global energy crisis, accompanied by environmental issues, is driving society to the need for a green transition. In this concept, lignocellulosic biomass should play the role of the renewable and non-edible source of carbon atoms [1,2]; therefore, its conversion into chemicals, materials, and fuels is widely studied [3,4]. Unlike cellulose, lignin has an irregular structure represented by phenolic fragments that makes lignin upgrading into valuable products difficult. Due to the high oxygen content reaching up to 40 wt.% [5], hydrodeoxygenation (HDO) via H_2 hydrogenation is considered as one of the most promising and effective methods of lignin transformations [6,7]. However, this method has important disadvantages from practical and environmental points of view. First, H_2 is known for its corrosive activity and explosiveness, and it requires high-pressure equipment presenting a significant safety hazard. Second, the low solubility of molecular hydrogen in almost all organic compounds results in low rates of hydrogenation in the liquid phase. At the same time, the use of organic solvents prevents biomass thermal decomposition and increases the effectiveness of the whole process [8].

Alcohols can be easily produced from renewable sources; moreover, they demonstrate the excellent H-donor activity in transfer hydrogenation (TH). For example, alcohols were used as H-donors in the TH of phenol-like compounds [9–11] and other bio-based molecules [12,13]. Sometimes alcohols demonstrate superior H-donor activity compared to H_2 [1,14]; however, as a rule, the use of organic H-donors requires harder conditions, including supercritical fluids [15,16]. In addition to donor activity, the supercritical state of alcohols also promotes the destruction of complex compounds such as lignin, polymers, or inorganic oxides [17–19]. Thus, supercritical alcohols are considered as a promising reducing agents for H_2-free hydrogenation.

Biomass catalytic conversion is aimed at obtaining a huge number of products, and heterogeneous catalysts are considered as an essential part of the process. One of the promising methods of the catalyst preparation is supercritical antisolvent (SAS) coprecipitation, which demonstrated promising results in the preparation of metal nanoparticles [10,20,21]. This method is based on high supersaturation of a precursor solution with supercritical fluid, in the role of which CO_2 is often used. The formation of a CO_2-solvent binary system results in a sharp decrease in dissolving power and the fast precipitation of catalyst precursors [22], which can be coprecipitated simultaneously with oxide sols, forming a catalyst support [21]. One of the key advantages of the SAS coprecipitation method is that it allows metal in a catalyst to reach up to 90% while maintaining the high dispersion of the metal particles [23,24]. Moreover, CO_2 and a solvent of the precursors can be easily separated and used repeatedly in a way that complies with the principals of green chemistry. In our earlier studies, the high-loaded nickel-based catalysts prepared according SAS coprecipitation demonstrated excellent results in the TH of anisole [10,25].

Considering the potential of supercritical fluids in catalyst preparation and H_2-free hydrogenation, this study aims to demonstrate the performance of the high-loaded Ni-based catalysts in the transfer hydrogenation of phenolic compounds, such as guaiacol and dimethoxybenzenes, for the first time. For this, the catalysts were synthesized accordingly the supercritical antisolvent coprecipitation method using sc-CO_2 as an antisolvent. The obtained catalysts containing 36–40 wt.% of nickel were used in the TH of guaiacol and dimethoxybenzenes modeling the products of lignin degradation. The transformations were carried out in sc-2-PrOH, which played the role of a solvent and an H-donor. Special attention was paid to the effect of copper addition to nickel and the comparison of guaiacol and dimethoxybenezene positional isomers in terms of their reactivity and the selectivity of their transformations.

2. Results

2.1. Catalyst Properties

The properties of the synthesized catalysts are described in detail in our previous study [25] and in the Supplementary Materials. Briefly, the results of XRF demonstrate that the Ni-SiO_2 catalyst contains 36.4 wt.% of Ni, while NiCu-SiO_2 contains 40.7 wt.% and 4.0 wt.% of nickel and copper, respectively (Table S1). XRD shows that Ni^0 and Cu^0 form the monophasic particles, which means the crystallite size does not exceed 5.5 nm, while the mean size of Ni^0 crystallites in Ni-SiO_2 is 6.0 nm (Figure S1). The samples have similar areas of Ni atoms measured by CO adsorption–35 m^2/g and 32 m^2/g for Ni-SiO_2 and NCu-SiO_2, respectively. TEM microphotographs shows that the nanosized particles are agglomerated in larger structures up to several micrometers in size (Figure S2). EDX data show that Ni and Cu are evenly distributed.

2.2. Transfer Hydrogenation of Guaiacol

Both catalysts provide higher conversion in TH of guaiacol than the other phenolic compounds used in this study. Guaiacol has a free OH group, which facilitates the interaction between this substrate and the catalyst surface according to the literature data [9,26]. The conversion of guaiacol reaches 100% after 2 h, and Ni-SiO_2, 2-methoxycyclohexanol, and cyclohexanol are found to be the main products (Figure 1). The kinetic data show that 2-methylcyclohexanol formed in the reaction mixture transforms into cyclohexanol and cyclohexane. In our recent study [25], methoxycyclohexane formed after the aromatic ring saturation of anisole demonstrated stability under very similar conditions. Thus, we suggest that the presence of two oxygen atoms in the substrate as well as the free OH group can facilitate hydrodeoxygenation.

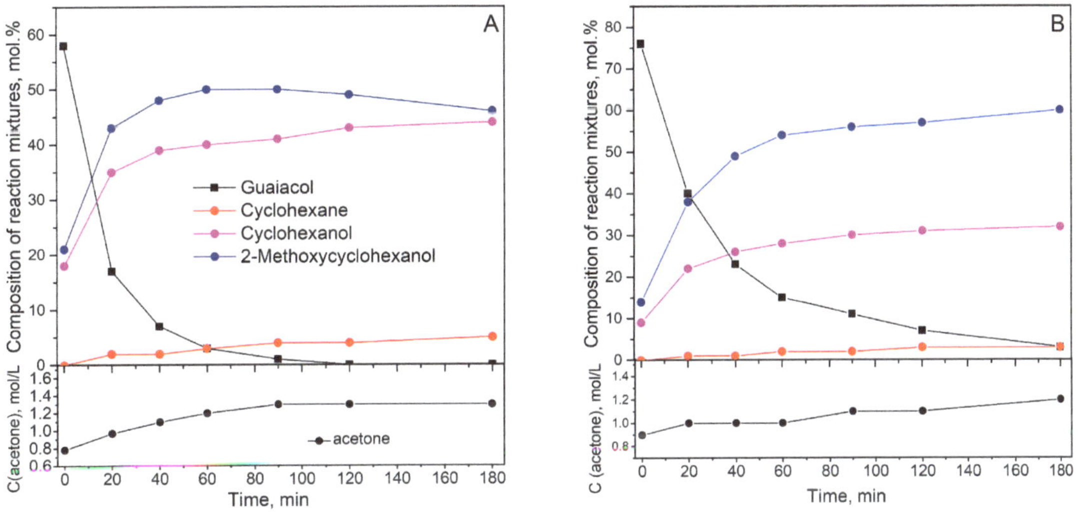

Figure 1. The composition of the reaction mixture in the course of guaiacol transformation over catalysts: (**A**)–Ni-SiO_2 and (**B**)–NiCu-SiO_2. 250 °C, 7.0–7.7 MPa, and m(catalyst) = 0.10–0.11 g.

The literature data show that the addition of Cu to Ni nanoparticles can improve the catalytic performance of the Ni-based catalysts. It has been demonstrated that Cu lowers the catalyst activation temperature [23,27] and contributes to the resistance of the Ni-Cu phase against oxidation [28,29]. Moreover, the presence of Cu can promote anisole HDO when H_2 is used [24,30]. However, in case of transfer hydrogenation, the Cu additives decrease the conversion of anisole as well as the rate of C-O bond cleavage [25]. These observations were confirmed in the present study because the use of NiCu-SiO_2 results in the lower guaiacol conversion compared to Ni-SiO_2 catalyst (Figure 1). At the same time, the cyclohexanol yield reaches only 32% vs. 44% after 3 h over NiCu-SiO_2 and Ni-SiO_2, respectively. Our previous studies demonstrated that this effect is based on acetone formation, which is suggested to adsorb on the catalyst surface, blocking active sites [25]. It is clearly seen in Figure S3 (see the Supplementary Materials) that the logarithmic dependence of guaiacol conversion on time is not linear.

2.3. *Transfer Hydrogenation of Dimethoxybenzenes*

The conversion of 1,2-dimethoxybenzene (1,2-DMB) achieves 83% and 88% for 3 h over Ni-SiO_2 and NiCu-SiO_2, respectively (Figure 2). In both cases, 1,2-dimethoxycyclohexane was found to be the main product, whereas the concentrations of other compounds did not exceed 10%. Thus, the oxygen-containing groups are almost uninvolved in the TH, in contrast to the results obtained for guaiacol. The difference can be related to the influence of OH group of guaiacol, which interacts with the catalyst facilitating the adsorption of the organic molecule. This promotes the transformations of the oxygen-containing groups on the catalyst surface. It is important to notice that the Cu addition results in the higher rate constant (Table 1) as well as the specific catalytic activity (Table 2), compared to the Cu-free catalyst. This fact is in good accordance with the known literature data [14,26,31] demonstrating the positive Cu influence on HDO under TH conditions.

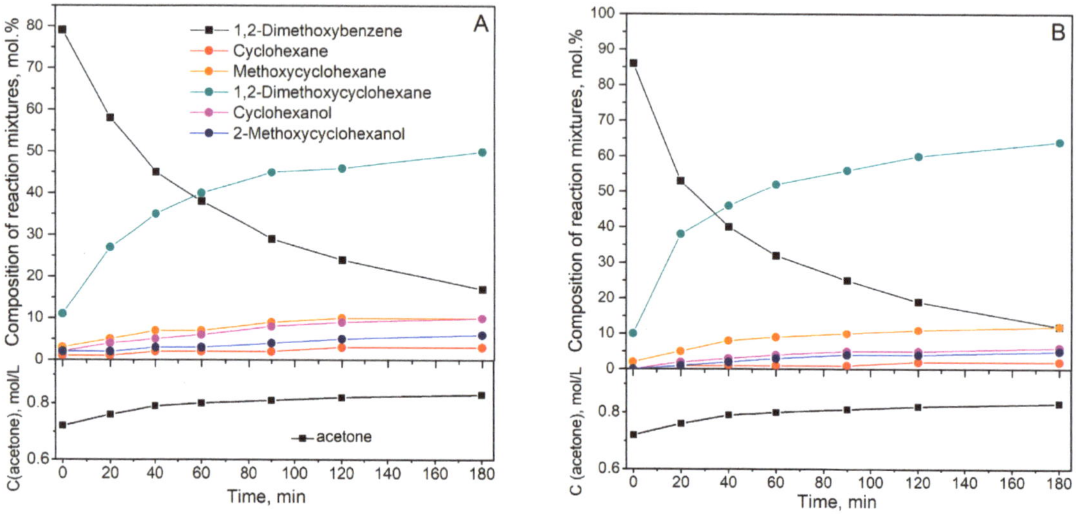

Figure 2. The composition of the reaction mixture in the course of 1,2-dimethoxybenzene (1,2-DMB) transformation over (**A**)–Ni-SiO_2 and (**B**)–NiCu-SiO_2. 250 °C, 7.0–7.7 MPa, and m(catalyst) = 0.10–0.11 g.

Table 1. Rate constants of phenolic compound transformations in TH with 2-PrOH, 250 °C, and 7.0–7.7 MPa. All constants were calculated taking into account the negative influence of acetone.

Catalyst	$k \times 10^{-2}$, min^{-1}			
	Guaiacol	1,2-DMB	1,3-DMB	1,4-DMB
Ni-SiO_2	5.9	1.3	0.65	1.3
NiCu-SiO_2	2.6	1.6	0.67	0.71

Table 2. Specific activity of the catalysts in TH of phenolic compounds at 250 °C.

Catalyst		$k/S_{CO} \times 10^{-3}$, $min^{-1} \times m^{-2}$			
	S_{CO}, m^2/g	Guaiacol	1,2-DMB	1,3-DMB	1,4-DMB
Ni-SiO_2	35	15	3.4	1.7	3.4
NiCu-SiO_2	32	7.4	4.5	1.9	2.0

Figure 3 shows the data obtained in the experiments with 1,3-dimethoxybenzene (1,3-DMB). After 3 h at 250 °C, the conversion of the initial substrate is relatively low and reaches 44% and 52% over the Ni-SiO_2 and NiCu-SiO_2 catalyst, respectively. In terms of the rate constants and the specific catalytic activity, the Cu addition has a limited positive effect (Tables 1 and 2). The qualitative content of the products has the significant differences compared to 1,2-DMB. Methoxycyclohexane and cyclohexanol were found to be the main products; at the same time, oxygen-free cyclohexane is presented in small amounts. Thus, the saturation of the 1,3-DMB benzene ring under TH conditions is always accompanied by C-O cleavage that is not typical for other phenolic compounds used in this study. Unfortunately, reductive transformations of 1,3-DMB are not widely studied in the literature; however, the results of the conventional hydrogenation of 1,3-DMB over silica-supported Pt and Pd [32] catalysts are quite similar to the data presented here.

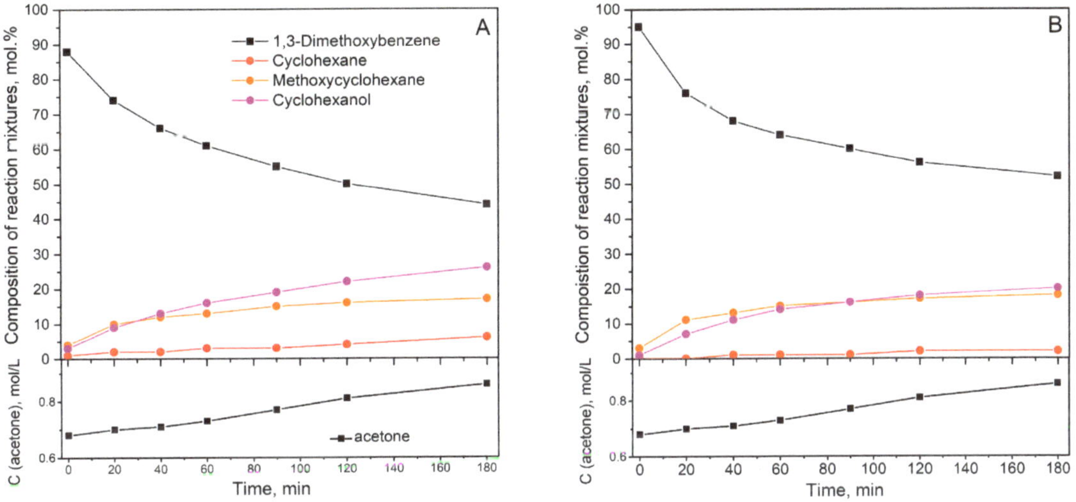

Figure 3. The composition of the reaction mixture in the course of 1,3-dimethoxybenzene (1,3-DMB) transformation over catalysts: (**A**)–Ni-SiO_2 and (**B**)–NiCu-SiO_2. 250 °C, 7.0–7.7 MPa, and m(catalyst) = 0.10–0.11 g.

In contrast to other position isomers, Cu negatively affects both the rate constant of 1,4-DMB consumption and the specific catalytic activity (Tables 1 and 2). This fact is mirrored in the conversion of 1,4-DMB, which reaches 84% and 49% over the Ni-SiO_2 and NiCu-SiO_2 catalysts, respectively (Figure 4). Thus, it is clearly seen that the influence of Cu on rate constants changes from positive for 1,2-DMB to almost neutral and negative for 1,3-DMB and 1,4-DMB, respectively. We assume that this effect can be related to the more effective adsorption of 1,2-DMB promoted by Cu. For example, De Castro et al. [26] explain the low reactivity of 3-methoxyphenol and 4-methoxyphenol in TH in terms of the adsorption on the catalyst surface. The results of the FTIR study provided by the authors demonstrated that after dissociation of the OH group, tilted structures are formed. The oxygen-containing groups of 2-methoxycyclohexanol are closer to the Ni surface compared to other derivatives, which causes the faster transformation of 2-methoxycyclohexanol. This fact, along with the known promoting effect of Cu on the adsorption of oxygen-containing compounds [31,33], proves our observations.

The transfer hydrogenation of 1,4-DMB results in the formation 1,4-dimethoxycyclohexane and 4-methoxycyclohexanol (Figure 4). Small amounts of methoxycyclohexane, cyclohexanol, and cyclohexane have been also found in the reaction mixtures. Figure 4 shows that the Cu addition affects the yields of the products negatively; however, the selectivity remains almost unchanged. As seen, the C_{arom}-O bonds are not significantly involved in the reductive transformations in contrast to the aromatic rings. At the same time, the formation of 4-methoxycyclohexanol indicates the noticeable activity of both catalysts in the CH_3-O bond cleavage.

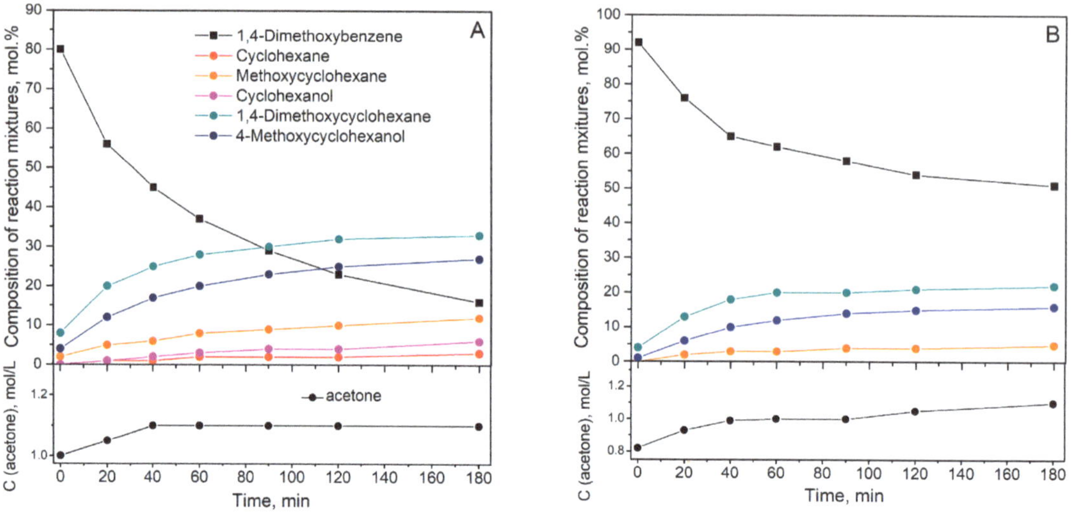

Figure 4. The composition of the reaction mixture in the course of 1,4-dimethoxybenzene (1,4-DMB) transformation over catalysts: (**A**)–Ni-SiO_2 and (**B**)–NiCu-SiO_2. 250 °C, 7.0–7.7 MPa, and m(catalyst) = 0.10–0.11 g.

2.4. Kinetic Studies

As it was mentioned above, the first-order kinetic model as well as any other models based on the higher-order kinetic equations of the elementary reaction do not fit the experimental data (see Supplementary Materials). In our recent study devoted to anisole transformations in the TH [25], the negative influence of acetone was demonstrated. In the kinetic calculations, it was mirrored as a negative term $k_{ac} \times C_{ac}$ (Equation (3)), decreasing the rate constant. This term includes k_{ac}, which is technically the quasi-equilibrium constant showing the ratio between acetone adsorbed on the catalysts surface and in the solution. The same kinetic model applied to guaiacol and dimethoxybenzenes describes the dependence of conversion on time quite well (see the Supplementary Materials).

The calculated rate constants (Table 1) show that the transformation of guaiacol over Ni-SiO_2 occurs at least 4.5 times faster than for dimethoxybenzenes. The Cu addition has a negative effect on the guaiacol conversion; therefore, the difference in the kinetic constants over NiCu-SiO_2 is not so high. However, in the case of 1,2-DMB copper increases the rate constant from 3.4×10^{-3} min^{-1} m^{-2} to 4.5×10^{-3} min^{-1} m^{-2}. Additionally, the limited positive effect of the copper addition on the rate constant was found for 1,3-DMB. To calculate the specific catalytic activity, the rate constants were normalized to the Ni^0 surface area (Table 2). Due to the close values of the Ni^0 area, the dependences observed for the rate constants are also valid for the specific catalytic activity.

The comparison of the data obtained in this study [25] shows that the specific catalytic activity in anisole transformation reaches 9.0×10^{-3} min^{-1} m^{-2} and 3.8×10^{-3} min^{-1} m^{-2} for the Ni-SiO_2 and NiCu-SiO_2 catalysts, respectively. Thus, the influence of copper on anisole TH is close to those observed for guaiacol and 1,4-DMB. At the same time, the specific catalytic activity of Ni-SiO_2 and NiCu-SiO_2 in the TH of anisole remains relatively high compared to that of dimethoxybenzenes.

Thus, the high-loaded Ni-based catalysts synthesized using coprecipitation in supercritical CO_2 demonstrated excellent performance in the TH of the phenolic compounds. The significant influence of the structure of the phenolic compound on the catalytic activity was shown. In particular, the transformations of guaiacol were found to be the fastest due to the presence of the non-methylated OH group. In contrast to other dimethylbenzene position isomers, 1,3-DMB actively loses oxygen atoms, transforming into methoxycyclohexane and cyclohexanol; however, the rate of its transformations remains the lowest compared to other isomers. There are many studies showing the positive effect of the Cu addition on the rate of the reductive transformations. This work demonstrates that in the case of TH of phenolic compounds, Cu can affect the rate constants in the different ways.

3. Materials and Methods

3.1. Materials

The following materials were used: 2-PrOH (≥99.8, EKOS-1), MeOH (J.T. Barker, Phillipsburg, NJ, USA, HPLC Gradient Grade), 1,2-dimethoxybenzene (99%, Sigma-Aldrich, Burlington, MA, USA), 1,3-dimethoxybenzene (≥98%, Sigma-Aldrich), 1,4-demethoxybenzene (99%, Sigma-Aldrich), guaiacol (≥98%, Sigma-Aldrich), dodecane (≥99%, Sigma-Aldrich), tetraethoxysilane (TEOS, 98%, Acros Organics, Geel, Belgium), $Ni(OAc)_2{\cdot}4H_2O$, (99% extra, Acros Organics), $Cu(OAc)_2{\cdot}H_2O$, (98%, Sigma Aldrich), and CO_2 (99.8%, Promgazservis, Orenburg, Russia).

3.1.1. Catalyst Preparation and Characterization

The catalysts were synthesized according to the original supercritical antisolvent co-precipitation method using the SAS-50 setup (Waters, Milford, MA, USA). The obtained solid particles were characterized by XRD, XRF, TEM, EDX, and CO adsorption. The detailed procedure of the catalyst synthesis and characterization is described in our previous publication [25]. Two samples, Ni-SiO_2 and NiCu-SiO_2, were synthesized.

3.1.2. Batch Experiments

Before application in the treatment of the phenolic compounds, the Ni-based catalysts were reduced in H_2 flow (30 L/h) at 400 °C for NiCu-SiO_2 and at 450 °C for Ni-SiO_2, to obtain the metal nanoparticles. The completeness of the reduction process was controlled by XRD. After 45 min under H_2 flow, the catalysts were cooled down to room temperature. Then, H_2 was replaced with Ar, and catalyst was placed under 2-PrOH, avoiding contact with air.

The catalytic experiments were carried out in the batch reactor (285 mL, AISI 316 L). The system was purged with Ar and then charged by the reaction mixture, including a catalyst (0.10–0.11 g); 2-PrOH (110 mL); phenolic compound (30 mmol, 3.72–4.14 g); and dodecane (0.30 g), which was used as an internal standard. After charging, the reactor was purged with Ar under stirring (mechanical agitator MagneDrive®, Erie, PA, USA, 800 rpm), closed, and then heated up to 250 °C for 30–35 min. During the experiment, the pressure was 7.0–7.7 MPa, and no H_2 or any other substances were added to the reaction mixture. The probes were collected just after reaching the target temperature, and then after 20, 40, 60, 90, 120, and 180 min of the reaction. The conversion of the initial substrate and yield of the products was calculated according to the following equations:

$$Conversion = \left(1 - \frac{C_{ph}^0 - C_{ph}}{C_{ph}^0}\right) \times 100\% \quad (1)$$

$$Yield = \frac{C_{pr}}{\sum C_{pr}} \times 100\% \quad (2)$$

where C_{ph}^0 and C_{ph} represent molar concentrations of the phenolic compounds (guaiacol and dimethoxybenzenes) in the reaction mixture at an initial and a certain time, and C_{pr} represents a concentration of a certain product at a certain time.

3.1.3. Kinetic Calculations

The rate constants describing the consumption of phenolic compounds were calculated according to Equation (3):

$$\frac{dC_{ph}}{dt} = -(k - k_{ac} \times C_{ac}) \times C_{ph} \quad (3)$$

where C_{ph} is a concentration of the phenolic compound in the solution at a certain time, C_{ac} is a concentration of acetone in the solution at a certain time, k is the kinetic constant of phenolic compound consumption, and k_{ac} is a coefficient showing the proportion between acetone in the solution and on the catalyst surface.

3.1.4. Product Analysis

The liquid products were analyzed using the Shimadzu GCMS-QP2010 SE spectrometer (Kyoto, Japan), equipped with an autosampler. The GsBP-INOWAX capillary chromatographic column (crosslinked polyethylene glycol) was used (length 30 m, internal diameter 0.32 mm, and stationary phase thickness 0.25 μm). The column was conditioned at 55 °C for 3 min, heated up to 200 °C at a rate of 15 °C per minute, and then heated up to 250 °C at a rate of 25 °C per minute. The evaporator temperature was 270 °C, and helium was used as a carrier gas. The products were identified using the peak retention time and the mass spectrum of the substance, which were compared with the corresponding data of the pure compounds or with the data from the NIST and Wiley electronic mass spectral libraries. Conversion of the initial phenolic compound and the yield of the products were evaluated by the internal standard method using dodecane.

To determine the qualitative composition of the formed gases, the reactor was cooled to 40 °C when the pressure dropped to 1.3–2.0 MPa. Then, the gas was sampled from the reactor using a 150 mL syringe. Before the analysis, a gas sample in the syringe was diluted 20 times with air. The chromatographic analysis was preformed using a Chromos GC 1000 (Chromos, Nizhny, Russia) equipped with a chromatography column (length 2 m, internal diameter 3 mm, and stationary phase—NaX zeolite) and a thermal conductivity detector. Ar was used as a carrier gas, and the temperature mode of column conditioning was as follows: 30 °C for 3 min, programmed heating up to 120 °C at a rate 24 °C/min, and then for 5 min at 120 °C.

4. Conclusions

This study shows that high-loaded Ni and Ni-Cu catalysts demonstrate promising results in transformations of phenolic compounds and 2-PrOH under supercritical conditions. The obtained kinetic constants show the high reactivity of guaiacol compared to dimethoxybenzenes and the negative influence of Cu addition on the reaction rate. At the same time, Cu promotes transformations of 1,2-dimethoxybenzene, increasing the rate constant from 1.3×10^{-2} min^{-1} to 1.6×10^{-2} min^{-1}. Guaiacol and 1,3-dimethoxybenzene actively lose oxygen atoms, transforming into cyclohexanol and methoxycyclohexane, whereas the main products of 1,2-and 1,4-dimethoxybenzene transfer hydrogenation are corresponding dimethoxycyclohexanes. Thus, the copper addition and the structure of phenolic compounds significantly affect the rate and selectivity of transfer hydrogenation. This study expands the understanding of the fundamental aspects of H_2-free hydrogenation.

Supplementary Materials: The following supporting information can be downloaded at https://www.mdpi.com/article/10.3390/catal12121655/s1, Figure S1: XRD data obtained for the catalysts after activation in H_2. A—Ni-SiO_2, B—NiCu-SiO_2; Table S1: Content of Ni and Cu in the Ni-Cu catalysts measured by XRF. D—mean crystallite size, a—lattice parameter, measured by XRD, and S_{CO}—surface area, measured by CO adsorption.; Figure S2: TEM and EDX-mapping pictures obtained for A—Ni-SiO_2, B—NiCu-SiO_2; Figure S3: Kinetic data for the experiments with guaiacol over A—Ni-SiO_2 and NiCu-SiO_2. The first-order kinetic model was used; the dots—experimental data, and the lines—calculated using Equation (3); Figure S4: Kinetic data for the experiments with 1,2-DMB over A—Ni-SiO_2 and NiCu-SiO_2. The first-order kinetic model was used; the dots—experimental data, and the lines—calculated using Equation (3); Figure S5: Kinetic data for the experiments with 1,3-DMB over A—Ni-SiO_2 and NiCu-SiO_2. The first-order kinetic model was used; the dots—experimental data, and the lines—calculated using Equation (3); Figure S6: Kinetic data for the experiments with 1,4-DMB over A–Ni-SiO_2 and NiCu-SiO_2. The first-order kinetic model was used— the dots–experimental data, and the lines—calculated using Equation (3).

Author Contributions: Conceptualization, methodology, validation, formal analysis, investigation, writing—original draft preparation, writing—review and editing, and funding acquisition, A.P. and N.N.; and writing—original draft preparation, writing—review and editing, supervision, project administration, and funding acquisition, O.M. All authors have read and agreed to the published version of the manuscript.

Funding: This research was funded by Russian Science Foundation grant number 21-73-00049, (https://rscf.ru/project/21-73-00049/ (accessed on 27 July 2021)).

Data Availability Statement: Not applicable.

Conflicts of Interest: The authors declare no conflict of interest.

References

1. Gilkey, M.J.; Xu, B. Heterogeneous catalytic transfer hydrogenation as an effective pathway in biomass upgrading. *ACS Catal.* **2016**, *6*, 1420–1436. [CrossRef]
2. Thanigaivel, S.; Priya, A.K.; Dutta, K.; Rajendran, S.; Sekar, K.; Jalil, A.A.; Soto-Moscoso, M. Role of nanotechnology for the conversion of lignocellulosic biomass into biopotent energy: A biorefinery approach for waste to value-added products. *Fuel* **2022**, *322*, 124236. [CrossRef]
3. Ke, L.; Wu, Q.; Zhou, N.; Xiong, J.; Yang, Q.; Zhang, L.; Wang, Y.; Dai, L.; Zou, R.; Liu, Y.; et al. Lignocellulosic biomass pyrolysis for aromatic hydrocarbons production: Pre and in-process enhancement methods. *Renew. Sustain. Energy Rev.* **2022**, *165*, 112607. [CrossRef]
4. Gnanasekaran, L.; Priya, A.K.; Thanigaivel, S.; Hoang, T.K.A.; Soto-Moscoso, M. The conversion of biomass to fuels via cutting-edge technologies: Explorations from natural utilization systems. *Fuel* **2023**, *331*, 125668. [CrossRef]
5. Scholze, B.; Meier, D. Characterization of the water-insoluble fraction from pyrolysis oil (pyrolytic lignin). Part I. PY-GC/MS, FTIR, and functional groups. *J. Anal. Appl. Pyrolysis* **2001**, *60*, 41–54. [CrossRef]
6. Xu, J.; Zhu, P.; El Azab, I.H.; Bin Xu, B.; Guo, Z.; Elnaggar, A.Y.; Mersal, G.A.M.; Liu, X.; Zhi, Y.; Lin, Z.; et al. An efficient bifunctional Ni-Nb_2O_5 Nanocatalysts for the hydrodeoxygenation of anisole. *Chin. J. Chem. Eng.* **2022**, *49*, 187–197. [CrossRef]
7. Shafaghat, H.; Rezaei, P.S.; Ashri Wan Daud, W.M. Effective parameters on selective catalytic hydrodeoxygenation of phenolic compounds of pyrolysis bio-oil to high-value hydrocarbons. *RSC Adv.* **2015**, *5*, 103999–104042. [CrossRef]
8. Shuai, L.; Luterbacher, J. Organic Solvent effects in biomass conversion reactions. *ChemSusChem* **2016**, *9*, 133–155. [CrossRef]
9. Shafaghat, H.; Tsang, Y.F.; Jeon, J.K.; Kim, J.M.; Kim, Y.; Kim, S.; Park, Y.K. In-situ hydrogenation of bio-oil/bio-oil phenolic compounds with secondary alcohols over a synthesized mesoporous Ni/CeO_2 catalyst. *Chem. Eng. J.* **2020**, *382*, 122912. [CrossRef]
10. Philippov, A.A.; Nesterov, N.N.; Pakharukova, V.P.; Martyanov, O.N. High-loaded ni-based catalysts obtained via supercritical antisolvent coprecipitation in transfer hydrogenation of anisole: Influence of the support. *Appl. Catal. A Gen.* **2022**, *643*, 118792. [CrossRef]
11. Park, Y.K.; Ha, J.M.; Oh, S.; Lee, J. Bio-oil upgrading through hydrogen transfer reactions in supercritical solvents. *Chem. Eng. J.* **2021**, *404*, 126527. [CrossRef]
12. Yfanti, V.L.; Lemonidou, A.A. Effect of hydrogen donor on glycerol hydrodeoxygenation to 1,2-propanediol. *Catal. Today* **2020**, *355*, 727–736. [CrossRef]
13. Chen, H.; Xu, Q.; Zhang, D.; Liu, W.; Liu, X.; Yin, D. Highly efficient synthesis of γ-valerolactone by catalytic conversion of biomass-derived levulinate esters over support-free mesoporous Ni. *Renew. Energy* **2021**, *163*, 1023–1032. [CrossRef]
14. Reddy Kannapu, H.P.; Mullen, C.A.; Elkasabi, Y.; Boateng, A.A. Catalytic transfer hydrogenation for stabilization of bio-oil oxygenates: Reduction of p-cresol and furfural over bimetallic Ni-Cu catalysts using isopropanol. *Fuel Process. Technol.* **2015**, *137*, 220–228. [CrossRef]
15. Philippov, A.A.; Chibiryaev, A.M.; Martyanov, O.N. Base-free transfer hydrogenation of menthone by sub- and supercritical alcohols. *J. Supercrit. Fluids* **2019**, *145*, 162–168. [CrossRef]
16. Shafaghat, H.; Kim, J.M.; Lee, I.G.; Jae, J.; Jung, S.C.; Park, Y.K. Catalytic hydrodeoxygenation of crude bio-oil in supercritical methanol using supported nickel catalysts. *Renew. Energy* **2019**, *144*, 159–166. [CrossRef]
17. Alekseev, E.S.; Alentiev, A.Y.; Belova, A.S.; Bogdan, V.I.; Bogdan, T.V.; Bystrova, A.V.; Gafarova, E.R.; Golubeva, E.N.; Grebenik, E.A.; Gromov, O.I.; et al. Supercritical fluids in chemistry. *Russ. Chem. Rev.* **2020**, *89*, 1337–1427. [CrossRef]
18. Kim, J.Y.; Park, J.; Kim, U.J.; Choi, J.W. Conversion of lignin to phenol-rich oil fraction under supercritical alcohols in the presence of metal catalysts. *Energy Fuels* **2015**, *29*, 5154–5163. [CrossRef]
19. Goto, M. Chemical recycling of plastics using sub- and supercritical fluids. *J. Supercrit. Fluids* **2009**, *47*, 500–507. [CrossRef]

20. Tang, Z.-R.; Edwards, J.K.; Bartley, J.K.; Taylor, S.H.; Carley, A.F.; Herzing, A.A.; Kiely, C.J.; Hutchings, G.J. Nanocrystalline cerium oxide produced by supercritical antisolvent precipitation as a support for high-activity gold catalysts. *J. Catal.* **2007**, *249*, 208–219. [CrossRef]
21. Nesterov, N.S.; Pakharukova, V.P.; Martyanov, O.N. Water as a cosolvent—Effective tool to avoid phase separation in bimetallic Ni-Cu catalysts obtained via supercritical antisolvent approach. *J. Supercrit. Fluids* **2017**, *130*, 133–139. [CrossRef]
22. Reverchon, E. Supercritical antisolvent precipitation of micro- and nano-particles. *J. Supercrit. Fluids* **1999**, *15*, 1–21. [CrossRef]
23. Nesterov, N.S.; Paharukova, V.P.; Yakovlev, V.A.; Martyanov, O.N. The facile synthesis of Ni-Cu catalysts stabilized in SiO_2 Framework via a supercritical antisolvent approach. *J. Supercrit. Fluids* **2016**, *112*, 119–127. [CrossRef]
24. Nesterov, N.S.; Smirnov, A.A.; Pakharukova, V.P.; Yakovlev, V.A.; Martyanov, O.N. Advanced green approaches for the synthesis of nicu-containing catalysts for the hydrodeoxygenation of anisole. *Catal. Today* **2021**, *379*, 262–271. [CrossRef]
25. Philippov, A.; Nesterov, N.; Pakharukova, V.; Kozhevnikov, I.; Martyanov, O. Advanced high-loaded Ni—Cu catalysts in transfer hydrogenation of anisole: Unexpected effect of Cu addition. *Catalysts* **2022**, *12*, 1307. [CrossRef]
26. De Castro, I.B.D.; Graça, I.; Rodríguez-García, L.; Kennema, M.; Rinaldi, R.; Meemken, F. Elucidating the reactivity of methoxyphenol positional isomers towards hydrogen-transfer reactions by ATR-IR spectroscopy of the liquid-solid interface of RANEY® Ni. *Catal. Sci. Technol.* **2018**, *8*, 3107–3114. [CrossRef]
27. Viar, N.; Requies, J.M.; Agirre, I.; Iriondo, A.; Gil-Calvo, M.; Arias, P.L. Ni-Cu bimetallic catalytic system for producing 5-hydroxymethylfurfural-derived value-added biofuels. *ACS Sustain. Chem. Eng.* **2020**, *8*, 11183–11193. [CrossRef]
28. Ganesan, M.; Liu, C.C.; Pandiyarajan, S.; Lee, C.T.; Chuang, H.C. Post-supercritical CO_2 electrodeposition approach for Ni-Cu alloy fabrication: An innovative eco-friendly strategy for high-performance corrosion resistance with durability. *Appl. Surf. Sci.* **2022**, *577*, 151955. [CrossRef]
29. Sequeira, C.A.C.; Cardoso, D.S.P.; Amaral, L.; Šljukić, B.; Santos, D.M.F. On the performance of commercially available corrosion-resistant nickel alloys: A review. *Corros. Rev.* **2016**, *34*, 187–200. [CrossRef]
30. Guo, Q.; Wu, M.; Wang, K.; Zhang, L.; Xu, X. Catalytic hydrodeoxygenation of algae bio-oil over bimetallic Ni-Cu/ZrO_2 catalysts. *Ind. Eng. Chem. Res.* **2015**, *54*, 890–899. [CrossRef]
31. Gao, B.; Zhang, J.; Yang, J.-H. Bimetallic Cu-Ni/MCM-41 catalyst for efficiently selective transfer hydrogenation of furfural into furfural alcohol. *Mol. Catal.* **2022**, *517*, 112065. [CrossRef]
32. Szczyglewska, P.; Feliczak-Guzik, A.; Nowak, I. Ordered mesoporous silica-supported metal catalysts for hydrodeoxygenation of anisole derivatives. *Microporous Mesoporous Mater.* **2021**, *312*, 110691. [CrossRef]
33. Ardiyanti, A.R.; Khromova, S.A.; Venderbosch, R.H.; Yakovlev, V.A.; Heeres, H.J. Catalytic hydrotreatment of fast-pyrolysis oil using non-sulfided bimetallic Ni-Cu catalysts on a δ-Al_2O_3 support. *Appl. Catal. B Environ.* **2012**, *117–118*, 105–117. [CrossRef]

catalysts

MDPI

Article

Effect of Cr on a Ni-Catalyst Supported on Sibunite in Bicyclohexyl Dehydrogenation in Hydrogen Storage Application

Leonid M. Kustov [1,2,3,*] and Alexander N. Kalenchuk [1,2]

1 Chemistry Department, Moscow State University, Leninskie Gory 1, Bldg. 3, 119992 Moscow, Russia
2 N.D. Zelinsky Institute of Organic Chemistry, Russian Academy of Sciences, Leninsky Prospect 47, 119991 Moscow, Russia
3 Laboratory of Nanochemistry and Ecology, National University of Science and Technology MISiS, Leninsky Prospect 4, 119991 Moscow, Russia
* Correspondence: lmk@ioc.ac.ru

Abstract: A comparison of the activity of mono- and bimetallic Ni-Cr/C catalysts deposited on a carbon carrier (sibunite) in the bicyclohexyl dehydrogenation reaction as a stage of hydrogen evolution in hydrogen storage systems is carried out. The interaction of Ni and Cr supported onto the carbon carrier—sibunite in bimetallic NiCr systems affects the change in the parameters of the crystal lattice of nickel, compared with the FCC lattice of Ni, as shown by the methods of XPS, TPR, XRD, high-resolution TEM and electron diffraction.

Keywords: hydrogen storage; dehydrogenation; bicyclohexyl; Pt-Ni-Cr supported catalysts; Ni-Cr alloys

1. Introduction

Recently, catalysis has become an increasingly important tool aimed at creating modern materials and technologies. For example, the successful replacement of lithium-ion batteries with cheaper methanol fuel cells largely depends on the development of new catalysts for the electrooxidation of methanol. It is known that the most active in this reaction, as well as in many other reactions, are supported systems based on noble metals, especially, platinum [1–3]. At the same time, these metals are prone to rapid CO poisoning in the methanol medium. The problem is solved by developing bi- and trimetallic catalysts, which not only increase resistance to the deactivating action of CO, but also allow one to reach a better performance of catalysts with a lower loading of precious metals. To increase the catalytic activity, transition metals (Fe, Ni, Co, Ag and others) are usually used as promoting additives [4–7]. The second group consists of metals that improve the corrosion stability of the catalysts (Cr, V, Mn) [8]. In particular, an increase in the conversion and selectivity in the methanol electrooxidation reaction with reduction of production costs is reported for bi- and trimetallic catalysts, such as PtCo, PtNi, PtFe, PtPb [9,10], PtNiCr and PtCoCr [11–13]. At the same time, the greatest resistance to CO poisoning is achieved when platinum is doped with ruthenium [14,15], including the addition (dilution) of third-base metals in the case of trimetallic systems, such as PtRuFe [16], PtRuNi [17], PtRuCo [18], PtRuW [19] and PtRuMo [20].

An increase in the activity of PtNiCr trimetallic catalytic systems supported on the carbon carrier, sibunite, was also found in [21] compared with mono- and binary systems applied in the bicyclohexyl dehydrogenation reaction. The reaction is crucial for hydrogen storage and hydrogen release systems based on the use of liquid organic hydrogen carriers (LOHC) as an alternative fuel system without the so-called carbon footprint instead of traditional hydrocarbon raw materials, since hydrogen oxidation does not produce CO_x gases that pollute the atmosphere. In this reaction, the high activity of PtNiCr/C trimetallic catalysts, as in the methanol electrooxidation reaction, is presumably achieved due to the stabilizing effect of the NiCr binary system on platinum [7]. In relation to hydrogen storage systems, the development of catalysts is limited by the need to carry out conjugated hydrogenation–dehydrogenation reactions without the formation of by-products of hydrogenolysis and cracking, which destroy the substrate and reduce the lifetime of hydrogen storage and release systems as a whole. This limits the choice of carriers and active metals; therefore, for the purposes of hydrogen storage, the study of the stimulating effect of promoting additives on the activity of Pt catalysts is of great interest. The aim of this study is to study the effect of the interaction of modifying base metals (Cr, Ni) on the activity of platinum in the bicyclohexyl dehydrogenation reaction.

2. Results

The morphology, average size of Ni and Cr particles and their distribution in bimetallic systems deposited on oxidized sibunite (C) and consisting of a mixture of Ni (3 wt.%) and Cr (1.5 wt.%) were studied using transmission electron microscopy (TEM). Figure 1 and Figure S1 (Supplementary Materials, SM) show an example of a micrograph of a catalyst (3Ni-1.5Cr)/C in a light (a) and dark field (b). It can be seen from Figure 1 that small particles of a regular shape are located on top of some large particles on the surface of the carbon carrier. The average size of large particles is ~20–30 nm and that of smaller ones is ~3–5 nm. The EDX data for this system, together with the analysis of interplane distances by the high-resolution TEM, show that the particles contain atoms of Ni and Cr metals coated with an oxide shell (NiO, Cr_2O_3). At the same time, a comparison of micrographs of TEM of a number of bimetallic Ni-Cr/C catalysts with their monometallic analogues showed that large particles belong to nickel and smaller ones are formed by chromium.

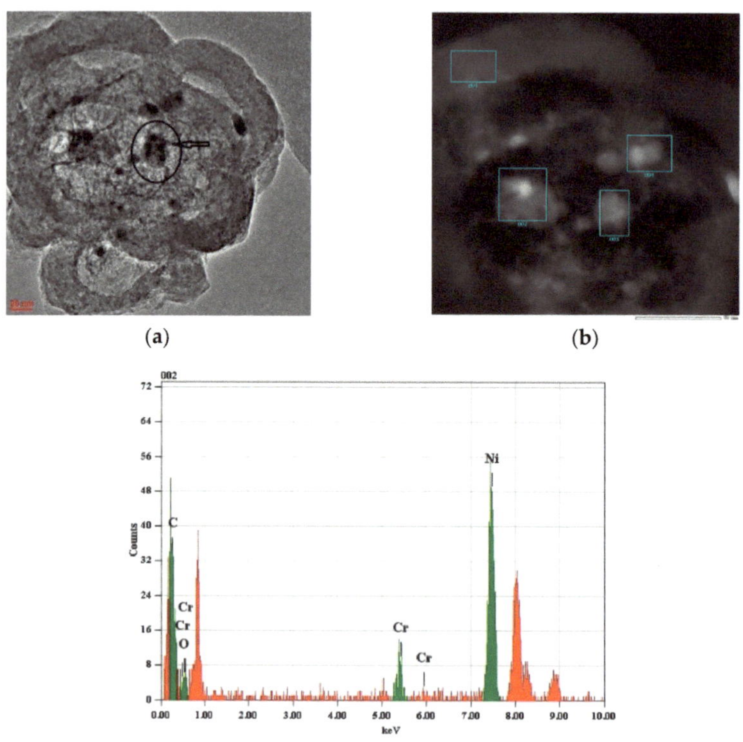

Figure 1. Catalyst (3Ni-1.5Cr)/C: micrograph in light (**a**) and dark field (**b**) with EDX (**c**).

Comparison with other investigated Ni-Cr systems indicates that the order of the application of metals does not have a noticeable effect on the change in the average size and distribution of particles on the surface of the sibunite. At the same time, the presence of Cr_xNi_{1-x} solid solutions with different Cr/Ni ratios, as well as small amounts of $NiCrO_4$ with a rutile structure and $NiCr_2O_4$ with a spinel structure, is determined at the points of contact of the two metals at interplanar distances. Interestingly, the formation of the latter, together with a decrease in the degree of oxidation of the catalyst surface due to the alloying of nickel with chromium, is indicated as a possible reason for an increase in the activity of the PtNiCr system in the electrooxidation of methanol compared with a monometallic Pt catalyst [7,8]. It was shown [11–13] that the substitution of chromium for Mn and Co also affects the activity of the Ni-Mn and Ni-Co binary systems; however, the effect is less pronounced.

In order to identify the forms of nickel and chromium in the catalysts under study, the corresponding samples were examined using X-ray photoelectron spectroscopy (XPS). The comparison showed that the XPS spectra of all the analyzed systems, regardless of the order of application of metals, are identical and have no fundamental differences among themselves. In particular, only a doublet of broad lines with a poorly expressed structure and with a binding energy of the Cr $2p_{3/2}$ component of about 577.0 eV, which is typical for trivalent chromium compounds, is observed in the Cr 2p spectra of all the samples studied [22]. In the Ni 2p spectra of the studied samples, a small shoulder with a binding energy of about 852.7 eV is observed, which indicates the presence of metallic nickel [23]. At the same time, the main contribution to the Ni 2p spectrum is made by the oxidized state of nickel, which corresponds to the Ni $2p_{3/2}$ component with a binding energy of about 855.8 eV and an intense shake-up satellite, which is characteristic of divalent nickel [23]. The result of deconvolution of the spectra by the example of the catalyst (3Ni-3Cr)/C is shown in Figure 2 and in Table 1. The proportions of nickel in various states in this catalyst are indicated in comparison with the most characteristic data for other studied Ni systems.

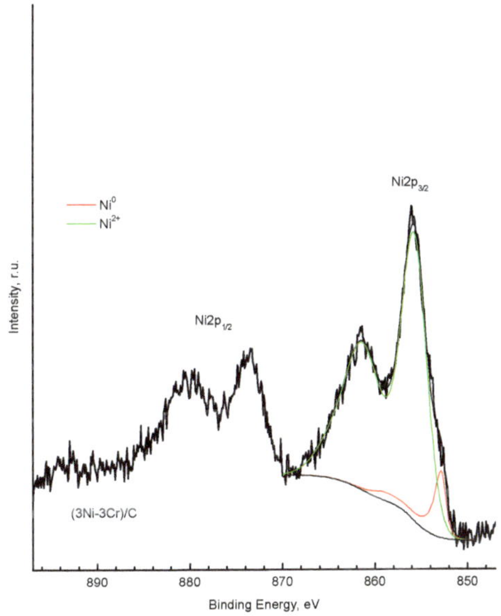

Figure 2. X-ray photoelectron spectra of Ni 2p electrons for (3Ni-3Cr)/C catalyst.

Table 1. The content of nickel in various degrees of oxidation on the surface of the studied samples according to the XPS data.

Catalyst	Ni 2p			
	E_b, eV	Content, at.%	E_b, eV	Content, at.%
3Ni/C	852.5	53	855.8	47
(3Ni-1.5Cr)/C	852.5	27	855.4	73
(3Ni-3Cr)/C	852.7	8	855.4	92
1.5Cr/3Ni/C	852.7	18	854.8	82

A comparison of the data in Table 1 shows that, regardless of the order of application of metals in the presence of chromium, the content of non-oxidized nickel (Ni(0)) in the studied bimetallic NiCr/C systems compared to the monometallic catalyst 3Ni/C decreases and the more strongly, the higher the Cr content. At the same time, it can be seen that there is a peak of Ni $2p_{3/2}$ electrons with binding energies E_b = 852.7 eV in the spectra of catalysts with a low Ni^0 content, which is slightly higher than in the spectrum of 3Ni/C (852.5 eV) This may indicate that metallic nickel acquires some positive charge in these samples. This is also evidenced by the binding energy characterizing the second state of nickel in NiCr/C systems (Ni^{2+}). The value of the E_b for this peak in the spectrum of the sample of 3Ni/C is equal to 855.8 eV, which is higher than that for the samples containing chromium. The difference in the electronic state of the components seems to indicate the presence of an interaction between these two metals. The effect is most pronounced for the 1.5Cr/3Ni/C catalyst, in which chromium is deposited on top of nickel.

The shape of the TPR profiles for the studied NiCr systems also shows a noticeable difference with individual monometallic catalysts. The TPR curves for the monometallic catalyst 3Ni/C [24] and 1.5Cr/C [25] demonstrate a number of characteristic peaks associated with the reduction processes of nickel and chromium. In contrast, in the NiCr systems, several processes occurring in a wide temperature range (T = 300–800 °C) are superimposed on the TPR profiles. This shape of the curves may be due to a variety of topochemical reactions responsible for the reduction of mixed Ni-Cr oxides, including the interaction between nickel and chromium noticed in the analysis of XPS spectra. At the same time, a low-temperature peak is highlighted in Figure 3, the position of which on the TPR curve for NiCr systems with different metal deposition, the order differs and is T = 138 °C for the catalyst (3Ni-1.5Cr)/C, T = 146 °C for 1.5Cr/3Ni/C and T = 170 °C for 3Ni/1.5Cr/C.

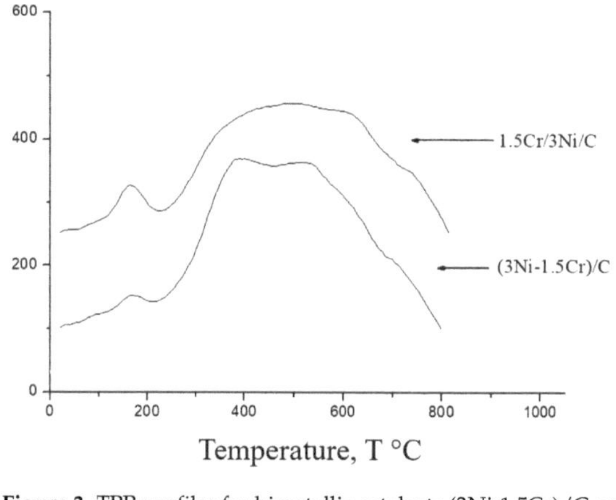

Figure 3. TPR profiles for bimetallic catalysts (3Ni-1.5Cr)/C and 1.5Cr/3Ni/C.

It is noteworthy that the position of each of these peaks decreases in comparison with both the Ni/C catalyst (T = 190 °C) and Cr/C catalyst (T = 265 °C) [25]. The intensity of the peaks for the bimetallic systems significantly exceeds those for monometallic catalysts. The authors [26,27] associated the increase in the intensity of TPR peaks with a decrease in the fraction of amorphous particles of metal oxides, including those produced due to the formation of the Ni-Cr alloy or bimetallic particles. It was also established that chromium doping of the Ni/MgO-La_2O_3 system leads to the modification of a part of the catalyst sites, with the formation of alloys with nickel and possibly the reduction of the size of crystallites [28]. Similar results were obtained for the Cu-Ni/Al_2O_3 system, where the addition of copper contributed to an increase in the stability of amorphous nanoalloy catalysts and a decrease in the Ni reduction temperature [29].

Comparison of monometallic Ni/C and bimetallic NiCr/C systems by the XRD method shows that the diffractograms of all the studied catalysts have peaks at 2θ = 25.7 and 43.4°, corresponding to the reflections of (002) and (101) of the carbon sibunite carrier (Figure 4). The maxima at 2θ = 44.4 and 51.8° are attributed to the crystallographic planes (111) and (200) of the FCC lattice of Ni (2θ = 44.505 and 51.86°; PDF 4-850). At the same time, the maximum of the (200) peak on the diffractogram of the monometallic catalyst 3Ni/C is shifted towards larger angles (2θ = 52.0°), as compared with bimetallic catalysts. The value of this shift corresponds to the lattice period a = 0.3514 (+0.0003) nm, which is lower than the table value for the FCC lattice of Ni (a = 0.3524 nm; PDF 4-850). Such an increase was associated [30] with the formation of metastable Ni_3C carbide, the enthalpy of formation of which is ΔH_{Ni3C} = 75.3kJ/mol. The enthalpy of formation of chromium carbides have large negative values ($\Delta H_{Cr3C2} = -9.5$ kJ/mol; $\Delta H_{Cr7C3} = -164.4$ kJ/mol; $\Delta H_{Cr23C6} = -341.4$ kJ/mol); therefore, Cr atoms interact more actively with Ni than with the carrier when added to the bimetallic system. The Ni lattice periods calculated by two reflexes (111) and (200) in bimetallic NiCr systems exceed those for both the FCC lattice of Ni and the monometallic catalyst 3Ni/C [21]. At the same time, the most intense reflex 110 of the Cr BCC lattice (2θ = 44.37°, PDF 6-694) overlaps with the indicated reflex (111) of the Ni FCC lattice and does not appear on diffractograms, apparently due to the low content of metallic Cr in the volume of the catalyst or its X-ray amorphous state. In the absence of the contribution of chromium to the position of peaks on the graphs, this excess indicates the formation of Cr-Ni solid solutions formed by the type of substitution of nickel atoms for chromium. The broadening of the (200) Ni lines on the diffractogram indicates that the CrNi-solid solution has a heterogeneous character, in which there are regions with different chromium contents (Cr_xNi_{1-x}).

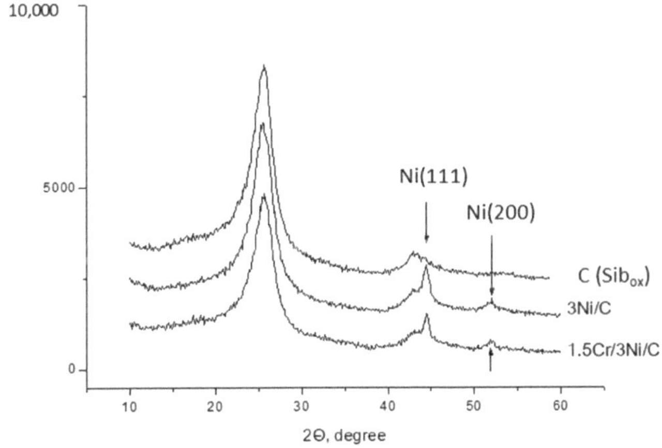

Figure 4. Diffractograms of the carbon carrier, 3Ni/C and 1.5Cr/3Ni/C catalysts.

The estimate of interplane distances of different sites obtained by the high-resolution TEM of the surface of the studied NiCr catalysts closely coincides with the data of the electron diffraction method. Two series of reflexes (111) and (200) are observed on electronograms of the NiCr catalysts: nickel (d(111) = 0.209 ± 0.006 nm and d(200) = 0.179 ± 0.002 nm) and chromium solid solution in nickel (d(111) = 0.215 ± 0.010 nm and d(200) = 0.187 ± 0.010 nm). The value d(111) of nickel in a two-component catalyst (3Ni-1.5Cr)/C calculated from the distance between adjacent parallel atomic planes is d = 0.2124 nm, which also exceeds the value of d(111) for the catalyst containing no chromium (d = 0.2091 nm). In accordance with the calculated parameters of the FCC lattice, the crystallographic structure of Ni-Cr is shown in Figure 5.

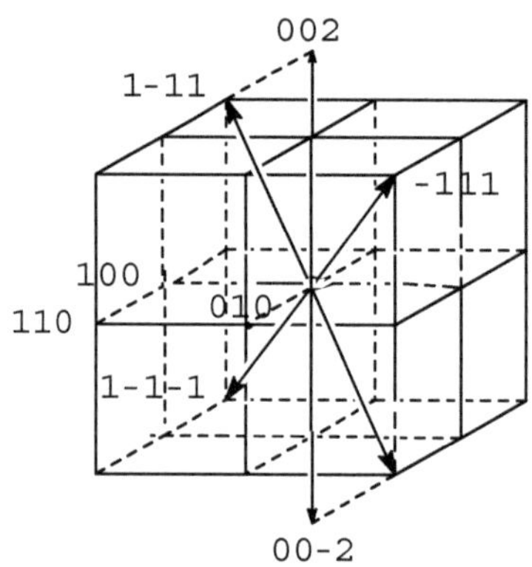

Figure 5. Crystallographic structure of the Ni-Cr phase.

The data obtained in studying the catalytic activity of catalysts in bicyclohexyl dehydrogenation showed that monometallic Ni and Cr catalysts exhibited a low activity in the selectivity formation of the final product, biphenyl ($C_{12}H_{10}$); thus, the rate of hydrogen evolution is low (Table 2). Both monometallic and bimetallic catalysts demonstrate a stable rate of hydrogen evolution and selectivities to the semi-hydrogenated and fully dehydrogenated products indicated in Table 2, without any noticeable loss for at least 8 h of operation.

Table 2. Catalytic properties of Ni-Cr/C catalysts of bicyclohexyl dehydrogenation (T = 320 °C, p = 1 atm, WHSV = 2.8 h^{-1}).

	3Ni/C	1.5Cr/C	3Ni/1.5Cr/C	(3Ni-1.5Cr)/C	1.5Cr/3Ni/C
Rate of hydrogen evolution, mol/h × 10^{-3}	1.863	2.760	1.021	1.542	1.014
S_{C12H10}, % *	35	60	48	49	47
S_{C12H16}, % *	65	40	52	51	53

* Selectivities to the semi-hydrogenated and fully dehydrogenated products.

It is seen from Table 2 that the rate of hydrogen evolution in the course of conversion of bicyclohexyl on the Cr catalysts was noticeably higher than on the Ni catalysts; in turn, it is much lower than that observed on the Pt catalysts [21]. However, the selectivity for the final product, biphenyl ($C_{12}H_{10}$), on the Cr catalysts was noticeably higher than on the Ni catalysts. In the case of nickel, the formation of trace amounts of cracking products [24] was also observed. Since the content of cracking products ranged from 50 to 200 ppm, their contribution was not taken into account when calculating the selectivity of the main reaction products. No by-products were found for the studied bimetallic NiCr systems, which is important for hydrogen storage systems. According to chromato-mass spectrometry, in all the cases, the reaction mixture in addition to fully unsaturated biphenyl, contained one intermediate cyclohexylbenzene. At the same time, the rate of bicyclohexyl conversion (hydrogen release) on NiCr systems, regardless of the order of introduction of metals, was even lower than on monometallic catalysts. This decrease correlates with a decrease in the content of non-oxidized nickel (see Table 1).

3. Materials and Methods

3.1. Preparation of Catalysts

Oxidized sibunite was used as a carrier (trademark C, Omsk, Russia), with an average granule diameter of 1.5–1.8 mm, a specific surface area of 243 m^2/g, an average pore size of 4.2 nm and a pore volume of 0.45 cm^3/g [31]. Monometallic Ni and Cr catalysts deposited on a carbon carrier were prepared by incipient wetness impregnation with aqueous solutions of the corresponding salts, nitrates $Ni(NO_3)_2$ $6H_2O$ (chemical grade purity; Khimmed, Russia) and $Cr(NO_3)_3$ $9H_2O$ (chemical grade purity; "Acros Organics"), and bimetallic catalysts—Ni-Cr/C—by impregnation of Ni/C and Cr/C or joint impregnation of the sibunite. After impregnation, the samples were successively dried for 24 h in air at room temperature and then, for 4 h at a temperature of 130 °C. Further, the samples were calcined at a temperature of 500 °C in a current of N_2 (99.9%, 50 mL/min). Pt-containing catalysts were prepared by incipient wetness impregnation of the carrier or Ni/C, Cr/C and Ni-Cr/C systems with an aqueous solution of H_2PtCl_6 $6H_2O$ (chemical grade purity; Khimmed, Russia); then, dried and calcined at a temperature of 350 °C in a flow of N_2 (99.9%, 50 mL/min) for 2 h.

3.2. Characterization of Catalysts

The surface morphology of the catalysts was studied by transmission electron microscopy (TEM) using a JEOL-2100F (Tokyo, Japan) electron microscope in light and dark field modes at an accelerating voltage of 200 kV. The charge state of metals and the composition of the catalyst surface were determined using X-ray photoelectron spectroscopy (XPS), with a Kratos Axis Ultra DLD device (Manchester, UK) with a monochromatic radiation Al Ka (hυ = 1486.6 eV, 150 W). The standard energy of the analyzer was 160 eV and 40 eV for high-resolution spectra.

The phase composition of the samples was studied by X-ray phase analysis (XRD) using an automatic diffractometer DRON-3 (Moscow, Russia), with Cu Ka radiation (λ = 1.5405 A) with a graphite monochromator in step-by-step scanning mode (step 0.1°, exposure per point—5 s). The main 2Θ interval was 20°–90°.

The dependence of the hydrogen absorption rate in the mode of thermoprogrammable reduction (TPR) of catalysts was recorded at the laboratory installation KL-1 (Moscow, Russia). The reduction was carried out with a mixture of gases of 5% H_2/Ar at a flow rate of 23 mL/min until a temperature of 850 °C. The linear heating rate of the detector was 10 °C/min.

Dehydrogenation of bicyclohexyl was carried out in a flow reactor without a carrier gas. Samples of all the studied catalysts with a weight of 1.85 g were placed in a steel reactor with an internal diameter of 10 mm and activated. Before starting the reaction, each portion of the new catalyst was precalcined in an inert atmosphere and then, reduced in a hydrogen flow. The substrate was fed at a rate of 6 mL/h (normal conditions, ρ = 0.864 g/cm^3) by a high-pressure pump HPP 5001. Commercial bicyclohexyl (99%, Acros Organics, Geel, Belgium, $C_{12}H_{22}$) was used as a substrate. Dehydrogenation was carried out at a temperature of 320 °C and atmospheric pressure for 4 h. Hydrogen and reaction products were separated. The hydrogen released during dehydrogenation was measured using a gas burette. To prevent the entrainment of organic substrates with a hydrogen flow, a system of cooling traps and membrane was used.

The reaction products were analyzed with a Crystallux 4000M chromatograph (Kazan, Russia) using a ZB-5 capillary column (Zebron, Phenomenex, Torrance, CA, USA) and a flame ionization detector of the FOCUS DSQ II chromato-mass spectrometer (Thermo Fisher Scientific, Waltham, MA, USA), with a TR-5ms capillary column. The analysis was performed in a programmable temperature mode of 70–220 °C at a heating rate of

6 °C/min. The purity of the hydrogen released was determined by gas chromatography with a thermal conductivity detector and a Porapak Q packed column. The conversion (X) of bicyclohexyl was calculated as the ratio of the change in the amount of bicyclohexyl before and after the reaction to the initial amount of bicyclohexyl. The selectivity (S) of the reaction products was determined as the ratio of the amount of one of the reaction products formed to the total amount.

4. Conclusions

Thus, the study of the composition and structure of the surface phases of mono- and bimetallic catalysts based on Ni and Cr deposited on a carbon carrier by methods of physico-chemical analysis showed the interaction of metals with each other. Due to the significant difference in the enthalpy of formation of nickel and chromium carbides when interacting with the carbon carrier, Cr atoms, when added to a bimetallic system, interact more actively with Ni than with the carrier. The data of XRD, TEM and electron diffraction indicate a change in the parameters of the nickel crystal lattice in bimetallic NiCr systems, compared with the FCC Ni lattice, which indicates, in turn, the formation of Cr-Ni solid solutions formed by the type of substitution of nickel atoms for chromium. However, this does not lead to a change in the conversion and selectivity in the bicyclohexyl dehydrogenation reaction on bimetallic NiCr catalysts, compared with the corresponding monometallic catalysts. At the same time, this circumstance leaves open the question of the high activity of trimetallic Pt catalysts deposited on the studied NiCr/C system [21,26]. Thus, the mechanism of increasing the activity of three metallic Pt catalysts deposited on the Ni-Cr/C system under study deserves additional study.

Supplementary Materials: The following supporting information can be downloaded at: https://www.mdpi.com/article/10.3390/catal12121506/s1, Figure S1: TEM pictures of the surface of catalysts (a) 1.5Cr/3Ni/C_{ox} and (b) 0.1Pt/1.5Cr/3Ni/C_{ox}; electron diffraction patterns of crystallites Cr-Ni (1, 2) and Pt (3) on the surface of the 0.1Pt/1.5Cr/3Ni/C_{ox} catalyst.

Author Contributions: Conceptualization, L.M.K.; methodology, A.N.K.; investigation, A.N.K.; writing—original draft preparation, A.N.K.; writing—review and editing, L.M.K.; supervision, L.M.K.; project administration, L.M.K.; funding acquisition, L.M.K. All authors have read and agreed to the published version of the manuscript.

Funding: The research was financially supported by the Ministry of Science and Higher Education of the Russian Federation (grant no. 075-15-2021-591).

Data Availability Statement: Data are available upon request from the authors.

Conflicts of Interest: The authors declare no conflict of interest.

References

1. Rodriguez, J.A.; Goodman, D.W. The nature of the metal-metal bond in bimetallic surfaces. *Science* **1992**, *257*, 897–903. [CrossRef] [PubMed]
2. Stakheev, A.Y.; Kustov, L.M. Effects of the support on the morphology and electronic properties of supported metal clusters: Modern concepts and progress in 1990s. *Appl. Catal. A Gen.* **1999**, *188*, 3–35. [CrossRef]
3. Kustov, L.M.; Kalenchuk, A.N.; Bogdan, V.I. Systems for accumulation, storage and release of hydrogen. *Rus. Chem. Rev.* **2020**, *89*, 897–916. [CrossRef]
4. Xiong, L.; Manthiram, A. Effect of atomic ordering on the catalytic activity of carbon supported PtM (M = Fe, Co, Ni, and Cu) alloys for oxygen reduction in PEMFCs. *J. Electrochem. Soc.* **2005**, *152*, 697–703. [CrossRef]
5. Baglio, V.; Stassi, A.; Blasi, A.D.; D'Urso, C.; Antonucci, V.; Arico, A.S. Investigation of bimetallic Pt-M/C as DMFC cathode catalysts. *Electrochim. Acta* **2007**, *53*, 1360–1364. [CrossRef]
6. Kaewsai, D.; Hunsom, M. Comparative study of the ORR activity and stability of Pt and PtM (M = Ni, Co, Cr, Pd) supported on polyaniline/carbon nanotubes in a PEM fuel cell. *Nanomaterials* **2018**, *8*, 299. [CrossRef] [PubMed]
7. Ting, H.; Kreidler, E.; Xiong, L.; Ding, E. Combinatorial screening and nano-synthesis of platinum binary alloys for oxygen electroreduction. *J. Power Source* **2007**, *165*, 87–91.

8. Rouibah, K.; Barama, A.; Benrabaa, R.; Guerrero-Caballero, J.; Kane, T.; Vannier, R.N.; Rubbens, A.; Lofberg, A. Dry reforming of methane on nickel-chrome, nickel-cobalt and nickel-manganese catalysts. *Int. J. Hydrogen Energy* **2017**, *42*, 29725–29734. [CrossRef]
9. Antolini, E.; Salgado, J.R.C.; Gonzalez, E.R. The methanol oxidation reaction on platinum alloys with the first row transition metals: The case of Pt–Co and Ni alloy electrocatalysts for DMFCs: A short review. *Appl. Catal. B Env.* **2006**, *63*, 137–149. [CrossRef]
10. Liu, Z.; Guo, B.; Tay, S.W.; Hong, L.; Zhan, X. Physical and electrochemical characterizations of PtPb/C catalyst prepared by pyrolysis of platinum(II) and lead(II) acetylacetonate. *J. Power Source* **2008**, *184*, 16–22. [CrossRef]
11. Jeon, M.K.; McGinn, P.J. Composition dependence of ternary Pt-Ni-Cr catalyst activity for the methanol electro-oxidation reaction. *J. Power Source* **2009**, *194*, 737–745. [CrossRef]
12. Jeon, M.K.; Zhang, Y.; McGinn, P.J. Effect of reduction conditions on electrocatalytic activity of a ternary PtNiCr/C catalyst for methanol electro-oxidation. *Electrochim. Acta* **2009**, *54*, 2837–2842. [CrossRef]
13. Cooper, J.S.; Jeon, M.K.; McGinn, P.J. Combinatorial screening of ternary Pt-Ni-Cr catalysts for methanol electro-oxidation. *Electrochem. Commun.* **2008**, *10*, 1545–1547. [CrossRef]
14. Petrii, O.A. Pt–Ru electrocatalysts for fuel cells: A representative review. *J. Solid State Electrochem.* **2008**, *12*, 609–642. [CrossRef]
15. Antolini, E. Platinum-based ternary catalysts for low temperature fuel cells: Part I. Preparation methods and structural characteristics. *Appl. Catal. B Environ.* **2007**, *74*, 324–336. [CrossRef]
16. Jeon, M.K.; Lee, K.R.; Daimon, H.; Nakahara, A.; Woo, S.J. $Pt_{45}Ru_{45}M_{10}$/C (M = Fe, Co, and Ni) catalysts for methanol electro-oxidation. *Catal. Today* **2008**, *132*, 123–126. [CrossRef]
17. Wang, Z.B.; Yin, G.P.; Shi, P.F.; Sun, Y.C. Novel Pt-Ru-Ni/C Catalysts for Methanol Electro-oxidation in Acid Medium. *Electrochem. Solid-State Lett.* **2006**, *9*, A13–A15. [CrossRef]
18. Strasser, P.; Fan, Q.; Devenney, M.; Weinberg, W.H.; Liu, P.; Norskov, J.K. High Throughput Experimental and Theoretical Predictive Screening of Materials—A Comparative Study of Search Strategies for New Fuel Cell Anode Catalysts. *J. Phys. Chem. B* **2003**, *107*, 11013–11021. [CrossRef]
19. Umeda, M.; Ojima, H.; Mohamedi, M.; Uchida, I. Methanol electrooxidation at Pt-Ru-W sputter deposited on Au substrate. *J. Power Source* **2004**, *136*, 10–15. [CrossRef]
20. Neto, A.O.; Franco, E.G.; Arico, E.; Linardi, M.; Gonzalez, E.R. Electro-oxidation of methanol and ethanol on Pt-Ru/C and Pt-Ru-Mo/C electrocatalysts prepared by Bönnemann's method. *J. Eur. Ceram. Soc.* **2003**, *23*, 2987–2992. [CrossRef]
21. Bogdan, V.I.; Kalenchuk, A.N.; Chernavsky, P.A.; Bogdan, T.V.; Mishanin, I.I.; Kustov, L.M. Synergistic effect of metal components of the low-loaded Pt-Ni-Cr/C catalyst in the bicyclohexyl dehydrogenation reaction. *Int. J. Hydrogen Energy* **2021**, *46*, 14532–14539. [CrossRef]
22. Biesinger, M.C.; Brown, C.; Mycroft, J.R.; Davidson, R.D.; McIntyre, N.S. X-ray photoelectron spectroscopy studies of chromium compounds. *Surf. Interface Anal.* **2004**, *36*, 1550–1563. [CrossRef]
23. Biesinger, M.C.; Lau, L.W.; Gerson, A.R.; Smart, R.S. The role of the Auger parameter in XPS studies of nickel metal, halides and oxides. *Phys. Chem. Chem. Phys.* **2012**, *14*, 2434–2442. [CrossRef] [PubMed]
24. Kalenchuk, A.N.; Leonov, A.V.; Bogdan, V.I.; Kustov, L.M. Dehydrogenation of bicyclohexyl over Ni/Oxidized Sibunite catalyst. *Rus. J. Phys. Chem. A* **2019**, *93*, 652–657. [CrossRef]
25. Kalenchuk, A.N.; Maslakov, K.I.; Bogdan, T.V.; Chernavsky, P.A.; Bogdan, V.I. Formation of the surface of Pt-Ni-Cr/C catalysts under reducing conditions. *Rus. Chem. Bul.* **2021**, *70*, 323–329. [CrossRef]
26. Carrero, A.; Calles, J.A.; García-Moreno, L.; Vizcaíno, A.J. Production of renewable hydrogen from glycerol steam reforming over bimetallic Ni-(Cu, Co, Cr) catalysts supported on SBA-15 Silica. *Catalysts* **2017**, *7*, 55. [CrossRef]
27. Wang, J.B.; Kuo, L.E.; Huang, T.J. Study of carbon dioxide reforming of methane over bimetallic Ni-Cr/yttria-doped ceria catalysts. *Appl. Catal. A Gen.* **2003**, *249*, 93–105. [CrossRef]
28. Garcia, L.; French, R.; Czernik, S.; Chornet, E. Catalytic steam reforming of bio-oils for the production of hydrogen: Effects of catalyst composition. *Appl. Catal. A Gen.* **2000**, *201*, 225–239. [CrossRef]
29. Tuan, L.A.; Luong, N.T.; Ishihara, K.N. Low-Temperature Catalytic Performance of Ni-Cu/Al_2O_3 Catalysts for Gasoline Reforming to Produce Hydrogen Applied in Spark Ignition Engines. *Catalysts* **2016**, *6*, 45. [CrossRef]
30. Portnoi, V.K.; Leonov, A.V.; Mudretsova, S.N.; Fedotov, S.A. Formation of nickel carbide of deformation treatment of Ni-C mixtures. *Phys. Met. Metallogr.* **2010**, *109*, 153–161. [CrossRef]
31. Kalenchuk, A.N.; Bogdan, V.I.; Dunaev, S.F.; Kustov, L.M. Effect of surface hydrophilization on Pt/Sibunite catalytic activity in bicyclohexyl dehydrogenation in hydrogen storage application. *Int. J. Hydrogen Energy* **2018**, *43*, 6191–6196. [CrossRef]

catalysts

MDPI

Article

Reductive Catalytic Fractionation of Spruce Wood over Ru/C Bifunctional Catalyst in the Medium of Ethanol and Molecular Hydrogen

Oxana P. Taran [1,2,*], Angelina V. Miroshnikova [1,2,*], Sergey V. Baryshnikov [1], Aleksandr S. Kazachenko [1,2], Andrey M. Skripnikov [1,2], Valentin V. Sychev [1,2], Yuriy N. Malyar [1,2] and Boris N. Kuznetsov [1,2]

[1] Institute of Chemistry and Chemical Technology SB RAS, Krasnoyarsk Science Center, Siberian Branch, Russian Academy of Sciences, Akademgorodok 50, Bld. 24, 660036 Krasnoyarsk, Russia
[2] Department of Non-Ferrous Metals and Materials Science, Siberian Federal University, pr. Svobodny 79, 660041 Krasnoyarsk, Russia
* Correspondence: taran.op@icct.krasn.ru (O.P.T.); miroshnikova.av@icct.krasn.ru (A.V.M.)

Abstract: Reductive catalytic fractionation (RCF) has emerged as an effective lignin-first biorefinery strategy to depolymerize lignin into tractable fragments in high yields. Herein, we propose the RCF of spruce wood over a Ru/C bifunctional catalyst in the medium of ethanol and molecular hydrogen to produce monomeric phenolic compounds from lignin, polyols from hemicelluloses, and microcrystalline cellulose. This contribution attempts to elucidate the role of the Ru/C bifunctional catalysts characteristics. The results clarify the particular effect of the carbon support acidity, catalyst grain size, content and dispersion of Ru on the effectiveness of lignin and hemicelluloses extraction and the yields of liquid and gaseous products. The most efficient catalysts for RCF of spruce wood, providing high yields of the monomeric phenols, glycols, and solid product with content of cellulose up to 90 wt%, bear 3 wt% of Ru with a dispersion of 0.94 based on an acidic oxidized graphite-like carbon support Sibunit®, and having a grain size of 56–94 µm. The Ru/C catalysts intensify the reactions of hydrodeoxygenation of liquid products from lignin. The main phenolic monomers are 4-propyl guaiacol, 4-propenyl guaiacol, and 4-propanol guaiacol. We explored the effect of the process temperature and time on the yield and composition of the liquid, solid, and gaseous products of spruce wood RCF. The optimal trade-off between the yields of phenolic monomers (30.0 wt%). polyols (18.6 wt%) and the solid product containing 84.4 wt% of cellulose is reached at 225 °C and 3 h over the most acidic Ru/C catalyst.

Keywords: spruce wood; fractionation; reduction; Ru/C catalysts; ethanol; hydrogen; liquid products; phenols; polyols; cellulose

1. Introduction

The depletion of fossil resources and CO_2 emissions negatively affecting the environment require new methods for the use of renewable plant feedstock, including agricultural waste, to ensure environmental safety and social and economic sustainability. In recent years, there has been increased interest in deep processing of renewable plant biomass due to the fact that the use of biomass, in contrast to fossil organic raw materials, does not change the CO_2 balance in the atmosphere [1–3]. Wood is a widespread and available resource for the production of various chemical products and biofuels. The main components of wood (cellulose, hemicelluloses, and lignin) form a complex biocomposite complicating chemical conversion of wood into high-demand products.

The most large-scale industrial process in the chemical conversion of wood is the production of cellulose [4]. In conventional pulp production processes, hemicelluloses and lignin are extracted as by-products; besides, technical lignins are highly condensed and contain sulfur that hinders their further processing. Hence, the development of new processes to ensure processing of all main components of lignocellulosic biomass into various chemical products and liquid biofuels is an important topic [5,6]. Among such processes is organosolv delignification of wood using organic solvents [7,8]. A lower degree of condensation of organo-soluble lignins than technical lignins and the absence of sulfur facilitates their catalytic processing into required phenolic and aromatic substances [5,9]. However, when organosolv lignins are isolated, the undesirable polymerization processes can occur to decrease their yield and reactivity. These circumstances make it urgent to develop methods of lignin depolymerization which eliminate the additional stage of its isolation from lignocellulosic material.

The most promising approaches to the complex conversion of wood are based on the methods of oxidative [10–12] and reductive [12–14] catalytic fractionation of wood biomass. Oxidative catalytic fractionation leads to the formation of valuable products: microcrystalline cellulose, aromatic and aliphatic acids, and vanillin. In particular, the reductive catalytic fractionation (RCF) of lignocellulosic biomass ensures depolymerization of lignin to form liquid hydrocarbons while preserving the main part of cellulose [4,6]. Hemicelluloses depolymerize partially to form soluble products. In this case, the products formed during the partial depolymerization of hemicelluloses can be separated from the lignin products by extraction with dichloromethane and water [15]. Processes of reductive depolymerization of lignin are achieved over metallic catalysts (Pd/C, Ni-Raney, Ni/C, and Ni/Al_2O_3), over the reductive hydrogen donor agent [16,17]. The hydrogen donor agents like aliphatic alcohols (methanol, ethanol, or isopropanol) [11–13] and formic acids [14,15] are used. Supported noble metals (Pt, Pd, and Ru) catalysts show highest activity in these processes, thus providing high yields of phenolic monomers and high-quality cellulose [13,18–20]. The nature of catalysts and solvents used, as well as the process conditions (temperature, time, etc.) determine the yield and composition of phenolic monomers from lignin [15].

The use of Ru/C, Pt/C, and Rh/C catalysts leads to propyl-substituted methoxyphenols S/G (propyl guaiacol/syringol) as the main monomer compounds in the liquid products of softwood and hardwood hydrogenation [21–23]. The Pd/C or Ni/C catalysts give propanol-substituted monomers (4-propanol guaiacol and syringol) [23,24]. Over the Ru/C catalyst, the main products of the RCF process are 4-propanol guaiacol in the butanol/water medium [18] and 4-propyl guaiacol in the methanol medium [23].

The resulting methoxyphenols can be used in various fields; in particular, 4-propyl guaiacol can serve as a component of epoxy resins and polycarbonates [25,26] and 4-propanolguaiacol and its derivatives with the antioxidant and anti-inflammatory activity have prospects for application in medicine [27]. Polyols, e.g., propylene glycol and ethylene glycol, are used to produce foams, elastomers, and adhesives, as well as pharmaceuticals, antifreezes, and solvents [28,29]. The cellulose product can be used to obtain high-demand chemicals (ethanol, esters, levulinic and lactic acids, 5-hydroxymethylfurfural and others) [30–32]. Microcrystalline cellulose finds application in the pharmaceutical, food, cosmetic, and perfume industries and in production of sorbents [33]. In addition, it can be a basis of nanocellulose materials and composites [34].

The use of bifunctional solid acidic catalysts with platinum metals results in intensification the lignin depolymerization and enhances the yield of monomeric phenol derivatives [13,35]. A bifunctional catalyst containing 3% of Ru on an acid-modified graphite-like carbon Sibunit® has been previously shown to exhibit the high activity in the hydrogenolysis of wood and ethanol lignin of birch [12,32,36] and abies [37]. However, the content and nature of acidic species on the surface of the carbon support can affect both the acidic properties of the catalyst and the deposition of metal particles on its surface. This issue has not been studied earlier.

Herein, we explored the influence of characteristics of the Ru/C bifunctional catalysts (the acidity of a carbon support, catalyst gain size, content and dispersion of Ru) as well as the conditions (temperature and time) of the process of spruce wood RCF on the yield, composition, and structure of the liquid, solid, and gaseous products. The conditions for the optimal balance between the yields of monomeric phenols and polyols as well as cellulose product have been found. One of the main forest forming tree species in Siberia and in the north of the European part of Russia—the spruce *Picea obováta* [38]—was chosen as the object of this study.

2. Results

2.1. Characterization of Ru/C Catalysts

We chose a commercial synthetic graphite-like carbon material of the Sibunit® family as the catalyst support. The large surface area, high mechanical strength and the mesoporous structures make Sibunit materials the promising liquid-phase catalysts. Original Sibunit grains (1.0–1.6 mm in size) and powder (56–94 μm in particle size) were used as supports. In order to create surface oxygenate species, Sibunit was oxidized with gas mixture saturated with water vapor ($O_2/N_2 = 1/5$) at 400, 450 and 500 °C.

Table 1 shows N_2 adsorption, TEM, and pH_{pzc} data on the oxidized carbon supports and Ru/C catalysts. After the oxidative treatment of almost all the carbon supports used, except the sample oxidized at 450 °C, the Brunauer–Emmett–Teller specific surface areas (S_{BET}) decreased. The deposition of ruthenium onto the carbon support surface also leads to a decrease in the specific surface area and pore volume (Table 1) due to blocking of some pores of the supports by particles of the active component [9,39].

Table 1. Characteristics of the carbon supports and ruthenium catalysts.

No.	Support/Catalysts	Label	S_{BET}, m^2/g	V_{pore}, cm^3/g	$\langle d \rangle_{pore}$, nm	pH_{pzc} [a]	Ru Particle Size, nm [b]				D_{Ru}
							d_{min}	d_{max}	$<d_l>$	$<d_s>$	
1	Sib-4 [c]	S	375	0.55	5.87	7.59	-	-	-	-	-
2	3%Ru/Sib-4 [c]	3RS	321	0.43	5.45	8.01	0.60	2.73	1.22 ± 0.01	1.48	0.88
3	Sib-4-granular [d]	Sg	364	0.51	5.66	7.66	-	-	-	-	-
4	3%Ru/Sib-4-granular	3RSg	273	0.32	4.77	8.05	0.76	3.46	1.42 ± 0.02	1.71	0.77
5	Sib-4-ox-400 [c]	S400	332	0.42	5.06	6.88	-	-	-	-	-
6	3%Ru/Sib-4-ox-400 [c]	3RS400	300	0.37	5.01	7.12	0.66	3.00	1.19 ± 0.01	1.40	0.94
7	Sib-4-ox-450 [c]	S450	380	0.53	5.66	5.33	-	-	-	-	-
8	1%Ru/Sib-4-ox-450 [c]	1RS450	368	0.52	4.80	6.06	0.52	1.79	1.06 ± 0.03	1.27	1.03
9	3%Ru/Sib-4-ox-450 [c]	3RS450	341	0.50	5.88	6.89	0.52	2.37	1.13 ± 0.01	1.39	0.94
10	Sib-4-ox-500-granular [d]	S500g	287	0.37	5.14	3.34	-	-	-	-	-
11	3%Sib-4-ox-500-granular [d]	3RS500g	233	0.28	4.80	6.44	0.69	3.14	1.30 ± 0.01	1.53	0.85

[a] pH_{pzc} is pH at the zero charge point (PZC); [b] d_{min} and d_{max} are the minimum and maximum particle size, $<d_l> = \Sigma d_i / N$ is the average particle size, $<d_S> = \Sigma d_i^3/\Sigma d_i^2$ is the weight average particle size, and D_{Ru} is the Ru dispersion; [c] the grains are 56–94 μm in size; [d] the grains are 1.0–1.6 mm in size.

The interaction of negatively charged surface functional groups of the oxidized support with ruthenium ions during the catalyst preparation prevents sintering of ruthenium particles in the course of the high temperature reduction of the catalyst [40]. Thus, the oxidative treatment of the support results in an increase in the content of sorption-active groups on the surface, which helps to reduce the size of metal particles (Figure 1).

The temperature elevation during the oxidative treatment of the carbon support leads to an increase in the support and the catalyst acidities (Table 1). The deposition of ruthenium onto the surface of the initial and oxidized supports raises pH_{pzc}. Note that a slightly alkaline reaction (pH) is characteristic of the catalyst slurry based on the nonoxidized Sibunit-4 support.

Thus, the bifunctional ruthenium catalysts bearing Ru nanoparticles can be prepared via deposition of $Ru(NO)(NO_3)_3$ on the surface of the oxidized Sibunit-4 carbon support. The prepared series of the Ru/C catalysts with different grain size, Ru content (1 and 3 wt%), dispersion (0.77–1.00), and acidity (pH 3.34–8.05) were tested for hydrogenation of spruce wood in an ethanol medium.

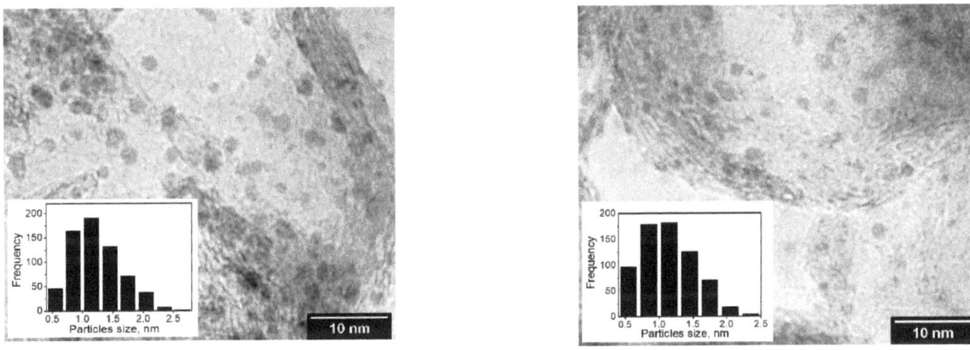

(A) **(B)**

Figure 1. High-resolution transmission electron microscopy images and Ru-particle side distribution for the Ru (3%)/Sib-4 (**A**) and Ru (3%)/Sib-4-ox-450 (**B**) catalysts.

2.2. Yield and Composition of the Spruce Wood RCF Products over the Ru/C Catalysts

Non-catalytic hydrogenation of spruce wood in ethanol at 225 °C and an initial H_2 pressure of 4 MPa for 3 h yields liquid products (23.8 wt%), gaseous products (4.4 wt%), and solid products (67.7 wt%) (Table 2). The aqueous phase contains up to 3.5 wt% of polyhydric alcohols including ethylene glycol, propylene glycol, 1,2-butanediol, and pentane-1,2,5-triol as products of decomposition of the carbohydrate constituent of the wood. The solid product contains 59.4 wt% of cellulose, 24.4 wt% of lignin, and 16.2 wt% of hemicelluloses (Table 2).

The yield of the liquid products of wood RCF over the 3RS450 catalyst increases to 36 wt% and the contents of phenolic monomers in the organic phase and polyols in the aqueous phase increase to 26 and 18.6 wt%, respectively. The cellulose content in the solid product increases to 84.4 wt% at a simultaneous decrease in the contents of hemicelluloses and lignin to 7.0 and 8.6 wt%, respectively (Table 2).

Table 2. The results of RCF of spruce wood at 225 °C for 3 h over different Ru catalysts.

Catalyst	Product Yield, wt%					Solid Product Composition, wt%		
	Solid	Liquid	Gaseous	Phenolic Monomers [c]	Polyols [d]	Hemicelluloses	Lignin	Cellulose
no	67.7	23.8	4.4	1.5	3.5	16.2	24.4	59.4
3RS450	49.5	36.0	10.6	26.0	18.6	7.0	8.6	84.4
3RS400	54.3	32.5	9.0	30.0	13.9	14.8	7.6	77.6
3RS500g [b]	54.7	27.5	12.2	12.0	11.3	15.8	9.3	74.9
3RS [a]	57.4	30.0	9.0	22.0	12.5	14.1	9.8	76.1
3RSg [a,b]	57.6	29.0	7.8	15.7	10.6	16.5	12.1	71.4

[a] Non-oxidized support; [b] a grain size of 1.0–1.6 mm; [c] the yield from the lignin weight in wood; [d] the yield from the polysaccharide weight in wood.

The gain size affects significantly the yield and composition of the liquid, solid, and gaseous products of spruce wood hydrogenation. The granular catalysts (1.0–1.6 mm fraction) are less efficient to the process of RCF of spruce wood than the powder catalysts (56–94 μm fraction). When used, the monomer and polyol yields are lower by a factor of 1.5–2 over the granular catalysts than over the powder catalysts, obviously, due to diffusion restrictions inherent in liquid-phase catalytic processes.

Figure 2 shows yields of the main RCF products obtained without catalysts and over different catalysts. The remaining monomer products were found in trace amounts. The maximum content (30 wt%) of monomer phenols was observed in the liquid products over the 3RS400 catalyst and the maximum content (18.6 wt%) of polyols over the 3RS450 catalyst. Notably, 4-propenyl guaiacol dominates among the phenolic products of the non-catalytic hydrogenation of spruce wood (Figure 2A). The main component of the phenol products in the RCF is 4-propyl guaiacol at a yield of 18.1 wt% over the 3RS450 catalyst and 23.9 wt% over the 3RS400 catalyst (Figure 2A). In the non-catalytic hydrogenation of wood, the yield of glycols is no more than 3.5 wt% (Figure 2B). In the liquid products of the RCF, the yield of glycols, mainly ethylene glycol and propylene glycol, rises sharply (by a factor of 2.9) (Figure 2B).

We observed a trend to an increase in the total yields of monomer methoxyphenols and polyols with the increasing dispersion of ruthenium in the catalysts. The higher Ru dispersion accounts for better active center accessibility to promote to the higher overall catalytic activity [41] (Figure 3).

Elemental analysis of the liquid products of spruce wood hydrogenation reveals intensification of the deoxygenation reactions over the ruthenium catalysts (Figure 4).

The liquid products of the spruce wood RCF at 225 °C contain a lower amount of oxygen and a higher amount of hydrogen and carbon compared to those in the initial wood; this leads one to conclude about more intensive deoxygenation reactions over the ruthenium catalysts. The minimum oxygen content and the maximum carbon content are observed in the liquid products of wood RCF over the 3RS400 and 3RS450 catalysts. According to the Van Krevelen diagram, the atomic ratio O/C of liquid products of non-catalytic and catalytic hydrogenation of spruce wood is much lower than in the original wood. The most significant decrease in O/C and the highest H/C ratio in liquid products were observed over the 3RS450 catalyst.

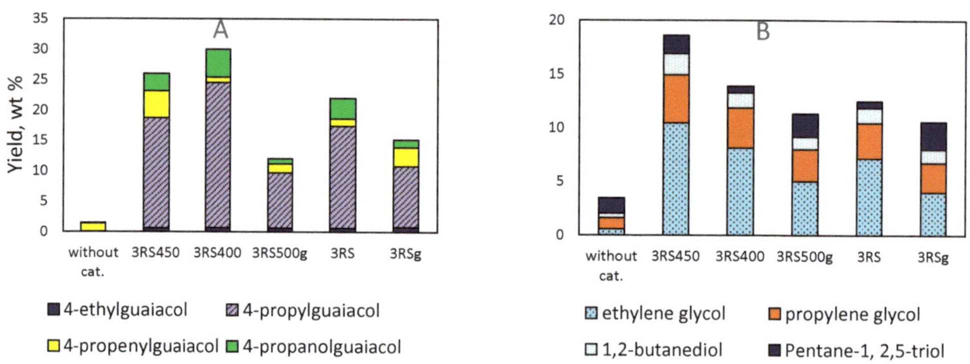

Figure 2. The yields of monomers from lignin (**A**) and polyols from polysaccharides (**B**) over the Ru catalysts.

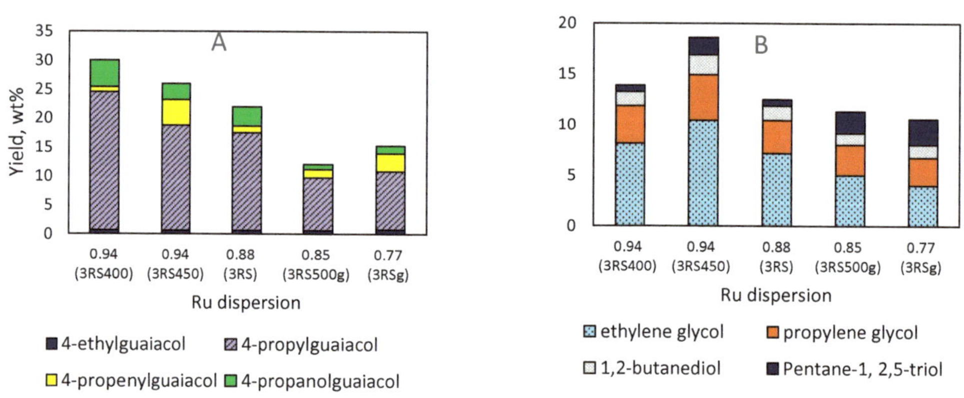

Figure 3. The influence of Ru dispersion in the catalyst on the yields of monomers from lignin and (**A**) polyols from polysaccharides (**B**).

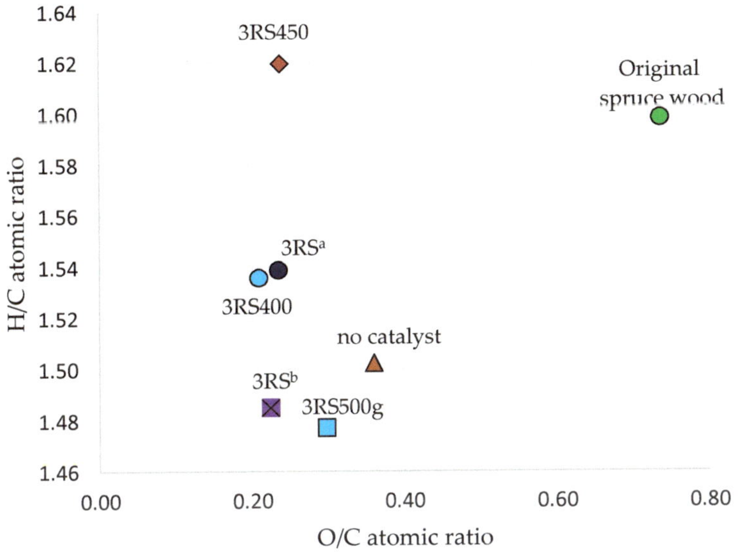

Figure 4. Van Krevelen diagram for the liquid products of spruce wood RCF at 225 °C for 3 h over different catalyst ([a] Original (non-oxidized) support; [b] the grain size is 1.0–1.6 mm).

Carbon oxides (4.4 wt%) and trace methane (less than 0.1 wt%) were detected in gaseous products of non-catalytic hydrogenation. The methane content increases significantly (up to 4.5 wt%) in the gaseous products over the ruthenium catalysts. Ruthenium is known to catalyze CO_2 conversion to methane. The usual reaction temperature range is from 200 to 450 °C depending on the catalyst, support and the reaction conditions. So, it is reasonable to assume such pathways in our case. However, the main pathway of methane formation is catalytic hydrocracking of aliphatic structural fragments of lignin [42].

We examined the effect of the ruthenium content in the catalyst on the yield and composition of the products of spruce wood RCF at 225 °C (Table 3).

Table 3. The effect of ruthenium content in the catalyst on the yield and composition of products of spruce wood RCF at 225 °C for 3 h.

Catalyst	Product Yield, wt%					Solid Residue Composition, wt%		
	Solid	Liquid	Gaseous	Phenolic Monomers [a]	Polyols [b]	Hemicelluloses	Lignin	Cellulose
no	67.7	23.8	4.4	1.5	3.5	16.2	24.4	59.4
1RS450	60.0	28.7	9.1	23.1	14.2	10.2	9.3	80.5
3RS450	49.5	36.0	10.6	26.0	18.6	7.0	8.6	84.4

[a] The yield from the lignin weight in wood; [b] the yield from the polysaccharide weight in wood.

Compared to the non-catalytic process, a 15-fold increase in the content of monomer phenolic compounds (Figure 5A) and a 4-fold increase in the content of polyols is observed in the spruce wood RCF over the catalyst containing 1 wt% of Ru. At the RCF of spruce wood, the contents of ethylene glycol and propylene glycol increase and the content of pentane-1,2,5-triol decreases (Figure 5B). At a higher Ru content in the catalyst (3 wt%), the yield of the liquid products of wood RCF increases up to 36.0 wt% and the content of phenolic monomers decreases to 26.0 wt% and the content of polyols to 18.6 wt%. While the yield of the solid product drops down, the cellulose content increases to 84.4 wt% (Table 3).

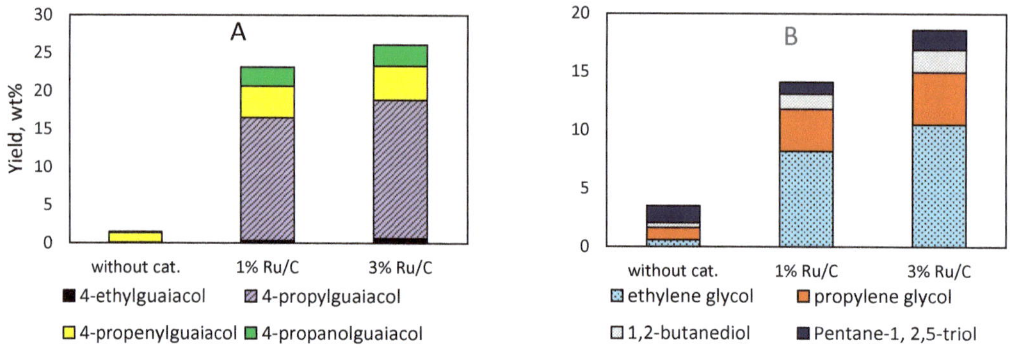

Figure 5. The effect of Ru content in the catalyst on the yield of phenolic monomers (**A**) and polyols (**B**) after spruce wood RCF at 225 °C for 3 h.

The influence of the metal content in the Ru/C catalyst on the molecular weight distribution (MWD) of the liquid products of spruce wood RCF at 225 °C for 3 h was studied using GPC (Figure 6).

Among the liquid products of the non-catalytic hydrogenation of spruce wood, there are fractions of organosolv lignin (M_W = 1000–10,000 g mol^{-1}) and products, mainly dimers, of its partial depolymerization. With the catalysts, the profile of the MWD curve of the liquid products changes significantly to indicate a change in their composition. The MWD curves of the liquid products of RCF contain an intense peak at M_W of ~150 g/mol corresponding to the monomer compounds. An increase in the amount of ruthenium in the Ru/C catalyst to 3 wt% leads to a decrease in the content of oligomers with 2–4 units (M_W ~ 400–600 g/mol) and an increase in the content of the monomer compounds in the liquid products (Figure 6, Table 4).

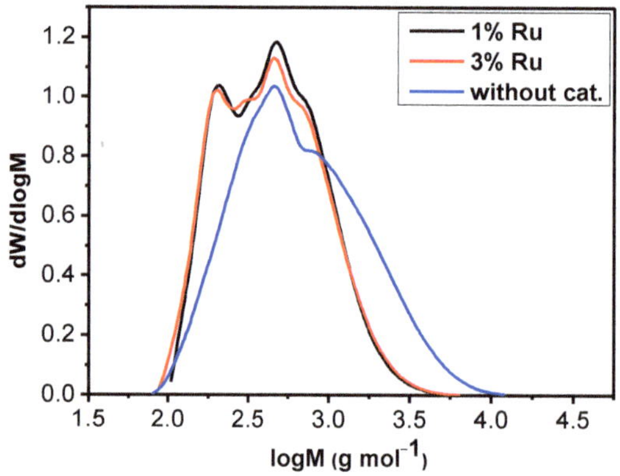

Figure 6. The effect of Ru content on the molecular weight distribution of the liquid products of spruce wood RCF at 225 °C for 3 h.

Table 4. Effect of the Ru content on molecular weight characteristics of the liquid products of spruce wood RCF at 225 °C for 3 h.

Catalyst	M_n [a], g mol^{-1}	M_w [b], g mol^{-1}	PD [c]
no	458	1001	2.186
3RS450	337	578	1.715
1RS450	351	580	1.652

[a] The number average molecular weight; [b] the weight average molecular weight; [c] the polydispersity M_w/M_n.

An increase in the ruthenium content in the catalyst from 1 to 3 wt% leads to a slight increase in the yield of gaseous carbon oxides with a noticeable rise in the yield of methane (Figure 7).

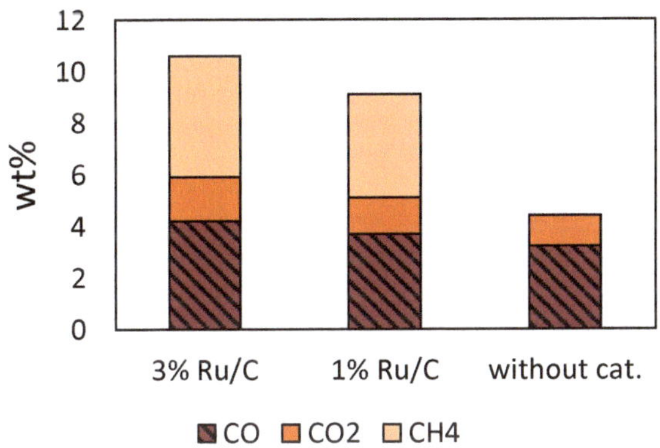

Figure 7. The effect of Ru content on the yield and composition of the gaseous products of spruce wood RCF at 225 °C for 3 h.

2.3. *Influence of Temperature on the Yield and Composition of the Products of Spruce Wood RCF over the 3RS450 Catalyst*

To study the effect of temperature on the yield and composition of the spruce wood RCF products, we chose a 3RS450 catalyst that provides the highest yield of the liquid products (Table 5).

Table 5. The effect of temperature on the yield and composition of the products of spruce wood RCF (3 h) over the 3RS450 catalyst.

Temperature, °C	Product Yield, wt%					Solid Residue Composition, wt%		
	Solid	Liquid	Gaseous	Phenolic Monomers [a]	Polyols [b]	Hemicelluloses	Lignin	Cellulose
200	69.0	19.6	6.7	15.6	10.2	18.6	14.5	56.7
225	49.5	36.0	10.6	26.0	18.6	7.0	8.6	84.4
250	41.0	32.0	18.1	35.0	19.1	4.4	5.2	90.4

[a] The yield from the lignin weight in wood; [b] the yield from the polysaccharides weight in wood.

The liquid product yield increases from 19.6 to 36 wt% over this catalyst as the process temperature rises from 200 to 225 °C. The temperature rises up to 250 °C leads to a decrease in the yield of the liquid products down to 32 wt% at a simultaneous increase in the contents of phenolic monomers and polyols. The solid product yield decreases from 69 to 41 wt% with the hydrogenation temperature rise from 200 to 250 °C (Table 5).

The process temperature elevation from 200 to 250 °C leads to an increase in the cellulose content in the solid product of wood RCF from 56.7 to 90.4 wt%, while the contents of lignin and hemicelluloses decrease from 13.2 and 17.6 to 5.2 and 4.4 wt%, respectively (Table 5). The content of phenolic monomers in the liquid products also increases from 15.6 to 35 wt% (Figure 8A). We observed an increase in the content of 4-ethylguaiacol (1), 4-propylguaiacol (2), and 4-propenylguaiacol (3) and reduction of the content of 4-propanolguaiacol (4) with the hydrogenation temperature rising from 200 to 225 °C. 4-Propanolguaiacol appears to convert into 4-propylguaiacol upon elimination of the γ-OH group with the rise of the RCF temperature [43].

The most noticeable increase (from 10.2 to 18.6 wt%) of the polyol yields is observed when the temperature rises from 200 to 225 °C; the further temperature elevation to 250 °C does not change significantly this yield (Figure 8B).

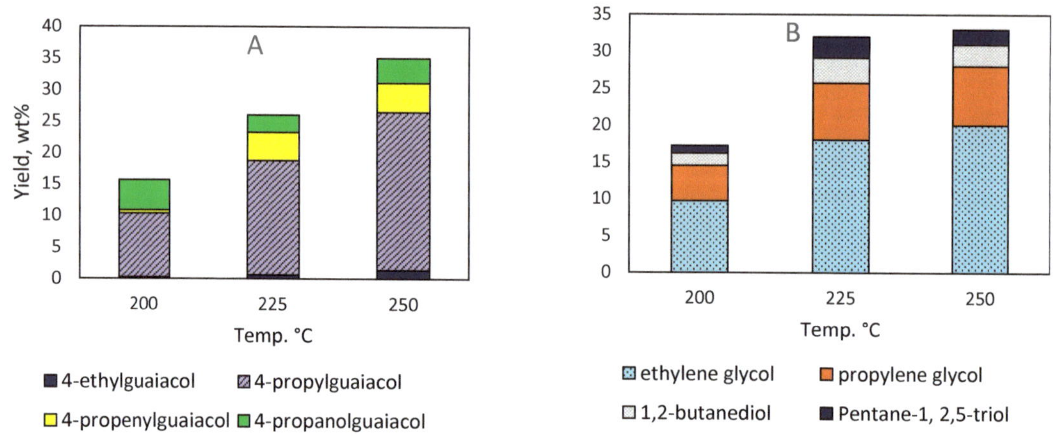

Figure 8. The effect of temperature of spruce wood RCF over the 3RS450 catalyst on the yield of phenolic monomers (**A**) and polyols (**B**).

From GPC data, the rise of the wood RCF temperature from 200 to 250 °C facilitates the reduction of the average molecular weight Mw of the liquid products from 578 to 502 g/mol (Table 6). The intensity of the MWD curves of the liquid products at the region corresponding to the monomeric compounds with MW ≈ 150 g/mol increases with the rise of the RCF temperature. In addition, a curve shoulder assigned to oligomeric fragments of lignin depolymerization narrows noticeably. However, at 250 °C, the intensity increases at the region of dimers with MW = 350–400 g/mol to indicate the contribution of the repolymerization reactions. The polydispersity PD of the liquid products passes through a maximum at 225 °C to increase from 1.614 to 1.715 and then decreases to 1.679 at 250 °C (Table 6, Figure 9).

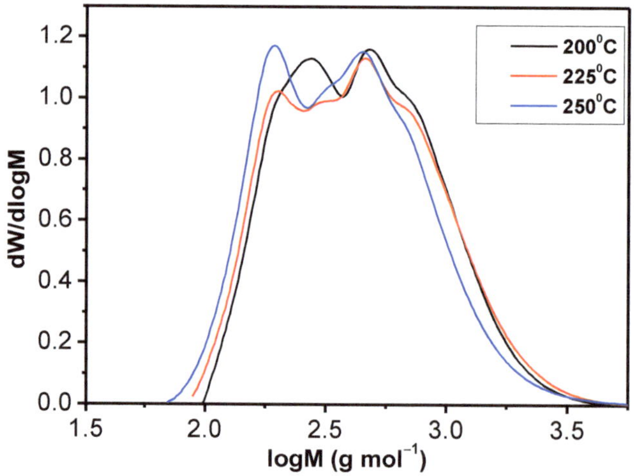

Figure 9. The effect of temperature of spruce wood RCF for 3 h over the 3RS450 catalyst on the molecular weight distribution of the liquid products.

Table 6. Effect of temperature of spruce wood RCF over the 3RS450 catalyst for 3 h on the molecular weight characteristics of the liquid products.

Temperature, °C	M_n [a], g mol^{-1}	M_w [b], g mol^{-1}	PD [c]
200	350	565	1.614
225	337	578	1.715
250	299	502	1.679

[a] The number average molecular weight; [b] the weight average molecular weight; [c] the polydispersity M_w/M_n.

2.4. The Yield and Composition of the Products of Spruce Wood RCF over the 3RS450 Catalyst Depending on Time

We studied the effect of the time of RCF of spruce wood at 225 °C over the 3RS450 catalyst on the product yield. As the process time lengthens from 3 to 4.5–6 h, the yields of the liquid and solid products decrease from 36 to 29 wt% and from 49.5 to 46 wt%, respectively, and the yield of the gaseous products increases from 10.6 up to 14.3 wt%; whereas the lignin and hemicelluloses contents in the solid product decrease from 8.6 to 6.8 wt% and from 7.0 to 6.0 wt%, respectively, while the cellulose content increases from 84.4 to 87.2 wt% (Table 7).

Table 7. The effect of time of spruce wood RCF at 225 °C over the 3RS450 catalyst on the products yield and composition.

Reaction Time, h	Product Yield, wt%					Solid Product Composition, wt%		
	Solid	Liquid	Gaseous	Phenolic Monomers [a]	Polyols [b]	Hemicelluloses	Lignin	Cellulose
3.0	49.5	36.0	10.6	26.0	18.6	7.0	8.6	84.4
4.5	48.2	31.0	11.7	22.6	15.5	6.2	7.6	86.2
6.0	46.0	28.8	14.3	21.4	13.6	6.0	6.8	87.2

[a] The yield from the lignin weight in wood; [b] the yield from the polysaccharide weight in wood.

The yields of monomer methoxyphenols in the liquid products decreases progressively from 26 to 21.4 wt% with lengthening of the spruce wood RCF time from 3 to 6 h. It seems like the yield of these compounds decreases due to the contribution of condensation reactions of the monomer compounds. The increasing process time affects also the composition of phenolic monomers: the yields of 4-propenylguaiacol (3) and 4-propylguaiacol (2) decrease and 4-propanolguaiacol (4) increase (Figure 10A).

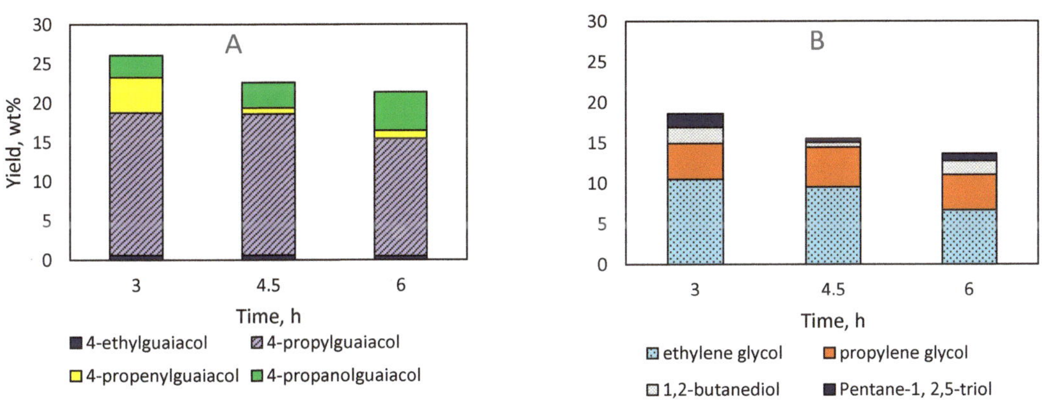

Figure 10. The effect of time of spruce wood RCF on the yield of phenolic monomers (**A**) and polyols (**B**) over the 3RS450 catalyst at 225 °C.

The yield of glycols, mainly ethylene glycol and propylene glycol, in the liquid products decreases more sharply than the yields of phenolic monomers (from 18.6 to 13.7 wt%) with the increasing RCF time (Figure 10B). An increase in the yield of the gaseous products with the spruce wood RCF time can be accounted for by the contribution of the hemicelluloses hydrocracking reaction to produce CO [44] (Figure 11).

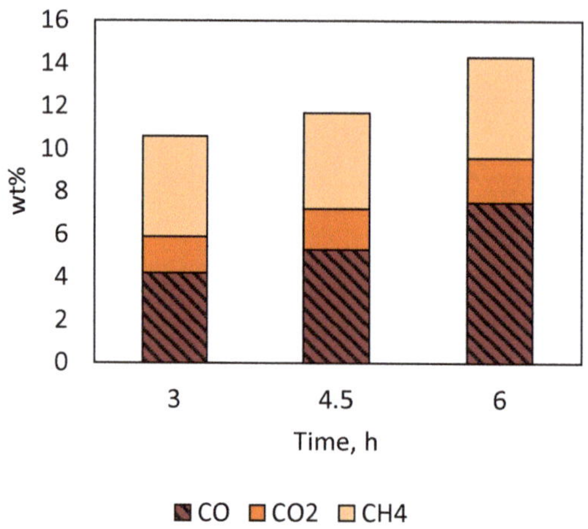

Figure 11. The effect of time of spruce wood RCF over the Ru/C catalyst at 225 °C on the yield and composition of the gaseous products.

The GPC data reveal a narrowed high-molecular shoulder on the MWD curve of the liquid products and decreasing polydispersity PD (from 1.715 to 1.623); this occurs due to a decrease in the average molecular weight M_w of the products from 578 to 526 g/mol at a longer RCF time. The intensity of the peaks corresponding to the mono- and dimeric products of the lignin transformation also reduces to indicate a decrease in their yields (Table 8, Figure 12).

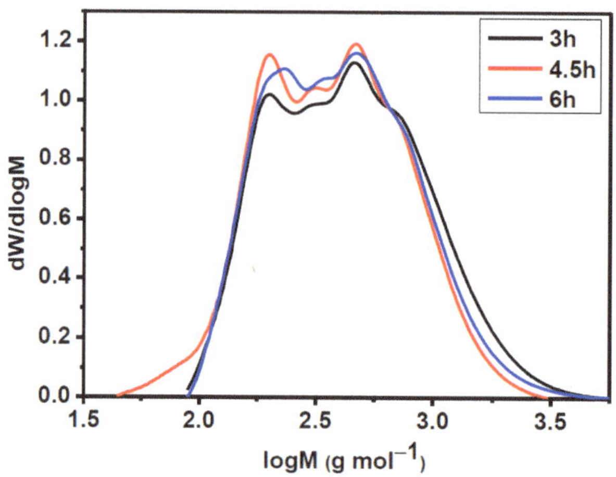

Figure 12. The effect of time of spruce wood RCF over the Ru/C catalyst at 225 °C for 3 h on the molecular weight distribution of the liquid products.

Table 8. The effect of time of spruce wood RSF over the 3RS450 catalyst at 225 °C for 3 h on the molecular weight characteristics of the liquid products.

Process Time, h	M_n [a], g mol^{-1}	M_w [b], g mol^{-1}	PD [c]
3.0	337	578	1.715
4.5	329	545	1.626
6.0	324	526	1.623

[a] The number average molecular weight; [b] the weight average molecular weight; [c] the polydispersity M_w/M_n.

2.5. Effect of Ru/C Catalyst Acidity on the Yield and Composition of the Spruce Wood RCF Products

Acidic conditions are known to promote hydrolysis of ether bonds. The most prominent event in acid-catalyzed lignin chemistry is the cleavage of b-O-4 ether bonds. Nevertheless, acidic media affect the lignin structure by facilitating both depolymerisation (i.e., acidolysis) and repolymerisation [4]. In order to check these impacts, we used bifunctional powder catalysts based on acidic supports.

The acid-base properties of the ruthenium catalysts were changed by oxidizing the carbon support Sibunit at 400 and 450 °C. The acidity of the catalysts determined from the pH_{pzc} (Table 9) vary from 6.89 to 8.01. The highest yields of the liquid products (36.0 and 32.5 wt%, respectively) were obtained over the 3RS450 and 3RS400 bifunctional catalysts at pH_{pzc} values of 7.12 and 6.89. The yield of the solid product decreases progressively from 57.4 to 49.5 wt% with decreasing pH_{pzc}. The solid products contain less lignin and hemicelluloses and more cellulose, in contrast to the case of the monofunctional ruthenium catalyst with the weak-alkaline properties (pH 8.01) (Table 9).

Table 9. The effect of pH_{pzc} value of the Ru catalysts on the yield and composition of products of spruce wood RCF at 225 °C for 3 h.

Catalyst	pH_{pzc}	Product Yield, wt%					Solid Product Composition, wt%		
		Solid	Liquid	Gaseous	Phenolic Monomers [b]	Polyols [c]	Hemicelluloses	Lignin	Cellulose
3RS450	6.89	49.5	36.0	10.6	26.0	18.6	7.0	8.6	84.4
3RS400	7.12	54.3	32.5	9.0	30.0	13.9	14.8	7.6	77.6
3RS [a]	8.01	57.4	30.0	9.0	22.0	12.5	14.1	9.8	76.1

[a] The original nonoxidized support; [b] the yield from the lignin weight in wood; [c] the yield from the polysaccharide weight in wood.

The yields of phenolic monomers and polyols (Figure 13A,B) are essentially affected by the catalyst acidity. On oxidation of the carbon support at 400 °C followed by pH_{pzc} decrease to 7.12, the yield of the monomer phenolic compounds increases from 22 to 30 wt%. Oxidation at higher temperature (450 °C) results in a decrease in pH_{pzc} to 6.89 and a decrease in the yield of these compounds to 26 wt%. The yield of polyols increases with a decrease in the catalyst pH_{pzc}; it reaches 18.6 wt% over the most acidic catalyst 3RS450 with pH_{pzc} 6.89. The increase in the yield of polyols seems to cause an increase in the efficiency of the glycosidic bond hydrolysis in hemicelluloses and/or amorphous cellulose with increasing acidity of the catalyst. [45]. (Table 9).

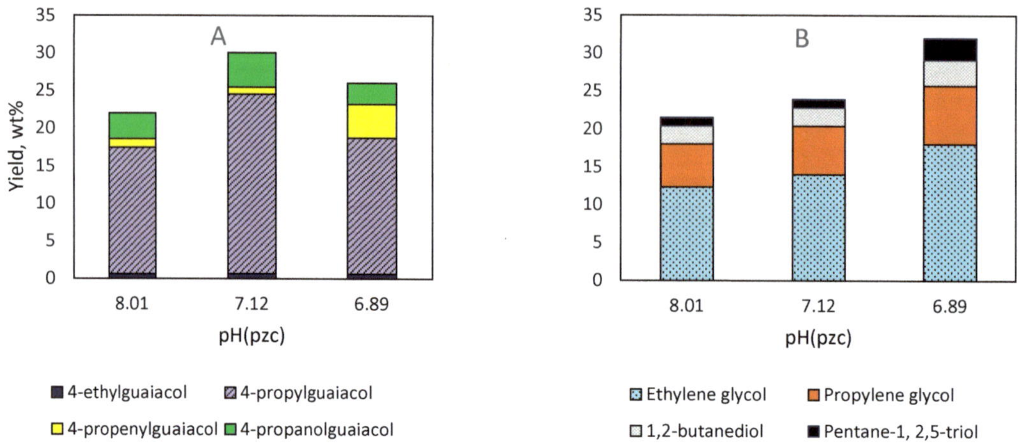

Figure 13. The effect of $pH_{(pzc)}$ of the catalysts on the yields of phenolic monomers (**A**) and polyols (**B**) at spruce wood RCF at 225 °C for 3 h.

2.6. Characterization of the Solids Obtained from RCF of Spruce Wood

The structure of the solid products of spruce wood hydrogenation was studied using infrared (IR) spectroscopy and X-ray diffraction (XRD) techniques. IR spectra of the initial wood and the solid products of its hydrogenation are shown in Figure 14. The IR spectrum of the initial spruce wood is a sum of absorption bands of its main structural components and includes the bands characteristic of bonds between cellulose, lignin, and hemicellulose macromolecules (Figure 14) [46]. The absorption band at 1737 cm^{-1} is assigned to stretching vibrations of the C=O group in the uronic acid ester monomer of hemicelluloses [47]. In the solid products of the RCF, the intensity at 1737 cm^{-1} almost disappears. The absorption bands at 1606 and 1510 cm^{-1} in the IR spectra of spruce wood and the solid products of its RCF correspond to skeletal vibrations of syringyl and guaiacyl aromatic rings [48]. These less intense bands, as well as band at 1269 cm^{-1} (C-O stretching of guaiacyl unit) [48], in the solid RCF products argue for a decrease in the lignin content.

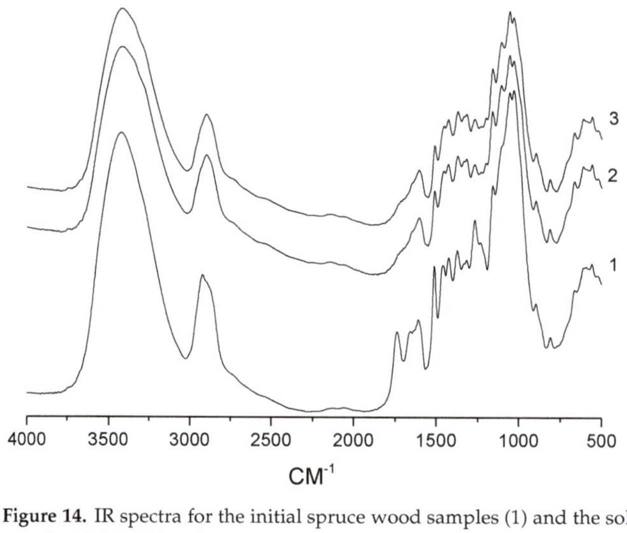

Figure 14. IR spectra for the initial spruce wood samples (1) and the solid products of wood RCF over the 3RS450 (2) and 3RS400 (3) catalysts.

The absorption bands at 1500–900 cm^{-1} are assigned to vibrations of C–H bonds in the methyl and methylene groups, C–O and O–H bonds, the glycosidic bond, and the cellulose glucopyranose ring [49].

Figure 15 shows XRD patterns of the samples of the initial spruce wood and solid products of its hydrogenation.

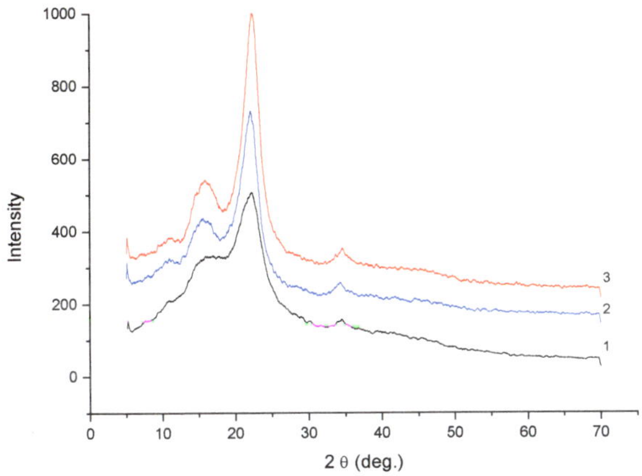

Figure 15. X-ray diffraction patterns for the initial spruce wood sample (1) and solid products of RCF over the 3RS400 (2) and 3RS450 (3) catalysts.

XRD patterns of all the samples under study contain two intense peaks with maxima at 2θ angles of 22.2° and 15.6° (Figure 15). They correspond to reflections of atoms at the (002) plane and superimposed reflections of atoms at the (101) and $(10\bar{1})$ planes of the cellulose crystal lattice [50]. A pronounced peak at a 2θ diffraction angle of ~22.2° is a criterion for the cellulose crystallinity and characterizes the fraction of densely packed cellulose molecules [51].

The crystallinity index of the initial spruce wood is 0.47. The solid products of RCF of the wood have the higher crystallinity indices as compared with those of the initial spruce wood. This relates to the removal of a significant part of amorphous carbohydrates during the wood hydrogenation (Table 10).

Table 10. Crystallinity indices for the spruce wood samples and solid products of wood RCF.

Sample	Crystallinity Index of the Solid Product
Spruce wood	0.47
Catalyzed solid hydrogenation products (RS400)	0.71
Catalyzed solid hydrogenation products (RS450)	0.78

The highest crystallinity index (0.78) is characteristic of the solid product of the catalytic hydrogenation of spruce wood over the more acidic RS450 bifunctional catalyst. The composition and structure of the solid product corresponds to the parameters of commercial microcrystalline cellulose with a crystallinity index of 0.75 [52].

2.7. Optimal Conditions for the Reductive Catalytic Fractionation of Spruce Wood

To be an effective biomass processing strategy for fuels and commodity chemical production, RCF should provide high levels of lignin and hemicelluloses extraction as well as depolymerization but minimize cellulose dissolution and decomposition. A trade-off between these parameters exists based on the data shown in the present study. A scenario aimed at achieving an optimal balance between the removal of lignin and the preservation of all carbohydrates in the pulp seems difficult to realize due to the fact that lignin and hemicellulose are intertwined in the cell wall.

The conditions for the RCF process providing the optimal trade-off between the yields of phenols, polyols, and cellulose form spruce wood are determined (Figure 16).

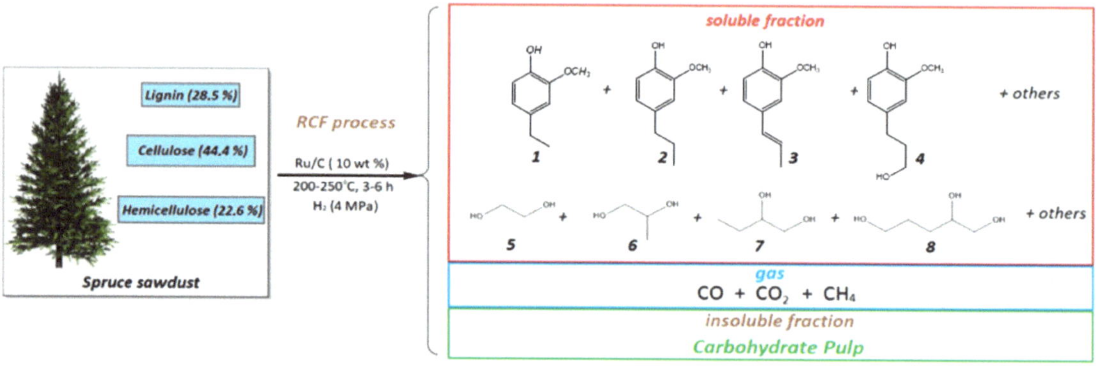

Figure 16. The main products of the reductive catalytic fractionation of spruce wood.

Trends in the influence of various process parameters (temperature, time, content and dispersion of ruthenium, catalyst acidity, catalyst particle size) on the yield and composition of liquid, solid, and gaseous products of spruce wood RCF have been revealed. The yield of monomeric phenols and polyols in the liquid products increases with the hydrogenation temperature elevation; an increase in the process time over 3 h reduces their yield. Thus, the highest yield of phenolic monomers and polyols are achieved at the process temperature of 250 °C and a process time of 3 h. The yields of phenolic and polyolic monomers increase with increasing content and dispersion of ruthenium in the catalyst as well as with increasing catalyst acidity.

Summarizing the results, we can conclude: the most efficient spruce wood RCF catalysts, that provide the highest yield of monomeric phenols and polyols, are the 3RS400 and 3RS450 bifunctional catalysts, which contain 3 wt% of Ru with a dispersion of 0.94, possess high acidity (pH 7.12 and 6.89, respectively), and have a particle size of 56–94 μm.

Figure 17 shows the Sankey diagram for the material balance of the spruce wood RCF process (3 g loaded into the reactor) calculated for the experiment over the 3RS450 catalyst (10 wt% of the wood weight). The initial pre-deresined spruce wood sample contained 1.407 g of cellulose, 0.869 g of lignin, and 0.719 g of hemicelluloses. The reductive catalytic fractionation in ethanol at 225 °C for 3 h yielded 1.470 g of microcrystalline cellulose (49.0% of the wood weight), 1.049 g of the liquid products (36.0% of the wood weight), and 0.318 g of gases (10.6% of the wood weight). After extraction of the liquid products with dichloroethane and water, 0.413 g of the aqueous phase containing 0.392 g of glycols (18.6% of the polysaccharide weight in the wood) and 0.636 g of the dichloromethane phase containing 0.233 g of phenolic monomers (26.0% of the lignin weight in the wood) were obtained (Figure 17). The total loss of these substances during the extraction and pyrogenetic water was 0.163 g (5.4% of the wood weight).

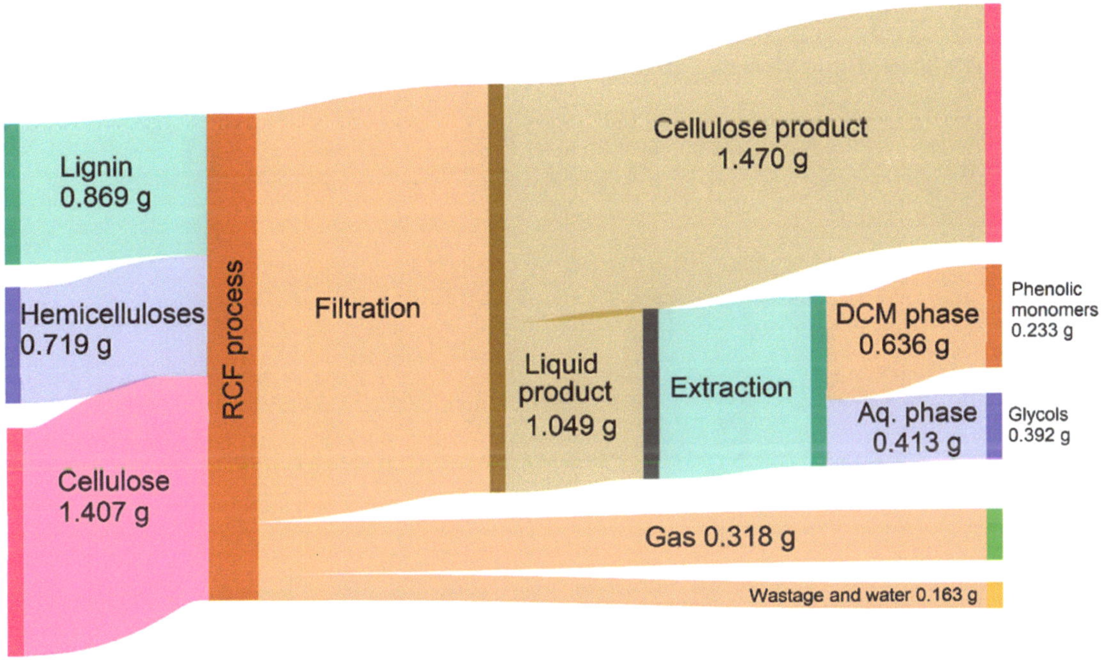

Figure 17. Sankey diagram for the material balance of the reductive catalytic fractionation of spruce wood over the 3RS450 catalyst.

Thus, the optimum trade-off between the yields of the liquid products enriched in phenolic monomers and polyols is obtained at the spruce wood hydrogenation temperature of 225 °C and the time of 3 h over the the highly-acidic 3% Ru/C bifunctional catalyst.

3. Materials and Methods

3.1. Materials and Reagents

Siberian spruce (*Pícea obováta*) wood (Krasnoyarsk region) used in this study contained cellulose (44.4), lignin (28.6), hemicelluloses (22.6), extractives (3.8), and ash (0.6%) from the weight of absolutely dry wood. The wood was ground into particles smaller than 1 mm in size, successively deresined with petroleum ether and acetone by a conventional ANSI/ASTM D 1105 method, and then dried at 24 °C to a constant moisture content (4.3 wt%).

In this study, we used AO RFK ethyl alcohol (95%), $Ru(NO)(NO_3)_3$ (31.3% of Ru, Alfa Aesar, Ward Hill, MA, USA), −N, O bistrifluoroacetamide (BSTFA + TMCS 99:1, Sigma-Aldrich, Burlington, MA, USA), and ionol (>99%, Sigma-Aldrich, Burlington, MA, USA).

3.2. Preparation and Study of the Ru/C Catalysts

Carbon supports containing acid sites were prepared by oxidizing a Sibunit commercial mesoporous carbon material with an oxygen–argon mixture containing 20 vol% of O_2 in the presence of water vapor (saturation at 90 °C, vapor pressure 70.1 kPa) at temperatures of 400, 450, and 500 °C for 2 h using the technique described elsewhere [53].

One percent Ru/C and 3% Ru/C catalysts were prepared by incipient wetness impregnation of a carbon support with the $Ru(NO)(NO3)_3$ aqueous solution followed by drying at room temperature for 2–3 h and then at 60 °C for 12 h. The active component was reduced in flowing hydrogen (30 mL/min) at 300 °C and the rate of 1 °C/min for 2 h. After cooling down to room temperature in the hydrogen atmosphere, the catalysts were passivated using 1% O_2 in argon as described in [54].

The texture characteristics of the samples were determined from the N_2 adsorption isotherms using a Micromeritics ASAP-2020 Plus analyzer (US) at 77 K. The size distribution of ruthenium particles in the ruthenium catalysts was determined using a Hitachi HT7700 transmission electron microscope (Toranomon Minato-Ku, Japan, 2014) at the accelerating voltage of 110 kV and resolution of 2 Å. The particle size distribution histograms were obtained by the statistical (500–800 particles) processing of electron microscopy images. The linear-average ($<d_l>$) and weight-average ($<d_s>$) diameters of deposited particles were calculated using the formulas [55]:

$$<d_l> = \Sigma d_i/N, <d_s> = \Sigma d_i^3/\Sigma d_i^2,$$

where d_i is the diameter of the deposited particle and N is the total number of particles.

The ruthenium dispersion D_{Ru} in the catalysts was calculated as:

$$D_{Ru} = 6 \cdot \frac{M_{Ru}}{a_{Ru} \cdot \rho \cdot N_0 \cdot \langle d_s \rangle} \quad (1)$$

where M_{Ru} = 0.101 kg/mol is the ruthenium molar mass, ρ = 12,410 kg/m^3 is the density of ruthenium metal, a_{Ru} = 6.13; 10^{-20} m^2 is the average effective area of a metal atom on the surface, N_0 is the Avogadro number, and d_s is the weight-average ruthenium particle diameter [55].

The acid-base properties of the catalysts were investigated by finding the point of zero charge (pzc) by the Sörensen–de Bruin method [41]. This method was shown [53] to be suitable for comparing the acid properties of carbon materials. The results obtained by this method correlate well with the results of XPS and acid-base titration [53], and the technique does not require expensive instruments or lengthy experiments compared to other methods. Ten mL of distilled water was poured in a potentiometric cell and the test sample was successively added as small (0.01-g) portions in specified (5–10 min) intervals under continuous stirring with a magnetic stirrer until the constant glass electrode potential was reached [41].

3.3. *Catalytic Hydrogenation of Spruce Wood*

The process of hydrogenation of crushed deresined wood was studied in a 300 mL ChemRe SYStem R-201 autoclave (Gwanyang–dong, Korea) (see Figure 18). Fifty mL of ethanol, 5.0 g of sawdust, and 0.5 g of the ruthenium catalyst were loaded into the reactor. Then, the autoclave was hermetically sealed, purged with argon to remove air, and filled with hydrogen to a pressure of 4 MPa. The hydrogenation was carried out at 200, 225, and 250 °C for 3 h under continuous stirring at the speed of 800 rpm. The operating pressure in the reactor ranged from 8.6 to 9.5 MPa.

After completion of the reaction and cooling of the reaction mixture to room temperature, gaseous products were collected in a gasometer, their volume was measured, and the composition was determined by gas chromatography (GC). The mixture of the liquid and solid products was quantitatively discharged from the autoclave by washing with ethanol and separated by filtration; then, the solid product was washed with ethanol until the solvent discoloration and dried to a constant weight at 80 °C. Ethanol was removed from the liquid product by distillation on a rotary evaporator and the liquid product was brought to a constant weight by drying under vacuum (1 mmHg) at room temperature. The yields of the liquid (a_1), solid (a_2), and gaseous (a_3) products and the spruce wood conversion (Y_1) were calculated using the equations:

$$a_1 = \frac{m_l(\text{g})}{m_{init}(\text{g})} \times 100\% \quad (2)$$

$$a_2 = \frac{m_s(\text{g}) - m_{cat}(\text{g})}{m_{init}(\text{g})} \times 100\% \quad (3)$$

$$a_3 = \frac{m_g(g)}{m_{init}(g)} \times 100\% \quad (4)$$

$$Y_1 = \frac{m_{init}(g) - m_s(g)}{m_{init}(g)} \quad (5)$$

where m_l (g) is the liquid product weight, m_{init} (g) the weight of the initial wood, m_s (g) the solid product weight, m_{cat} (g) is the catalyst weight, and m_g (g) the gaseous products weight.

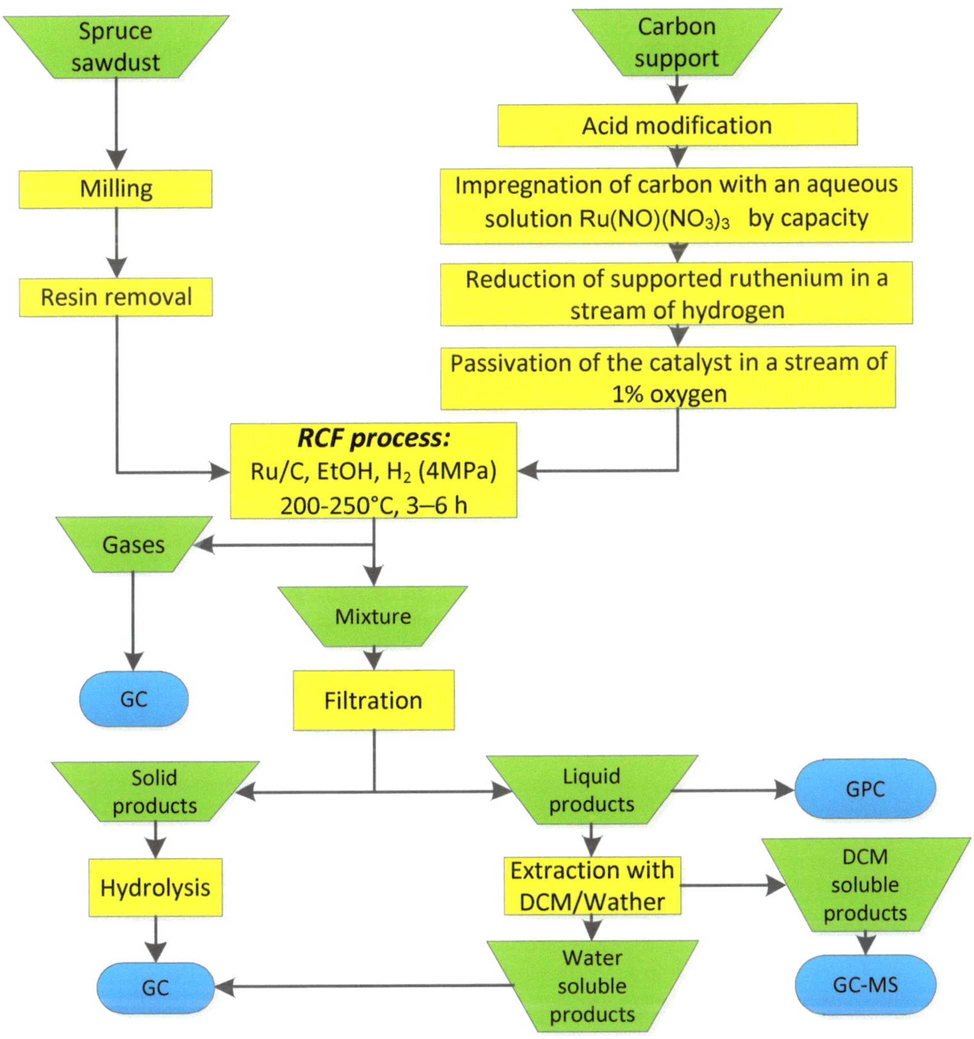

Figure 18. Overview of process flow diagram of catalytic hydrogenation of spruce wood.

3.4. Study of the Spuce Wood RCF Products

The ethanol-soluble liquid products of wood hydrogenation were extracted with a mixture of dichloromethane and water to separate into the aqueous and organic fractions (see Figure 17). The aqueous fraction was silylated by evaporating 1 mL of the solution to dryness and adding 100 µL of pyridine and 100 µL of –N,O-bis(trimethylsilyl)trifluoroacetamide (BSTFA); then, 10 µg of the internal standard (ionol) was introduced and kept at 70 °C for

1 h. After that, the organic and aqueous phases were analyzed by gas chromatography–mass spectrometry (GC–MS) using an Agilent 7890A chromatograph with an Agilent 7000A Triple Quad selective mass detector and an HP-5MS capillary column (30 m) at temperatures programmed within 40–250 °C. The compounds were identified using the NIST MS Search 2.0 database. Propyl guaiacol, propenyl guaiacol, ethyl guaiacol (Sigma-Aldrich) were used as standard substances for quantifying the yield of monomers from lignin. The response factor for each standard compound was defined relative to the internal standard [56]. Phenanthrene was used as an internal standard. Ethylene glycol, propylene glycol, 1,2-butanediol and glycerol (Acros Organics, Fair Lawn, NJ, USA) were used as standards for the quantitative determination of products in the aqueous phase.

The molecular weight distribution of the liquid wood RCF products, weight average molecular weight M_w, number average molecular weight M_n, and polydispersity PD of the liquid product samples were determined by gel permeation chromatography (GPC) using an Agilent 1260 Infinity II Multi-Detector GPC/SEC System with the triple detection using a refractometer, a viscometer, and light scattering. The mixtures were separated in a PLgel Mixed-E column using tetrahydrofuran stabilized with 250 ppm BHT as a mobile phase. The column was calibrated using Agilent polydisperse polystyrene standards (Santa Clara, USA). The eluent feed rate was 1 mL/min and the volume of the injected sample was 100 μL. Before the analysis, the samples were dissolved in THF (1 mg/mL) and filtered through a 0.45-μm Millipore PTFE membrane filter. The data obtained were collected and processed using the Agilent GPC/SEC MDS software ver. 2.2.

The composition of the gaseous products was determined by GC using a Chromatec Crystal 2000 M chromatograph (Russia) with a thermal conductivity detector in a helium flow at the flow rate of 15 mL/min and a detector temperature of 170 °C. The CO and CH_4 analysis was carried out in a NaX zeolite column (3 m × 2 mm) in the isothermal mode at 60 °C. The analysis of CO_2 and hydrocarbon gases was carried out in a Porapak Q column in the following mode: 60 °C for 1 min and then elevation of temperature to 180 °C at the rate of 10 °C/min.

The solid wood product was analyzed for hemicelluloses, cellulose, and lignin contents. The residual lignin content was determined by hydrolysis with 72% H_2SO_4 at 98 °C [57]. The composition and concentration of monosaccharides in the solution obtained by hemicelluloses hydrolysis with 4% sulfuric acid were determined by GC [57]. The cellulose content was calculated from the difference between the wood weight (or the solid residue) and the hemicelluloses and lignin contents.

The GC study was carried out with a Varian-450 GC gas chromatograph (Varian, Inc., Palo Alto, CA, USA), a flame ionization detector, and a VF-624ms capillary column with a length of 30 m, an inner diameter of 0.32 mm, and helium as a gas carrier at an injector temperature of 250 °C. Before the analysis, the solution was derivatized to produce trimethylsilyl derivatives according to the procedure described elsewhere [58]. Sorbitol was used as an internal standard. The peaks were identified from retention times of the tautomeric forms of monosaccharides.

The elemental compositions of wood and the products of its conversion were determined using a CHNSO VARIO EL CUBE analyzer (Elementar, Germany).

X-ray diffraction (XRD) analysis was carried out using a PANalyticalX'Pert Pro (PANalytical, EA Almelo, The Netherlands) spectrometer with CuKα radiation (λ = 0.54 nm). The analysis was performed in the angle 4 range of 2θ = 5–70° in 0.1° intervals on the powder sample in a 2.5-cm diameter cuvette. The crystallinity index was calculated as the ratio between the crystalline peak intensities I_{002}–I_{AM} and the total intensity I_{002} after subtraction of the background signal:

$$CI = \frac{I_{002} - I_{AM}}{I_{002}}, \tag{6}$$

where I_{002} is the height of peak 002 and I_{AM} is the height of the minimum between peaks 002 and 101 [51].

4. Conclusions

Reductive catalytic fractionation (RCF) of lignocellulose offers a biorefinery technology which integrates both lignin and carbohydrates valorization. During this process, lignin is solvolytically extracted and simultaneously depolymerized via hydrogenolysis. In this study, we propose the RCF of spruce wood over a bifunctional catalyst Ru/C in the medium of ethanol and molecular hydrogen to produce monomer phenolic compounds from lignin, polyols from hemicelluloses, and microcrystalline cellulose.

Tle influence of the Ru/C bifunctional catalyst characteristics was assessed. The ruthenium catalysts were characterized by transmission electron microscopy. The solid, liquid, and gaseous products of the reductive fractionation of spruce wood biomass were characterized by IR spectroscopy, XRD analysis, GC-MS, GPC, GC, N_2 adsorption methods and by elemental and chemical analysis. The results show the particular effect of the carbon support acidity, catalyst grain size, content and dispersion of Ru on the effectiveness of lignin and hemicelluloses extraction and the yields of liquid and gaseous products. The most efficient catalysts for RCF of spruce wood providing high yields of monomer phenols, glycols, and solid product with content of cellulose up to 90 wt% bear 3 wt% of Ru with a dispersion of 0.94 based on an acidic oxidized graphite-like carbon support Sibunit®, and having a grain size of 56–94 μm.

Author Contributions: Conceptualization, O.P.T. and B.N.K.; methodology, A.V.M., A.S.K., A.M.S. and S.V.B.; formal analysis A.M.S., V.V.S., A.V.M., Y.N.M. and S.V.B.; investigation, A.V.M., A.S.K., V.V.S. and A.M.S.; data curation A.S.K., A.V.M. and Y.N.M.; writing—original draft preparation, A.V.M. and S.V.B.; writing—review and editing, O.P.T. and B.N.K.; visualization, A.V.M. and V.V.S.; supervision, O.P.T. and B.N.K.; project administration, O.P.T.; funding acquisition, O.P.T. All authors have read and agreed to the published version of the manuscript.

Funding: This study was supported by the Russian Science Foundation, project no. 21-73-20269, https://rscf.ru/project/21-73-20269/ (accessed on 1 November 2022).

Data Availability Statement: Data available on request from the authors.

Acknowledgments: This study was conducted using the equipment of the Krasnoyarsk Regional Centre for Collective Use, Krasnoyarsk Scientific Center, Siberian Branch of the Russian Academy of Sciences.

Conflicts of Interest: The authors declare that they have no conflict of interest.

References

1. Luterbacher, J.S.; Martin Alonso, D.; Dumesic, J.A. Targeted chemical upgrading of lignocellulosic biomass to platform molecules. *Green Chem.* **2014**, *16*, 4816–4838. [CrossRef]
2. Gallezot, P. Conversion of biomass to selected chemical products. *Chem. Soc. Rev.* **2012**, *41*, 1538–1558. [CrossRef] [PubMed]
3. Dusselier, M.; Mascal, M.; Sels, B.F. Top Chemical Opportunities from Carbohydrate Biomass: A Chemist's View of the Biorefinery. In *Selective Catalysis for Renewable Feedstocks and Chemicals*; Nicholas, K.M., Ed.; Springer International Publishing: Cham, Switzerland, 2014; pp. 1–40.
4. Schutyser, W.; Renders, T.; Van den Bosch, S.; Koelewijn, S.F.; Beckham, G.T.; Sels, B.F. Chemicals from lignin: An interplay of lignocellulose fractionation, depolymerisation, and upgrading. *Chem. Soc. Rev.* **2018**, *47*, 852–908. [CrossRef] [PubMed]
5. Kuznetsov, B.N.; Chesnokov, N.V.; Sudakova, I.G.; Garyntseva, N.V.; Kuznetsova, S.A.; Malyar, Y.N.; Yakovlev, V.A.; Djakovitch, L. Green catalytic processing of native and organosolv lignins. *Catal. Today* **2018**, *309*, 18–30. [CrossRef]
6. Renders, T.; Van den Bossche, G.; Vangeel, T.; Van Aelst, K.; Sels, B. Reductive catalytic fractionation: State of the art of the lignin-first biorefinery. *Curr. Opin. Biotechnol.* **2019**, *56*, 193–201. [CrossRef] [PubMed]
7. Johansson, A.; Aaltonen, O.; Ylinen, P. Organosolv pulping—Methods and pulp properties. *Biomass* **1987**, *13*, 45–65. [CrossRef]
8. Zhang, Z.; Harrison, M.D.; Rackemann, D.W.; Doherty, W.O.S.; O'Hara, I.M. Organosolv pretreatment of plant biomass for enhanced enzymatic saccharification. *Green Chem.* **2016**, *18*, 360–381. [CrossRef]
9. da Silva Fonseca, B.C.; Araújo, L.S.; da Silva Pinheiro, B.; dos Santos, A.S.; Amaral-Labat, G.; Matsushima, J.T.; Baldan, M.R. Bio-based Carbon Electrochemically Decorated with Cu Nanoparticles: Green Synthesis and Electrochemical Performance. *Mater. Res.* **2022**, *25*. [CrossRef]
10. Kuznetsov, B.N.; Sudakova, I.G.; Garyntseva, N.V.; Kondrasenko, A.A.; Pestunov, A.V.; Djakovitch, L.; Pinel, C. Catalytic peroxide fractionation processes for the green biorefinery of wood. *React. Kinet. Mech. Catal.* **2019**, *126*, 717–735. [CrossRef]

11. Tarabanko, V.E.; Kaygorodov, K.L.; Skiba, E.A.; Tarabanko, N.E.; Chelbina, Y.V.; Baybakova, O.V.; Kuznetsov, B.N.; Djakovitch, L. Processing Pine Wood into Vanillin and Glucose by Sequential Catalytic Oxidation and Enzymatic Hydrolysis. *Wood Chem. Technol.* **2017**, *37*, 43–51. [CrossRef]
12. Chikunov, A.S.; Shashkov, M.V.; Pestunov, A.V.; Kazachenko, A.S.; Mishenko, T.I.; Taran, O.P. Hydrogenolysis of Birch Ethanol-Lignin in Supercritical Over Bifunctional Ru and Ni Catalysts Bifunctional Supported on Oxidized Carbon. *J. Sib. Fed. Univ. Chem.* **2018**, *11*, 131–150.
13. Van den Bosch, S.; Schutyser, W.; Vanholme, R.; Driessen, T.; Koelewijn, S.-F.; Renders, T.; De Meester, B.; Huijgen, W.; Dehaen, W.; Courtin, C.; et al. Reductive lignocellulose fractionation into soluble lignin-derived phenolic mono- and dimers and processable carbohydrate pulp. *Energy Environ. Sci.* **2015**, *8*, 1748–1763. [CrossRef]
14. Kuznetsov, B.N.; Sharypov, V.I.; Baryshnikov, S.V.; Miroshnikova, A.V.; Taran, O.P.; Yakovlev, V.A.; Lavrenov, A.V.; Djakovitch, L. Catalytic hydrogenolysis of native and organosolv lignins of aspen wood to liquid products in supercritical ethanol medium. *Catal. Today* **2021**, *379*, 114–123. [CrossRef]
15. Zhang, K.; Li, H.; Xiao, L.-P.; Wang, B.; Sun, R.-C.; Song, G. Sequential utilization of bamboo biomass through reductive catalytic fractionation of lignin. *Bioresour. Technol.* **2019**, *285*, 121335. [CrossRef]
16. Song, Q.; Wang, F.; Cai, J.; Wang, Y.; Zhang, J.; Yu, W.; Xu, J. Lignin depolymerization (LDP) in alcohol over nickel-based catalysts via a fragmentation–hydrogenolysis process. *Energy Environ. Sci.* **2013**, *6*, 994–1007. [CrossRef]
17. Galkin, M.V.; Samec, J.S.M. Selective Route to 2-Propenyl Aryls Directly from Wood by a Tandem Organosolv and Palladium-Catalysed Transfer Hydrogenolysis. *ChemSusChem* **2014**, *7*, 2154–2158. [CrossRef] [PubMed]
18. Renders, T.; Cooreman, E.; Van den Bosch, S.; Schutyser, W.; Koelewijn, S.-F.; Vangeel, T.; Deneyer, A.; Van den Bossche, G.; Courtin, C.; Sels, B. Catalytic lignocellulose biorefining in n-butanol/water: A one-pot approach toward phenolics, polyols, and cellulose. *Green Chem.* **2018**, *20*, 4607–4619. [CrossRef]
19. Parsell, T.; Yohe, S.; Degenstein, J.; Jarrell, T.; Klein, I.; Gencer, E.; Hewetson, B.; Hurt, M.; Kim, J.I.; Choudhari, H.; et al. A synergistic biorefinery based on catalytic conversion of lignin prior to cellulose starting from lignocellulosic biomass. *Green Chem.* **2015**, *17*, 1492–1499. [CrossRef]
20. Klein, I.; Marcum, C.; Kenttämaa, H.; Abu-Omar, M.M. Mechanistic investigation of the Zn/Pd/C catalyzed cleavage and hydrodeoxygenation of lignin. *Green Chem.* **2016**, *18*, 2399–2405. [CrossRef]
21. Li, C.; Zheng, M.; Wang, A.; Zhang, T. One-pot catalytic hydrocracking of raw woody biomass into chemicals over supported carbide catalysts: Simultaneous conversion of cellulose, hemicellulose and lignin. *Energy Environ. Sci.* **2012**, *5*, 6383–6390. [CrossRef]
22. Yan, N.; Zhao, C.; Dyson, P.J.; Wang, C.; Liu, L.-t.; Kou, Y. Selective Degradation of Wood Lignin over Noble-Metal Catalysts in a Two-Step Process. *ChemSusChem* **2008**, *1*, 626–629. [CrossRef] [PubMed]
23. Van den Bosch, S.; Schutyser, W.; Koelewijn, S.F.; Renders, T.; Courtin, C.M.; Sels, B.F. Tuning the lignin oil OH-content with Ru and Pd catalysts during lignin hydrogenolysis on birch wood. *Chem. Commun.* **2015**, *51*, 13158–13161. [CrossRef] [PubMed]
24. Chen, J.; Lu, F.; Si, X.; Nie, X.; Chen, J.; Rui, L.; Xu, J. High Yield Production of Natural Phenolic Alcohols from Woody Biomass Using a Nickel-Based Catalyst. *ChemSusChem* **2016**, *9*, 3353–3360. [CrossRef] [PubMed]
25. Zhao, S.; Abu-Omar, M.M. Renewable Epoxy Networks Derived from Lignin-Based Monomers: Effect of Cross-Linking Density. *ACS Sustain. Chem. Eng.* **2016**, *4*, 6082–6089. [CrossRef]
26. Koelewijn, S.F.; Van den Bosch, S.; Renders, T.; Schutyser, W.; Lagrain, B.; Smet, M.; Thomas, J.; Dehaen, W.; Van Puyvelde, P.; Witters, H.; et al. Sustainable bisphenols from renewable softwood lignin feedstock for polycarbonates and cyanate ester resins. *Green Chem.* **2017**, *19*, 2561–2570. [CrossRef]
27. Lakani, S.A.; Hedjazi, S.; Abdulkhani, A. Chemical analysis and antioxidant activities of bark extracts from four endemic species of Hyrcanian forests in Iran. *Holzforschung* **2019**, *73*, 287–294. [CrossRef]
28. Hu, S.; Luo, X.; Li, Y. Polyols and Polyurethanes from the Liquefaction of Lignocellulosic Biomass. *ChemSusChem* **2014**, *7*, 66–72. [CrossRef]
29. Sun, P.; Zhang, W.; Yu, X.; Zhang, J.; Xu, N.; Zhang, Z.; Liu, M.; Zhang, D.; Zhang, G.; Liu, Z.; et al. Hydrogenolysis of Glycerol to Propylene Glycol: Energy, Tech-Economic, and Environmental Studies. *Front. Chem.* **2022**, *9*, 778579. [CrossRef]
30. Kang, S.; Fu, J.; Zhang, G. From lignocellulosic biomass to levulinic acid: A review on acid-catalyzed hydrolysis. *Renew. Sustain. Energy Rev.* **2018**, *94*, 340–362. [CrossRef]
31. Li, G.; Liu, W.; Ye, C.; Li, X.; Si, C.-L. Chemocatalytic Conversion of Cellulose into Key Platform Chemicals. *Int. J. Polym. Sci.* **2018**, *2018*, 4723573. [CrossRef]
32. Hou, Q.; Qi, X.; Zhen, M.; Qian, H.; Nie, Y.; Bai, C.; Zhang, S.; Bai, X.; Ju, M. Biorefinery roadmap based on catalytic production and upgrading 5-hydroxymethylfurfural. *Green Chem.* **2021**, *23*, 119–231. [CrossRef]
33. Thoorens, G.; Krier, F.; Leclercq, B.; Carlin, B.; Evrard, B. Microcrystalline cellulose, a direct compression binder in a quality by design environment—A review. *Int. J. Pharm.* **2014**, *473*, 64–72. [CrossRef] [PubMed]
34. Shankar, S.; Rhim, J.-W. Preparation of nanocellulose from micro-crystalline cellulose: The effect on the performance and properties of agar-based composite films. *Carbohydr. Polym.* **2016**, *135*, 18–26. [CrossRef] [PubMed]
35. Ullah, N.; Odda, A.H.; Liang, K.; Kombo, M.A.; Sahar, S.; Ma, L.-B.; Fang, X.-X.; Xu, A.-W. Metal–acid nanoplate-supported ultrafine Ru nanoclusters for efficient catalytic fractionation of lignin into aromatic alcohols. *Green Chem.* **2019**, *21*, 2739–2751. [CrossRef]

36. Ayusheev, A.B.; Taran, O.P.; Afinogenova, I.I.; Mishchenko, T.I.; Shashkov, M.V.; Sashkina, K.A.; Semeikina, V.S.; Parkhomchuk, E.V.; Agabekov, V.E.; Parmon, V.N. Depolymerization of Birch-Wood Organosolv Lignin Over Solid Catalysts in Supercritical Ethanol. *J. Sib. Fed. Univ. Chem.* **2016**, *9*, 353–370. [CrossRef]
37. Kazachenko, A.S.; Baryshnikov, S.V.; Chudina, A.I.; Malyar, Y.N.; Sychev, V.V.; Taran, O.P.; Djakovitch, L.; Kuznetsov, B.N. Hydrogenation of abies wood and ethanollignin by molecular hydrogen in supercritical ethanol over bifunctional Ru/C catalyst. *Chem. Plant Raw Mater.* **2019**, *2*, 15–26. [CrossRef]
38. Shorohova, E.; Kuuluvainen, T.; Kangur, A.; Jõgiste, K. Natural stand structures, disturbance regimes and successional dynamics in the Eurasian boreal forests: A review with special reference to Russian studies. *Ann. For. Sci.* **2009**, *66*, 201. [CrossRef]
39. De Lucas-Consuegra, A.; Serrano-Ruiz, J.C.; Gutiérrez-Guerra, N.; Valverde, J.L. Low-Temperature Electrocatalytic Conversion of CO2 to Liquid Fuels: Effect of the Cu Particle Size. *Catalysts* **2018**, *8*, 340. [CrossRef]
40. Gallezot, P.; Chaumet, S.; Perrard, A.; Isnard, P. Catalytic Wet Air Oxidation of Acetic Acid on Carbon-Supported Ruthenium Catalysts. *J. Catal.* **1997**, *168*, 104–109. [CrossRef]
41. Carter, J.L.; Cusumano, J.A.; Sinfelt, J.H. Catalysis over Supported Metals. V. The Effect of Crystallite Size on the Catalytic Activity of Nickel. *J. Phys. Chem.* **1966**, *70*, 2257–2263. [CrossRef]
42. Kazachenko, A.S.; Tarabanko, V.E.; Miroshnikova, A.V.; Sychev, V.V.; Skripnikov, A.M.; Malyar, Y.N.; Mikhlin, Y.L.; Baryshnikov, S.V.; Taran, O.P. Reductive Catalytic Fractionation of Flax Shive over Ru/C Catalysts. *Catalysts* **2021**, *11*, 10042. [CrossRef]
43. Liu, X.; Li, H.; Xiao, L.-P.; Sun, R.-C.; Song, G. Chemodivergent hydrogenolysis of eucalyptus lignin with Ni@ZIF-8 catalyst. *Green Chem.* **2019**, *21*, 1498–1504. [CrossRef]
44. Boonyasuwat, S.; Omotoso, T.; Resasco, D.E.; Crossley, S.P. Conversion of Guaiacol over Supported Ru Catalysts. *Catal. Lett.* **2013**, *143*, 783–791. [CrossRef]
45. Sun, S.; Sun, S.; Cao, X.; Sun, R. The role of pretreatment in improving the enzymatic hydrolysis of lignocellulosic materials. *Bioresour. Technol.* **2016**, *199*, 49–58. [CrossRef]
46. Teacă, C.A.; Roşu, D.; Bodîrlău, R.; Roşu, L. Structural Changes in Wood under Artificial UV Light Irradiation Determined by FTIR Spectroscopy and Color Measurements—A Brief Review. *BioResources* **2013**, *8*, 1478–1507. [CrossRef]
47. Shi, J.-B.; Yang, Q.-L.; Lin, L.; Peng, L.-C. Fractionation and characterization of physicochemical and structural features of corn stalk hemicelluloses from yellow liquor of active oxygen cooking. *Ind. Crops Prod.* **2013**, *44*, 542–548. [CrossRef]
48. Cheng, S.; Huang, A.; Wang, S.; Zhang, Q. Effect of Different Heat Treatment Temperatures on the Chemical Composition and Structure of Chinese Fir Wood. *BioResources* **2016**, *11*, 4006–4016. [CrossRef]
49. Shi, J.; Xing, D.; Lia, J. FTIR Studies of the Changes in Wood Chemistry from Wood Forming Tissue under Inclined Treatment. *Energy Procedia* **2012**, *16*, 758–762. [CrossRef]
50. Šturcová, A.; His, I.; Apperley, D.C.; Sugiyama, J.; Jarvis, M.C. Structural Details of Crystalline Cellulose from Higher Plants. *Biomacromolecules* **2004**, *5*, 1333–1339. [CrossRef]
51. Park, S.; Baker, J.O.; Himmel, M.E.; Parilla, P.A.; Johnson, D.K. Cellulose crystallinity index: Measurement techniques and their impact on interpreting cellulase performance. *Biotechnol. Biofuels* **2010**, *3*, 10. [CrossRef]
52. Sudakova, I.G.; Garyntseva, N.V.; Chudina, A.I.; Kuznetsov, B.N. Experimental and Mathematical Optimization of the Peroxide Delignification of Larch Wood in the Presence of MnSO4 Catalyst. *Catal. Ind.* **2020**, *12*, 265–272. [CrossRef]
53. Taran, O.P.; Polyanskaya, E.M.; Ogorodnikova, O.L.; Descorme, C.; Besson, M.; Parmon, V.N. Sibunit-based catalytic materials for the deep oxidation of organic ecotoxicants in aqueous solution: I. Surface properties of the oxidized sibunit samples. *Catal. Ind.* **2010**, *2*, 381–386. [CrossRef]
54. Taran, O.P.; Descorme, C.; Polyanskaya, E.M.; Ayusheev, A.B.; Besson, M.; Parmon, V.N. Sibunit-based catalytic materials for the deep oxidation of organic ecotoxicants in aqueous solutions. III: Wet air oxidation of phenol over oxidized carbon and Rr/C catalysts. *Catal. Ind.* **2013**, *5*, 164–174. [CrossRef]
55. Minh, D.P.; Aubert, G.; Gallezot, P.; Besson, M. Degradation of olive oil mill effluents by catalytic wet air oxidation: 2-Oxidation of p-hydroxyphenylacetic and p-hydroxybenzoic acids over Pt and Ru supported catalysts. *Appl. Catal. B Environ.* **2007**, *73*, 236–246. [CrossRef]
56. Kim, J.-Y.; Oh, S.; Hwang, H.; Kim, U.-J.; Choi, J.W. Structural features and thermal degradation properties of various lignin macromolecules obtained from poplar wood (Populus albaglandulosa). *Polym. Degrad. Stab.* **2013**, *98*, 1671–1678. [CrossRef]
57. Sluiter, J.B.; Ruiz, R.O.; Scarlata, C.J.; Sluiter, A.D.; Templeton, D.W. Compositional Analysis of Lignocellulosic Feedstocks. 1. Review and Description of Methods. *J. Agric. Food Chem.* **2010**, *58*, 9043–9053. [CrossRef]
58. Ruiz-Matute, A.I.; Hernández-Hernández, O.; Rodríguez-Sánchez, S.; Sanz, M.L.; Martínez-Castro, I. Derivatization of carbohydrates for GC and GC–MS analyses. *J. Chromatogr. B* **2011**, *879*, 1226–1240. [CrossRef]

www.ingramcontent.com/pod-product-compliance
Lightning Source LLC
Chambersburg PA
CBHW041402210926
51850CB00013B/3155

MDPI AG
Grosspeteranlage 5
4052 Basel
Switzerland
Tel.: +41 61 683 77 34

Catalysts Editorial Office
E-mail: catalysts@mdpi.com
www.mdpi.com/journal/catalysts